Optimization of Design for Better Structural Capacity

Mourad Belgasmia
Setif 1 University, Algeria

A volume in the Advances in Civil and Industrial Engineering (ACIE) Book Series

IGI Global
DISSEMINATOR OF KNOWLEDGE

Published in the United States of America by
IGI Global
Engineering Science Reference (an imprint of IGI Global)
701 E. Chocolate Avenue
Hershey PA, USA 17033
Tel: 717-533-8845
Fax: 717-533-8661
E-mail: cust@igi-global.com
Web site: http://www.igi-global.com

Library of Congress Cataloging-in-Publication Data

Names: Belgasmia, Mourad, 1973- editor.
Title: Optimization of design for better structural capacity / Mourad Belgasmia, editor.
Description: Hershey, PA : Engineering Science Reference, an imprint of IGI Global, [2019] | Includes bibliographical references and index.
Identifiers: LCCN 2018014735| ISBN 9781522570592 (hardcover) | ISBN 9781522570608 (ebook)
Subjects: LCSH: Structural optimization.
Classification: LCC TA658.8 .O679 2019 | DDC 624.1/7713--dc23 LC record available at https://lccn.loc.gov/2018014735

This book is published in the IGI Global book series Advances in Civil and Industrial Engineering (ACIE) (ISSN: 2326-6139; eISSN: 2326-6155)

British Cataloguing in Publication Data
A Cataloguing in Publication record for this book is available from the British Library.

All work contributed to this book is new, previously-unpublished material. The views expressed in this book are those of the authors, but not necessarily of the publisher.

For electronic access to this publication, please contact: eresources@igi-global.com.

Advances in Civil and Industrial Engineering (ACIE) Book Series

Ioan Constantin Dima
University Valahia of Târgovişte, Romania

ISSN:2326-6139
EISSN:2326-6155

Mission

Private and public sector infrastructures begin to age, or require change in the face of developing technologies, the fields of civil and industrial engineering have become increasingly important as a method to mitigate and manage these changes. As governments and the public at large begin to grapple with climate change and growing populations, civil engineering has become more interdisciplinary and the need for publications that discuss the rapid changes and advancements in the field have become more in-demand. Additionally, private corporations and companies are facing similar changes and challenges, with the pressure for new and innovative methods being placed on those involved in industrial engineering.

The **Advances in Civil and Industrial Engineering (ACIE) Book Series** aims to present research and methodology that will provide solutions and discussions to meet such needs. The latest methodologies, applications, tools, and analysis will be published through the books included in **ACIE** in order to keep the available research in civil and industrial engineering as current and timely as possible.

Coverage

- Urban Engineering
- Earthquake engineering
- Quality Engineering
- Construction Engineering
- Coastal Engineering
- Production Planning and Control
- Optimization Techniques
- Operations research
- Structural Engineering
- Productivity

IGI Global is currently accepting manuscripts for publication within this series. To submit a proposal for a volume in this series, please contact our Acquisition Editors at Acquisitions@igi-global.com or visit: http://www.igi-global.com/publish/.

The Advances in Civil and Industrial Engineering (ACIE) Book Series (ISSN 2326-6139) is published by IGI Global, 701 E. Chocolate Avenue, Hershey, PA 17033-1240, USA, www.igi-global.com. This series is composed of titles available for purchase individually; each title is edited to be contextually exclusive from any other title within the series. For pricing and ordering information please visit http://www.igi-global.com/book-series/advances-civil-industrial-engineering/73673. Postmaster: Send all address changes to above address.

Titles in this Series

For a list of additional titles in this series, please visit: www.igi-global.com/book-series

Contemporary Strategies and Approaches in 3-D Information Modeling
Bimal Kumar (Glasgow Caledonian University, UK)
Engineering Science Reference • copyright 2018 • 313pp • H/C (ISBN: 9781522556251) • US $205.00 (our price)

New Approaches, Methods, and Tools in Urban E-Planning
Carlos Nunes Silva (University of Lisbon, Portugal)
Engineering Science Reference • copyright 2018 • 407pp • H/C (ISBN: 9781522559993) • US $205.00 (our price)

Designing Grid Cities for Optimized Urban Development and Planning
Guiseppe Carlone (Independent Researcher, Italy) Nicola Martinelli (Politecnico di Bari, Italy) and Francesco Rotondo (Politecnico di Bari, Italy)
Engineering Science Reference • copyright 2018 • 305pp • H/C (ISBN: 9781522536130) • US $210.00 (our price)

Design Solutions for nZEB Retrofit Buildings
Elzbieta Rynska (Warsaw University of Technology, Poland) Urszula Kozminska (Warsaw University of Technology, Poland) Kinga Zinowiec-Cieplik (Warsaw University of Technology, Poland) Joanna Rucinska (Warsaw University of Technology, Poland) and Barbara Szybinska-Matusiak (Norwegian University of Science and Technology, Norway)
Engineering Science Reference • copyright 2018 • 362pp • H/C (ISBN: 9781522541059) • US $225.00 (our price)

Handbook of Research on Perception-Driven Approaches to Urban Assessment and Design
Francesco Aletta (University of Sheffield, UK) and Jieling Xiao (Birmingham City University, UK)
Engineering Science Reference • copyright 2018 • 641pp • H/C (ISBN: 9781522536376) • US $295.00 (our price)

Convective Heat and Mass Transfer in the Free Atmosphere Emerging Research and Opportunities
Kaliyeva Kulyash (Lorraine University, France)
Engineering Science Reference • copyright 2018 • 127pp • H/C (ISBN: 9781522530503) • US $135.00 (our price)

Innovative Applications of Big Data in the Railway Industry
Shruti Kohli (University of Birmingham, UK) A.V. Senthil Kumar (Hindusthan College of Arts and Science, India) John M. Easton (University of Birmingham, UK) and Clive Roberts (University of Birmingham, UK)
Engineering Science Reference • copyright 2018 • 395pp • H/C (ISBN: 9781522531760) • US $235.00 (our price)

Dynamic Stability of Hydraulic Gates and Engineering for Flood Prevention
Noriaki Ishii (Osaka Electro-Communication University, Japan) Keiko Anami (Osaka Electro-Communication University, Japan) and Charles W. Knisely (Bucknell University, USA)
Engineering Science Reference • copyright 2018 • 660pp • H/C (ISBN: 9781522530794) • US $245.00 (our price)

IGI Global
DISSEMINATOR OF KNOWLEDGE

701 East Chocolate Avenue, Hershey, PA 17033, USA
Tel: 717-533-8845 x100 • Fax: 717-533-8661
E-Mail: cust@igi-global.com • www.igi-global.com

Dedicated to my Parents, Belgasmia Omar and Chabha.

Table of Contents

Detailed Table of Contents

The aim of this chapter is to confirm the possibility of using the simple diffusion model to predict the behavior of a building material during the application of drying process under variable operating conditions. This approach can be considered as a simulation of the effect of the variable climatic conditions on the building material. During this research, the thermo-physical properties of the tested material as well as the drying air are considered as variable and changing with the operating conditions. Accordingly, diffusion coefficient is determined experimentally and is considered as variable with the temperature and the humidity and represented as function of the wet bulb temperature. Two sorts of conditions are tested: constant flux and convective flux. Furthermore, two types of changes are also tested: sudden changes and progressive changes of the drying conditions. The results of the study are mainly represented by the drying curves or the drying kinetics.

This chapter shows, through the example of the addition of a plate and shell element to freeware FEM-object, an object-oriented (C++) finite element program, how object-oriented approaches, as opposed to procedural approaches, make finite element codes more compact, more modular, and versatile but mainly more easily expandable, in order to improve the continuity and the compatibility between software of research and industrial software. The fundamental traits of object-oriented programming are first briefly reviewed, and it is shown how such an approach simplifies the coding process. Then, the isotropic shell and orthotropic plate formulations used are given and the discretized equations developed. Finally, the necessary additions to the FEM-object code are reviewed. Numerical examples using the newly created plate membrane plate element are shown.

This chapter aims to develop a system identification methodology for determining structural parameters of linear dynamic systems, taking into consideration practical constraints such as insufficient sensors. Based on numerical analysis of measured responses (output) due to known excitations (input), structural parameters such as stiffness values are identified. If the values at the damaged state are compared with the identified values at the undamaged state, damage detection and quantification can be carried out. To retrieve second-order parameters from the identified state space model, various methodologies developed thus far impose different restrictions on the number of sensors and actuators employed. The restrictions are relaxed in this study by a proposed method called the condensed model identification and recovery (CMIR) method. To estimate individual stiffness coefficient from the condensed stiffness matrices, the genetic algorithms approach is presented to accomplish the required optimization problem.

Soil structure interaction can significantly affect the behavior of buildings subjected to seismic attacks, wind excitation, and other dynamic loading types. Different researches were developed in the last decade demonstrating the importance of taking account of soil properties and its effect in changing the behavior of the structures. It is common practice to analyze the structures assuming a fixed base, but this approach is not appropriate for the reason that neglecting the soil parameters such as the stiffness and the damping affect the behavior of the structure. Therefore, the nonlinear static approach provided the nonlinear response behavior of a structure for different types of soil. In this chapter, the authors will discuss some proposed methods in taking account of soil-structure interaction that must be considered from the very beginning of the design process and its impact on the structural behavior optimization by adding springs and dashpots to reproduce the soil behavior.

In the long history of mankind's existence, nature's forces have influenced human existence to a great extent. Of all natural disasters, the least understood and most destructive are earthquakes. Their claim of human lives and material losses constantly force people to search for better protection, still a great challenge for engineers and researchers worldwide. Although important progress has been done in understanding seismic activity and developing buildings technology, a better way of protecting buildings on large scale is still in search. The essential features of earthquake resistance structure are stable foundation design, regularity, ductility, adequate stiffness, redundancy, and ruggedness. The chapter focuses on increasing the knowledge dictum of earthquake resistant design and discusses the various sorts of issues and challenges. It also presents a wide view on optimization techniques that are required to be done in the latest technology currently in practice so as to achieve the optimum design techniques.

Foreword

I am pleased and honored to write a foreword for this book. This book gives the various optimization techniques for designing better structural capacity. The book is comprehensive and self-contained. It strikes an optimum balance between theory and practice. Application of heuristics algorithms has been also described.

This book provides a valuable window on the effect of various parameters on structural capacity. It provides a widely useful compilation of ideas, innovation approaches and practical strategies for designing better structural capacity.

It is my hope and expectation that this book will provide an effective learning experience and referenced resource for civil engineers. I hope that this book will become a primer for teachers, teacher educators, and professional developers, helping teachers across the country to learn, teach, and practice.

Pijush Samui
NIT Patna, India

Preface

Despite the evolution of methods and models as well as the algorithms, optimization is always of great importance, to overcome structural failures and guarantee a long life and go to tap hidden structural capacities The goal is to improve the capacity, to limit structural damage, to assess the structure dynamic response and to reduce damage caused different types of loads by optimizing the design, like pre-dimensioning dumpers and fire design optimizing both structural model and implementation by adding the dashpot and changing the manner of programming using the Object Oriented programming if implementation of codes is needed.

This book will be appropriate for Universities (researchers, and students), another part of this book can be in the contexts of industry.

This book is an attempt to present various computational techniques for application in structural engineering design. It consists of nine chapters, and aims to present the developments in Optimization of Design for Better behaviour of structure under different types of loads.

Chapter 1 of this book opens and confirms the possibility of using the simple diffusion model to predict the behavior of a building material during the application of drying process under variable operating conditions. This approach can be considered as a simulation of the effect of the variable climatic conditions on the building material. During this research, the thermo-physical properties of the tested material as well as the drying air, are considered as variable and changing with the operating conditions. Accordingly, diffusion coefficient is determined experimentally and is considered as variable with the temperature and the humidity and represented as function of the wet bulb temperature. Two sorts of conditions are tested: constant flux and convective flux. Furthermore, two types of changes are also tested; sudden changes and progressive changes of the drying conditions. The results of the study are mainly represented by the drying curves or the drying kinetics.

There is a lack of knowledge about the behavior of the different materials during the application of variable drying conditions, which can be meet in several industries as well as during solar drying. Moreover, the most developed models that describes the drying kinetics or the drying curves consider the application of constant drying conditions, which make the results given by these models not accurate, and a significant difference with the experimental results is usually observed. Moreover, the presented models are considered as complicated models that need specific experiments in order to determine the coefficients needed in those models.

Accordingly, two parts are presented in this chapter: the first part is linked to an experimental determination of the drying kinetics or the drying curves under variable drying conditions. The second part presents the functionality of using the simple diffusion model to predict these drying curves and in these particular variable operating conditions.

Chapter 2 shows, through the example of the addition of a plate and shell element to freeware FEM-Object, an Object-Oriented (C++) finite element program, how Object-Oriented approaches, as opposed to procedural approaches, make finite element codes more compact, more modular and versatile but mainly more easily expandable; in order to improve the continuity and the compatibility between software of research and industrial software.The fundamental traits of Object-Oriented programming are first briefly reviewed, and it is shown how such an approach simplifies the coding process. Then, the isotropic shell and orthotropic plate formulations used are given and the discretized equations developed. Finally, the necessary additions to the FEM-Object code are reviewed. Numerical examples using the newly created plate and shell element are shown. In the last decades, the finite element method has gained a large acceptance as a general tool for modeling and simulating physical systems. It has become, in fact, the most widely used technique applicable in fields such as solid mechanics, fluid mechanics, electromagnetics, etc. In recent years, easy maintenance, extensibility and flexibility of the finite element code, encouraged many researchers to change from procedural to object oriented programming in order to make the finite element software more flexible.

In Chapter 3 two types of Algorithms Genetic and Estimation Distribution Algorithm are reviewed and analysed. So several heuristics algorithms can be employed to solve single row layout in construction site planning. Firstly, this research builds Tabu Search to deal with the problem. Other heuristics methods which are Genetic Algorithm (GA) and Estimation Distribution Algorithm (EDA) are also developed against Tabu Search. A comparative study is performed to test the effectiveness and efficiency of the algorithms. The statistical test, ANOVA followed by t-test, compares the results of the three algorithms. Then, the pros and cons of using the algorithms are stated.

The arrangement of machines in construction site layout can improve in cost reduction and time savings during the construction process. In construction site planning, there is a case when the facilities need to be placed in a row. The case is also known as single row facility layout problem. The objective is to minimize the sum of the distance between facilities if given their traffic loads. Genetic Algorithm (GA) simulates natural evolution process which generates a group candidate solution with selection to the best candidates, crossover, and mutation. GA become a useful optimization technique to deal with many kinds of problems since it is flexible and only need pieces of knowledge. Estimation Distribution Algorithm (EDA) is an algorithm that mimics the statistical process of the sampling technique. EDA also uses randomization technique like GA in the beginning but choose the candidate solution based on the probabilistic model that defined before.

Chapter 4 describes the major items that need to be incorporated in the design of an offshore structure for fire resistance. Structural engineers in collaboration with safety engineers should perform the structural fire design of offshore structures. As we know that fire will always be a major threat to this type of structure (offshore structure) as oil and gas always passes through the installation. The design against accidental fire situation should be included in the structural design of offshore structures in collaboration with safety engineers. The design of offshore structures for fire safety involves considering fire as a load condition, assessment of fire resistance, use of fire protection materials, and so on. This chapter presents a methodology that will enable an engineer to design an offshore structure to resist fire. It aims to highlight the major requirements of design and to establish a common approach in carrying out the design.

Chapter 5 illustrates and shows essential concepts in wind engineering. Relying on some brief theoretical background about wind loads on structures, its nature and components, which will be presented in such an illustrative way that engineers and young researchers may understand with ease. Along with time series generation which can be seen as an important task in wind engineering. In the same manner, discussion and description of an energy dissipation system is presented in this chapter, historical use, advantages and mainly the design methods which can be seen as a main objective of this chapter; various attempts of an optimized design process will be shown and it will be concentrated on an effective and simple procedure. Finally, all the methodology discussed will be illustrated with some examples on an RC structure, along with results and comments that gives directions towards further research.

The aims of Chapter 6 is to develop a system identification methodology for determining structural parameters of linear dynamic system, taking into consideration of practical constraints such as insufficient sensors. Based on numerical analysis of measured responses (output) due to known excitations (input), structural parameters such as stiffness values are identified. If the values at the damaged state are compared with the identified values at the undamaged state, damage detection and quantification can be carried out. damage may even cause structures to collapse catastrophically, resulting in loss of lives and assets. The only way to safeguard the safety of human life and to reduce loss of wealth is to carry out regular monitoring for early detection of structural damage. It is therefore essential to detect the existence, location and extent of damage in the structure early and to carry out remedial work if necessary. To retrieve second-order parameters from the identified state space model, various methodologies developed thus far impose different restrictions on the number of sensors and actuators employed. The restrictions are relaxed in this study by a proposed method called the Condensed Model Identification and Recovery (CMIR) Method. To estimate individual stiffness coefficient from the condensed stiffness matrices, the genetic algorithms approach is presented to accomplish the required optimization problem.

Chapter 7 introduces the evaluation of dynamic response of reinforced concrete building under lateral seismic loads and especially to demonstrate how the structure is going to behave when a changing in the soil parameters and an addition of springs and dashpots in the base of the structure is done. summarizes some proposed methods in modeling soil-structure interaction effects on building structures, also, a brief history of the soil structure interaction and researches done in this field will be presented, and finally, the results of the case studied will be discussed the basic concepts. Soil structure interaction can significantly affect the behavior of buildings subjected to seismic attacks, wind excitation and other dynamic loading types. Different researches were developed in the last decade demonstrating the importance of taking account of soil properties and its effect in changing the behavior of the structures. It is common practice to analyze the structures assuming a fixed base but this approach is not appropriate for the reason that neglecting the soil parameters such as the stiffness and the damping affect the behavior of the structure. Therefore, the nonlinear static approach provided the nonlinear response behavior of a structure for different types of soil. In this chapter, the authors will discuss some proposed methods in taking account of soil-structure interaction that must be considered from the very beginning of the design process and its impact on the structural behavior optimization by adding springs and dashpots to reproduce the soil behavior.

Chapter 8 reviews the design issues and challenges of earthquake resistance and focuses on to increase the knowledge dictum of earthquake resistant design and discusses the various sorts of issues and challenges. It also presents a wide view on optimization techniques that are required to be done in the latest technology currently in practice so as to achieve the optimum design techniques. Because in the long history of mankind's existence, nature's forces have influenced human existence to a great extent. Of all natural disasters the least understood and most destructive are earthquakes. Their claim of human lives and material losses constantly forced people to search for better protection and it is still a great challenge for engineers and researchers worldwide. Although important progress has been done in understanding seismic activity and developing buildings technology, but still a better way of protecting buildings on large scale is still in search. The essentials features of earthquake resistance structure are stable foundation design, regularity, ductility, adequate stiffness, redundancy and ruggedness.

Chapter 9 focuses on optimized design foundation in which three currently practiced categories of foundation are briefly introduced which are shallow foundations, deep foundations and intermediate foundations related to reinforced or improved soils. For each category, the adequate type of foundation is decided on the basis of an optimized solution, e.g., cost effective and acceptable time of execution. Then, two illustrative Tunisian case histories are analyzed to explain, first, when the practiced foundation solution was inadequately chosen how a non-cost-effective solution can be avoided. Secondly why an unsuitable foundation solution can lead to the stopping of the structure functioning and, then, how to proceed for the design of retrofit solution to be executed for restarting the functioning of the structure.

This book provides a valuable window on the effect of various parameters on structural capacity. So all the chapters are written by more than 10 experts in the subjects with a hope that the book will give impetus for new methods and models as well as the algorithms for assessing performance and optimization of behaviour and design of structure. We are deeply indebted and wish to thank all the authors for their valuable contributions. We thank all reviewers for their helpful supports. A special thanks goes to the staff members of IGI Global.

I ardently believe that open, honest and true collaboration in research is one the best means at our disposal to improve our world.

Mourad Belgasmia
Setif 1 University, Algeria

Chapter 1
Using Diffusion Model for Prediction and Optimization of Drying Process of Building Material:
Simulation of Variable Environmental Conditions

Lyes Bennamoun
University of New Brunswick, Canada

ABSTRACT

The aim of this chapter is to confirm the possibility of using the simple diffusion model to predict the behavior of a building material during the application of drying process under variable operating conditions. This approach can be considered as a simulation of the effect of the variable climatic conditions on the building material. During this research, the thermo-physical properties of the tested material as well as the drying air are considered as variable and changing with the operating conditions. Accordingly, diffusion coefficient is determined experimentally and is considered as variable with the temperature and the humidity and represented as function of the wet bulb temperature. Two sorts of conditions are tested: constant flux and convective flux. Furthermore, two types of changes are also tested: sudden changes and progressive changes of the drying conditions. The results of the study are mainly represented by the drying curves or the drying kinetics.

INTRODUCTION

Drying is a process that has been always used by human. Liu et al. (2002) confirmed that China already used drying process six thousand years ago. Moyne and Roques (1986) reported that the United Kingdom, in 1727, extensively used wood drying for the shipyard. Drying consists on the removal of

DOI: 10.4018/978-1-5225-7059-2.ch001

the moisture content that a material is holding by the application of a source of heat. Drying process is used as a method of conservation and preservation of agro-alimentary products such as fish, corn, rice, potato and tomato with the objective of preserving valuable organoleptic characteristics which include odor, flavor, texture and color, as well as the nutritional quality (Bennamoun and Belhamri, 2003). It can also be an unavoidable process for the fabrication of the products and making it marketable such as wood, paper and building materials. The technological and scientific development has confirmed the importance of this process in different fields: biology, chemistry, pulp and paper, wood, textile, etc. However, drying is not an easy process and needs a real management of the whole steps of the process, in order to avoid deterioration of the final product or make it expensive. Indeed, it is established that non-studied drying conditions can lead to the deformation of the final product. It can conduct to the appearance of cracks as well (Dong et al., 2018, Hammouda and Mihoubi, 2017) and Bennamounet al. 2013a). In the field of fabrication of building materials, it is well known that application of random or un-controlled drying conditions to building materials can lead to a quick deterioration of those ones. Consequently, it is necessary to have the fundamental knowledge of the behavior of the material to be dried before performing the process itself. The common technique that represents drying process is the *drying curve* or more accurately the *drying kinetic*. The drying curve represents the variation of the moisture content with time and the drying kinetic is usually represented by the variation of the drying rate (frequently given by the following unit: $kg.m^{-2}.s^{-1}$) versus the moisture content. This representation is called *Krischer's curve*. This representation can be considered as the most representative of drying process, as the different drying phases are easily determined using this representation, as confirmed by Bennamoun et al. (2016) and Bennamoun et al. (2013b). It is important to mention that the behavior of the different existing materials during drying is not the same. van Brakel (1980) presented a classification of the different materials that can be dried and divided them into: capillary porous, hygroscopic porous and into non-porous materials, then he assembled the materials based on the shape of the drying kinetics in 16 categories, which include building materials.

Moreover, mathematical modeling and simulation, in particular the drying kinetics, are also important steps that can play a role into acquiring the necessary fundamental knowledge that can give at the end, the optimum conditions to be applied. Accordingly, different mathematical models were performed for building materials, such as the studies presented by Belhamri (2003), Belhamri and Fohr (1996), Fohr et al. (1989) and Ali Mohamed (1992). This last researcher (Ali Mohamed, 1992) applied a model known by Whitaker's model. This one is identified as a strong model that applies the heat and mass balance in the microscopic scale of the tested material. Therefore, the balance is applied for the liquid, gaseous and solid parts. Even, this model is presented as a strong model, it shows several constraints that can be summarized in the difficulty of obtaining the different used coefficients. Luikov presented another model that was widely used in building materials. Guimarães et al. (2017) compared experimental results to modeling results using Luikov model. This last applies two important parameters; the hydric diffusivity and the thermal diffusivity. The authors were able to present these two parameters as function of the moisture content of the concrete sample then representing the distribution of the moisture content as a function of the thickness of the sample and the drying time. The modeling results were in a good agreement with experimental ones. Krokida et al. (2002) used a mathematical model that shows the drying rate as function of the moisture content. Moreover, the drying parameters of the used model were function of the operating conditions, such as the temperature of the air, its velocity, and thickness of the tested sample. The authors applied this approach for several building materials and offered a compilation. Even, these sorts of models have its limitation based on the tested conditions, they can be very useful and practical to have a quick overview on the behavior of the material during the drying process and can give an idea about the necessary drying time.

Moreover, drying is considered as an intensive process as it consumes a huge amount of energy, which will be reflected on the price of the final product. In order to avoid this constraint, application of simulation and optimization of the operating conditions will definitively help to reach the objective of having a good building material with minimum expenses for its drying process. In this chapter, two main parts are presented. The first part is related to study the behavior of a building material during drying process with the application of different drying conditions (ie. constant and variable). Then, the second part is linked to the used of the simple diffusion model for the prediction of these drying kinetics and study how accurate are the results with the experimental ones.

MAIN FOCUS OF THE CHAPTER

There is a lack of knowledge about the behavior of the different materials during the application of variable drying conditions, which can be meet in several industries as well as during solar drying. Moreover, the most developed models that describes the drying kinetics or the drying curves consider the application of constant drying conditions, which make the results given by these models not accurate, and a significant difference with the experimental results can be observed. Moreover, the presented models are considered as complicated models that need specific experiments in order to determine the coefficients to be introduced in those models.

Accordingly, two parts are presented in this chapter: the first part is linked to an experimental determination of the drying kinetics or the drying curves under variable drying conditions. The second part presents the functionality of using the simple diffusion model to predict these drying curves in particular with the application of variable operating conditions.

Experimental Determination of Drying Curves or Kinetics Under Variable Drying Conditions

Bennamoun et al. (2009) presented an experimental study and built a loop that allows recording the variation of the mass of the sample versus drying time. The proposed loop was constituted of an electrical heater in order to heat the ambient air, a tranquilizing chamber that allows having a non-turbulent flow of the air, a fan, a control system of the humidity and eventually injection of vapor. Another part of the loop allows control of the temperature of the drying air. Moreover, an electronic balance was linked to a computer, which recorded the variation of the mass with time. The proposed loop is shown in Figure 1. Thermocouples are also used in order to measure the temperature of the air and the temperature of the sample.

The tested material was a sample of brick that has the following dimensions: 205×100×36 mm^3. As determined by Belhamri and Fohr (1996), the brick shows during drying a long constant drying phase, which was reported due to special porosity of the material. It had an initial porosity of 0.6 and a volumetric mass of 1250 kg/m^3. The initial moisture content was determined by inserting a sample in an oven under 105°C for 24 hours, which was equal to: 0.4 kg/kg (dry basis).

Based on Krischer's curve representation, Bennamoun et al. (2009) showed the variation of the drying kinetic, at constant temperatures of 40°C (curve a) and 54°C (curve b). Suddenly, the drying temperature

Figure 1. Scheme of a drying loop
(Bennamoun et al., 2009)

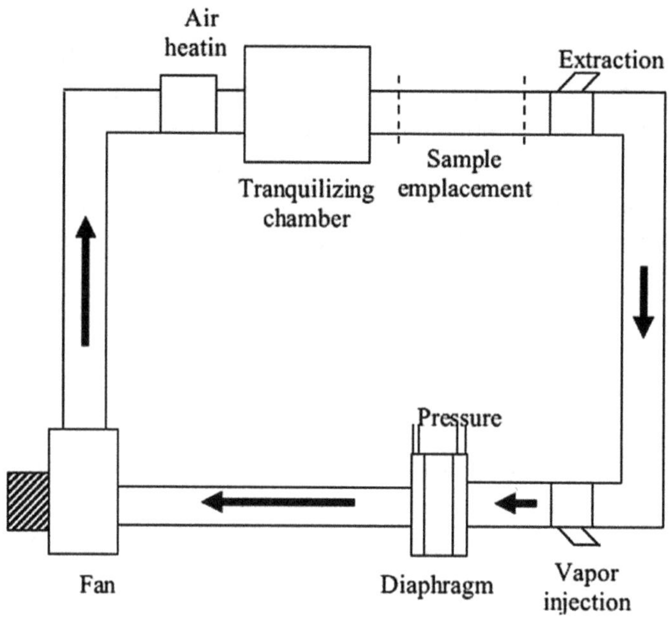

was changed from 40°C to 54°C. In this case, the drying kinetic shows an adaptation with observation of a certain time. Indeed, even the change in temperature was instant, the drying kinetic shows a non-instant modification to the new drying conditions represented by curve (c) of the Figure 2.

Figure 2 shows also a clear existence of the long constant drying phase called (C.D.R.P) represented by the constant drying rate as shown.

Rabha et al. (2017) studied solar drying of chili pepper with the presentation of the variation of the drying rate with time. As it is known, solar drying depends entirely on the solar radiation and the ambient conditions, accordingly, the drying rate curve showed intermittence during the low radiation periods, which makes difficult to distinguish the different drying phases. Similarly, Khama et al. (2016) exposed during solar drying of tomato the dependence of the drying rate and the drying curve on the solar radiation and the ambient operating conditions (i.e. air temperature, air humidity and air velocity) as shown in Figure 3. As it is seen most of solar drying studies are applied for food products and not for building materials because of the inconvenient of solar drying and its dependence on the weather conditions, which makes it not practical for industrialists. It the literature review, few studies linked to solar drying of building materials can be found. The work developed by Zhamalov (1989) and Jayamaha et al. (1996) are cited as examples. It is clear that these studies were performed long time ago and it is concluded that, point of view industrial application, it is not worth to study solar drying of building materials.

Using Diffusion Model for Prediction of Drying Under Variable Conditions

In one hand, Ali Mohamed (1992) used the model of Whitaker in order to predict the evolution of the drying kinetic with the moisture content during application of variable drying conditions. The comparison between the experimental and modeling results was favorable. However, the complication of the model

Figure 2. Drying kinetic of brick sample under different drying conditions and with sudden change
(Bennamoun et al., 2009)
C: is the moisture content of the tested sample in (kg/kg dry basis)

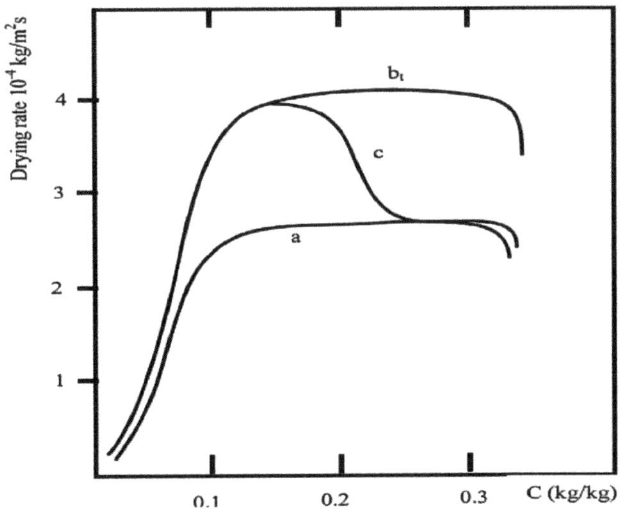

Figure 3. Variation of the moisture content and the drying rate of tomato during solar drying. Square represents the drying rate and diamond is the moisture content
(Khama et al., 2016)

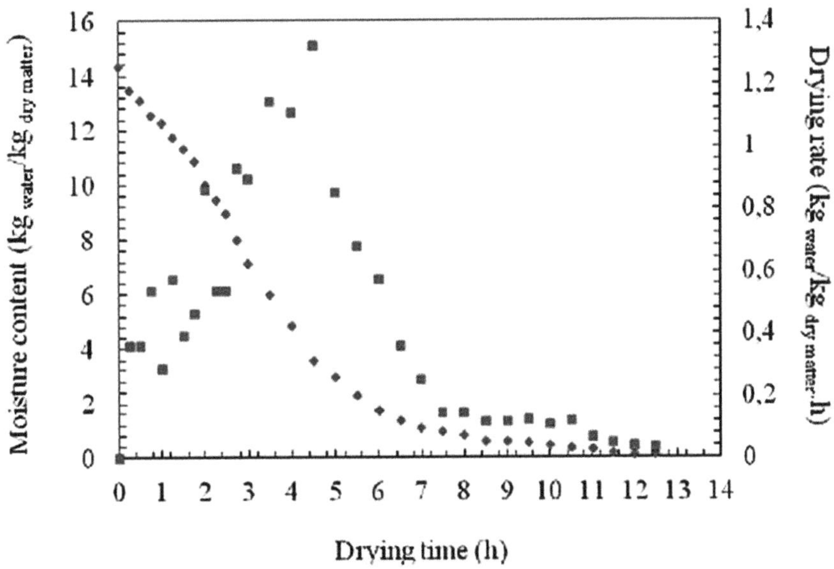

and the coefficients needed for the implementation of the model was an obstacle for having a practical model, easy for application. On the other hand, the studies done by Belhamri (1992) and Pel Landman and Kaasschieter (2002) done on the brick, were a proof of the possibility of application of diffusion model with a good prediction of the evolution of the moisture content with time.

Diffusion model is based on the simple Fick's law, written by Crank (1975) under the following form:

$$\frac{\partial C}{\partial T} = \nabla(D\nabla C) \quad (1)$$

where D is the coefficient of diffusion (m^2/s)

In drying and in order to distinguish the different drying phases, diffusion model is commonly written under the following form:

$$\frac{\partial C}{\partial t} = D_i \frac{\partial^2 C}{\partial y^2} \quad (2)$$

where 'i' can equal to 1 or 2. i=1 represents diffusion in the constant drying rate phase.

i=2 is diffusion during the falling drying rate phase.

Even the apparent simplicity of diffusion equation, finding an exact solution of this equation was not possible. Consequently, Crank (1975) proposed analytical solutions based on the initial and boundary conditions. It is important to mention that in our study; the flow of the air is considered unidirectional and thus the moisture content is varying with time and the thickness of the sample (y direction).

As shown in Figure 4, the proposed problem can be considered as symmetric, which allow writing using equation (2), the following boundary and initial conditions:

At the initial time the distribution of the moisture content within the material is considered to be uniform and constant:

$$t=0: C=C_0 \quad (3)$$

Figure 4. Flow of the air during drying of a brick sample
(Bennamoun et al., 2009)

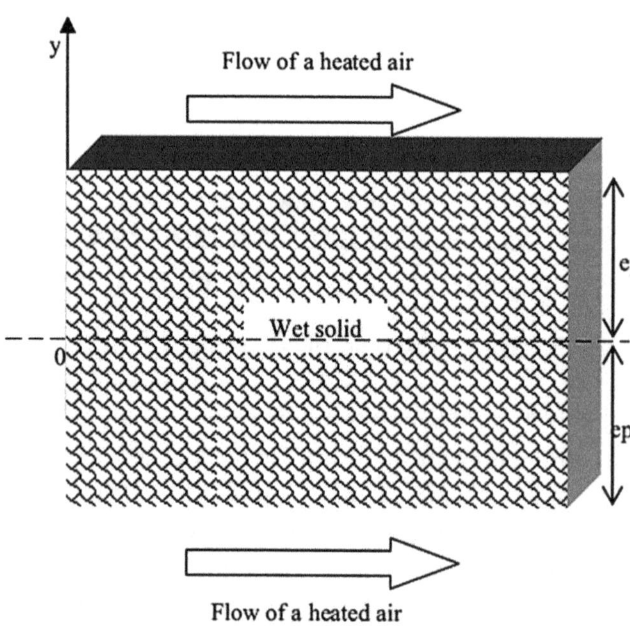

Also, the symmetry can be written mathematically using the following equation:

$$\left.\frac{\partial C}{\partial y}\right)_{y=0} = 0 \tag{4}$$

Moreover, there is a flow of the heated air at the surface of the sample, which can be translated mathematically to:

$$-\rho_S D_i \left.\frac{\partial C}{\partial y}\right)_{y=ep} = F \tag{5}$$

where ρ_s is the density of the material (kg/m^3)

'ep' is the thickness of the sample
'F' is the flux of the air.

In this study two cases will be modeled; constant flux and convective flux.

Crank (1975) presented the analytical solution of the diffusion equation for a plate shape, as it is represented in Figure 4 and the solution was written under the following form:

$$\frac{\left(C - C_0\right)\rho_S D_i}{F \cdot ep} = \frac{D_{i \cdot t}}{ep^2} + \frac{3y^2 - ep^2}{6ep^2} - \frac{2}{\pi^2}\sum_{n=1}^{\infty}\frac{\left(-1\right)^n}{n^2}\exp\left(\frac{-D_i n^2 \pi^2 t}{ep^2}\right)\cos\left(\frac{yn\pi}{ep}\right) \tag{6}$$

C_0 is the initial moisture content of the sample calculated in kg/kg (dry basis)

All the other parameters were defined previously.

Also, Crank (1975) presented the solution of the diffusion equation in the case of the application of a convective flux. The solution was written as follows:

$$\frac{C - C_0}{C_{eq} - C_0} = 1 - \sum_{n=1}^{\infty}\frac{2L\cos\left(\frac{\beta_n y}{ep}\right)\exp\left(\frac{-\beta_n^2 D_i t}{ep^2}\right)}{\left(\beta_n^2 + L^2 + L\right)\cos\left(\beta_n\right)} \tag{7}$$

Where:

C_{eq} is the equilibrium moisture content (kg/kg dry basis)
β_n are the roots of the following equation:

$$L = \beta\tan\left(\beta\right) \tag{8}$$

And 'L' is Biot number, which can be written as follows:

$$L = \frac{eph}{D_i} \quad (9)$$

Moreover, the convective flux takes the following form:

$$\frac{F}{\rho_S} = h\left(C - C_{eq}\right) \quad (10)$$

where 'h' is the convective mass transfer coefficient (m/s).

Knowing the coefficient of diffusion is important in order to complete solving numerically the equation of diffusion. In our case for the brick we used the values extracted experimentally by Bennamoun et al. (2009). Therefore, the coefficient of diffusion is presented as a function of the wet bulb temperature, using the following equations:

For the constant drying rate phase:

$$D_1 = \left(-0.021T_{wb}^2 + 1.098T_{wb} - 12.185\right)10^{-8} \quad (11)$$

And for the falling drying rate phase:

$$D_2 = \left(-0.003T_{wb}^2 + 0.197T_{wb} - 2.024\right)10^{-8} \quad (12)$$

T_{wb} is the wet bulb temperature and is function of the dry bulb temperature that can be recorded by a thermometer or a thermocouple and the humidity of the air. The temperature presented in equations (11) and (12) is in (°C) and the diffusion coefficient is presented in (m^2/s).

The values obtained experimentally are represented in the following Table1. As shown, the coefficient of diffusion is more important during the constant drying rate phase than the falling drying rate phase. This observation can be reported to the existence of important quantity of moisture during the constant drying rate phase, which is not the case for the falling one. Such observation was obtained by Chirife (1983), who studied convective drying of several materials including food.

Numerical Solution of Diffusion Equation

Numerical solution of diffusion equation is one of the important targets of this study. Indeed, this part will show how accurate is diffusion model to predict the changes in the drying curves at the application of variable operating conditions.

It is common and more appropriate to see the use of the dimensionless form of the diffusion equation. Accordingly, the following equations show the dimensionless form of the diffusion equation, initial condition and boundary conditions, respectively.

Table 1. Experimental values of diffusion coefficient for different values of temperatures and humidity

Dry Bulb Temperature (°C)	Relative Humidity (%)	$D_1 \times 10^8$ (m²/s)	$D_2 \times 10^9$ (m²/s)
27.5	51	1.20	7.16
40	27	1.82	9.76
50	28	1.87	7.82
50	31	1.91	8.20
50	16	1.87	10.70

(Bennamoun et al., 2009)

$$\frac{\partial C^*}{\partial t^*} = \frac{\partial^2 C^*}{\partial y^{*2}} \tag{13}$$

The initial condition becomes:

$$t^* = 0 : C^* = 1 \tag{14}$$

And the boundary conditions will be:

$$\left.\frac{\partial C^*}{\partial y^*}\right)_{y=0} = 0 \tag{15}$$

$$\left.\frac{\partial C^*}{\partial y^*}\right)_{y^*=1} = Sh(C^*) \tag{16}$$

Where: C* is the dimensionless moisture content and commonly can be written under the following form:

$$C^* = \frac{C - C_{cr}}{C_0 - C_{cr}} \tag{17}$$

C_{cr} is the critical moisture content, which represents the moisture content at the end of the constant drying rate phase. The term critical can be changed by equilibrium when going to the falling drying rate phase or during drying of food where there is no constant drying rate phase.

t* is the dimensionless time and can be given by (t* is also known by Fourier number):

$$t* = \frac{t \cdot D}{ep^2} \tag{18}$$

y* represents the dimensionless coordinate and commonly is written under the following form:

$$y^*=y/ep \tag{19}$$

Sh is the dimensionless Sherwood number. In our case, this number can be written:

$$Sh = 0.332\,\mathrm{Re}^{0.5}\,Sc^{0.33} \tag{20}$$

Re and Sc are Reynolds and Schmidt dimensionless numbers respectively. These numbers depend on the physical characteristics of the drying air (i.e. density, viscosity, temperature and humidity and other parameters). It is important to note that when solving diffusion equation, numerically, the variations of the characteristics of the drying air, presented in the appendix, were taken into consideration.

Discretization of the obtained equation (13) with application of the boundary and initial conditions (equations 14 to 16) lead to have a system of non-linear equations, that can be written in a matrix form, as follows (in case of the application of a constant flux):

$$\begin{bmatrix} \frac{1}{Fo}+\frac{2}{(\Delta y)^2} & -\frac{2}{(\Delta y)^2} & 0 & \cdots & 0 & 0 \\ -\frac{1}{(\Delta y)^2} & \frac{1}{Fo}+\frac{2}{(\Delta y)^2} & -\frac{1}{(\Delta y)^2} & 0 & \cdots & 0 \\ 0 & -\frac{1}{(\Delta y)^2} & \frac{1}{Fo}+\frac{2}{(\Delta y)^2} & -\frac{1}{(\Delta y)^2} & 0 & \cdots \\ \cdots & \cdots & \cdots & \cdots & \cdots & \cdots \\ \cdots & \cdots & \cdots & \cdots & \cdots & \frac{1}{Fo}+\frac{2}{(\Delta y)^2} \end{bmatrix} \begin{bmatrix} C(1) \\ C(2) \\ \vdots \\ \vdots \\ \vdots \end{bmatrix} = \begin{bmatrix} \frac{1}{Fo} \\ \frac{1}{Fo} \\ \vdots \\ \vdots \\ \vdots \end{bmatrix} \tag{21}$$

As it is shown the matrix contains several coefficients that equal to zero. Accordingly, and as cited by Gerald and Wheatly (1989), using Gauss-Seidel iterative method is more appropriate with a fast convergence of the iterations. Gauss-Seidel method can then be considered as more economical in terms of memory requirement of the used computer. This approach allowed us determination and distribution of the moisture content inside the tested sample.

Figure 5 shows the numerical solution of the equation of diffusion and accordingly distribution of the moisture content in the brick sample during drying with application of a constant flux.

It is clearly demonstrated that drying occurs with a non-homogeneous manner, as the surface, which is in contact with the operating conditions, dry more quickly than the core. In our example, it is shown that the surface represented by y*=1 needed just around 10800 seconds to reach a moisture range of 0.1-0.2 kg/kg dry basis. However, the moisture content of the core was in the range of 0.2-0.3 kg/kg dry basis. This last needed around 13000 seconds to go out of this range and around 18000 seconds to reach acceptable moisture content. This difference in moisture between the surface and core, if it is not well controlled, can lead to an inappropriate final product with an eventual appearance of the cracks making the final product not marketable.

In the next part, equation of diffusion is now used to simulate the behavior of the drying curves at the application of variable drying conditions. As the constant drying rate phase is the longest phase and the most important phase of brick drying. We will focus in our presentation on this phase

It is important to note that drying of building materials is particular brick is considered happening without alteration of the physical shape of the sample, in other terms there is no shrinkage. However, this can happen for other materials such as food. Thus, we do believe that it is important to show how equation of diffusion can be written when shrinkage is introduced. Bennamoun and Belhamri (2006) wrote equation of diffusion for shrinkable spherical material, taking the following form:

$$\frac{R^2}{D}\frac{\partial C^*}{\partial t} = \left(\frac{\partial^2 C^*}{\left(\partial r^* + r^* \frac{\partial R}{R}\right)^2} + \frac{2}{r^*} \frac{\partial C^*}{\partial r^* + r^* \frac{\partial R}{R}} \right) \tag{22}$$

We can see from this equation that radius of the sample is changing with time (∂R).
r* is considered as the dimensionless radius and equal to:

$$r^* = r/R \tag{23}$$

'R' is the total radius of the sample

At the time t=0: C*=1 (24)

At the center of the sample, the following condition was used:

$$r^* = 0 : \frac{\partial C^*}{\partial r^* + r^* \frac{\partial R}{R}} = 0 \tag{25}$$

Figure 5. Distribution of the moisture content during drying of brick sample

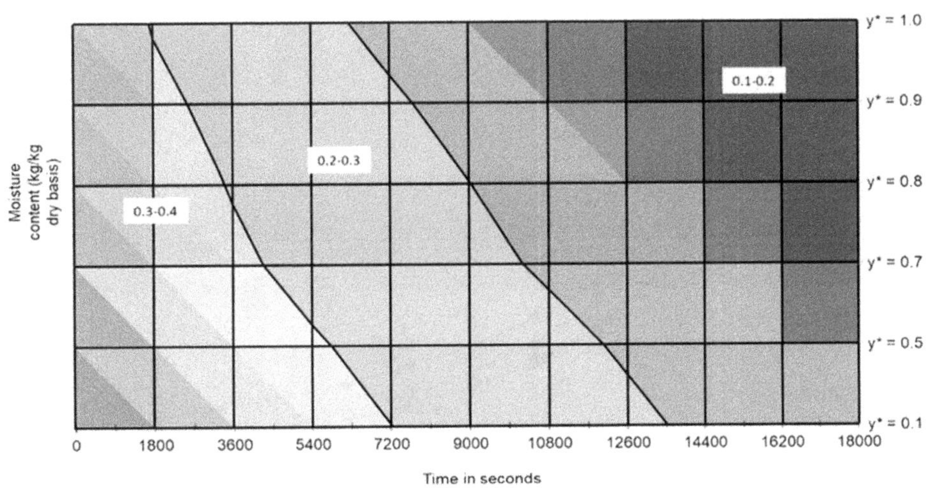

For this particular case, convective condition was applied at the surface of the sample, which bring us to the following equation:

$$r^* = 1 : -\frac{\partial C^*}{\partial r^* + r^* \frac{\partial R}{R}} = \frac{Sh}{2}\left(C^*\right) \tag{26}$$

Similarly, after discretization of the new equation of diffusion with introduction of the initial and boundary conditions, the authors (Bennamoun and Belhamri, 2006) were able to predict the drying curves with application of variable drying conditions. This study, in particular, was applied for solar drying which can be stated as the most representative example of the application of variable drying conditions.

Simulation of Variable Drying Conditions Using Equation of Diffusion

As it was mentioned previously, two different case will be simulated: application of a constant flux and application of a convective flux.

Application of a Constant Flux

Figure 6 shows the drying curves at two different temperatures of 40°C and 54°C, at the surface of the sample (y*=1). It shows that increasing the drying air temperature leads to the decrease in the drying time. Consequently, as shown in Figure 6, around 1200 seconds are needed for the surface of the sample to reach a moisture content of about 0.12 kg/kg dry basis at the application of a temperature of 54°C. This drying time increases to around 1800 seconds when applying a temperature of 40°C. The figure shows also that a sudden change and increase of temperature from 40°C to 54°C is made around 400 seconds. Accordingly, the drying curve starts changing its behavior and starts moving to the drying curve of 54°C, as shown in the figure. However, this last is reached at the time of around 1200 seconds, which means

Figure 6. Evolution of the moisture content of the surface of a brick sample at the application of different temperatures and sudden change in temperature

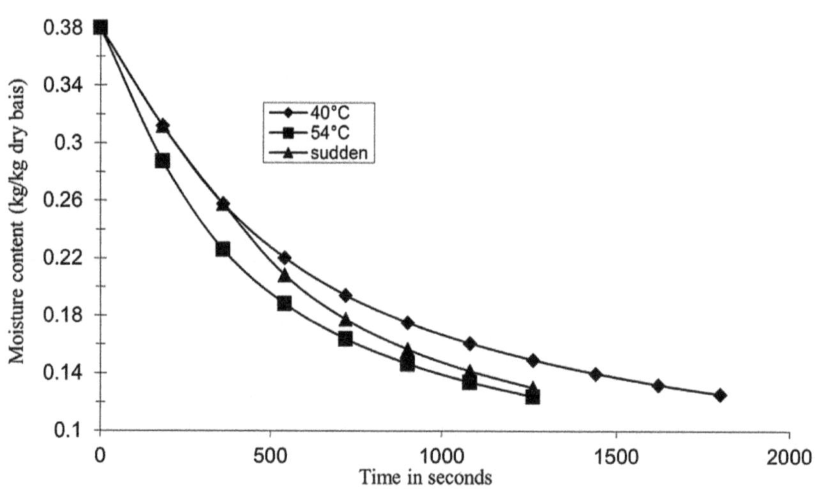

that there was a reaction time of about 800 seconds. Similar results were observed by Bennamoun et al. (2009) with the application of the same temperatures, but at the core of the sample. The reaction time in this case was higher and it took around 7200 seconds to reach the drying curve of 54°C.

A similar test is performed regarding the influence of the air velocity on the drying kinetic. It is found that the influence of the air velocity is less important than the temperature. However, when making a sudden change in the air velocity, comparable change in the drying curve, as the temperature, was observed, as shown in Figure 7. However, we do believe that better results can be obtained if we perform the calculation with a coefficient of diffusion that varies with the temperature of the air and its velocity.

Figure 8 shows another type of a graphical representation. It shows the use of the dimensionless parameters. In fact, in this figure the dimensionless moisture content described by Equation (17) is represented

Figure 7. Drying curves at fixed air velocity and at the application of a sudden change

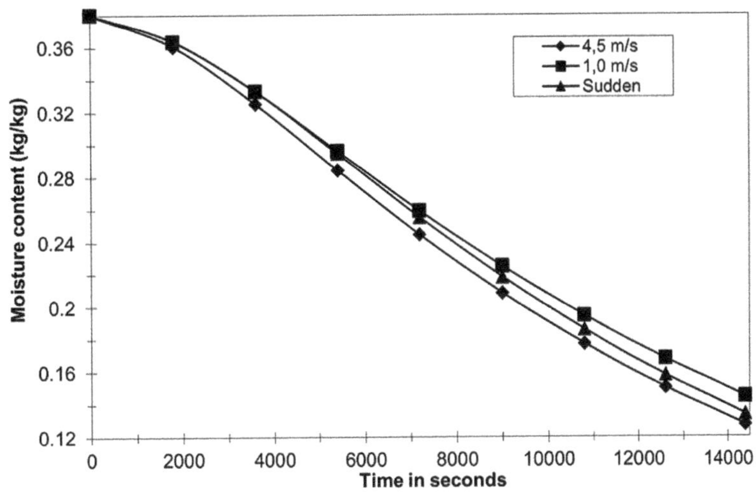

Figure 8. Drying curve using dimensionless parameters and application of progressive change in the temperature of the drying air

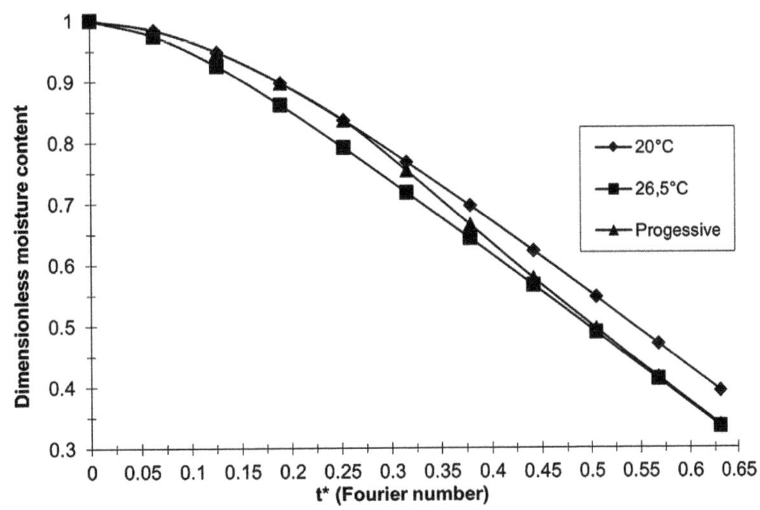

versus the dimensionless time or also called Fourier number, given by Equation (18). Figure 8 shows the drying curves of the brick sample at the application of a wet bulb temperature of 20°C and 26.5°C. Similar to what was deduced previously. The temperature of the drying air has an important influence on drying curves. Thus, increasing the temperature allows reducing drying time. Figure 8 shows also the variation of the drying curve at the application of a progressive change in the temperature by increasing the temperature by 3.5°C each 1000 second. The general behavior of the drying curve was comparable to the one at the application of a sudden change. There was a time of reaction. However, as obtained by Bennamoun et al. (2009), the reaction time to the new conditions was lower at the application of progressive changes than during sudden changes. We can report this observation that at the application of progressive changes, the material to be dried has more time to react to the new conditions

Application of a Convective Flux

Figure 9 shows the variation of the drying curves at the application of a convective flux. Comparing to constant flux, it is deduced that the effect of the drying air temperature has more influence during convective drying (as shown in Figure 9).

Figure 9 shows also the new behavior of the drying kinetic after a sudden change was made after of about 4000 seconds. So, even after 7000 seconds the curve at 54°C was not reached yet, which was not the case when a constant flux was applied (as shown in Figure 6 and Figure 8).

In this part, the same drying conditions shown in the previous Figure (i.e., Figure 9) was applied, then a sudden and a progressive change in the temperature was applied. Figure 10 shows the difference in applying these two changes. Confirming the results obtained previously, during the application of a constant flux; the drying curve with a progressive change was going faster to the new conditions than the

Figure 9. Drying curves of the brick sample at the application of convective conditions with different temperatures and with a sudden change in the temperature

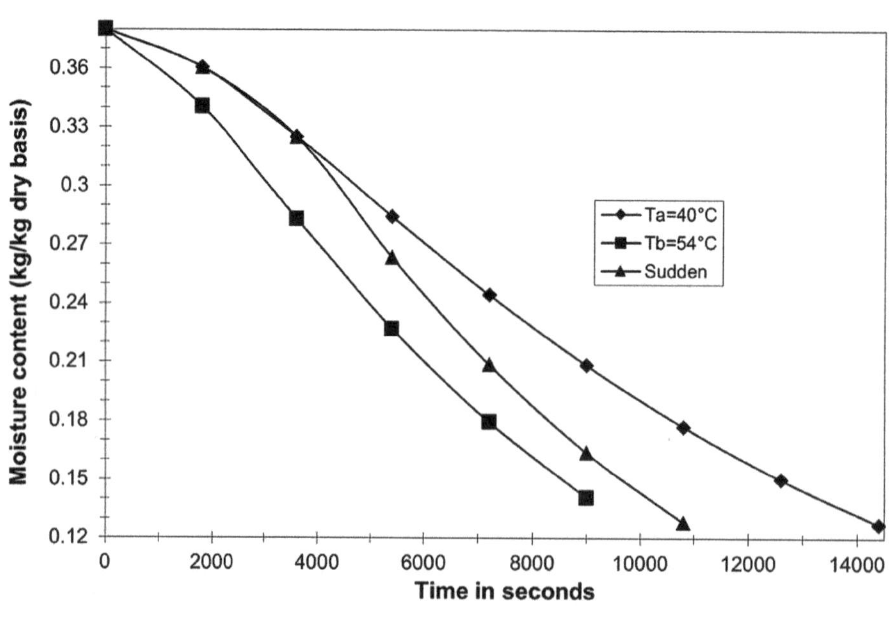

Figure 10. Variation of the drying curve at the application of a sudden change and a progressive change of the temperature

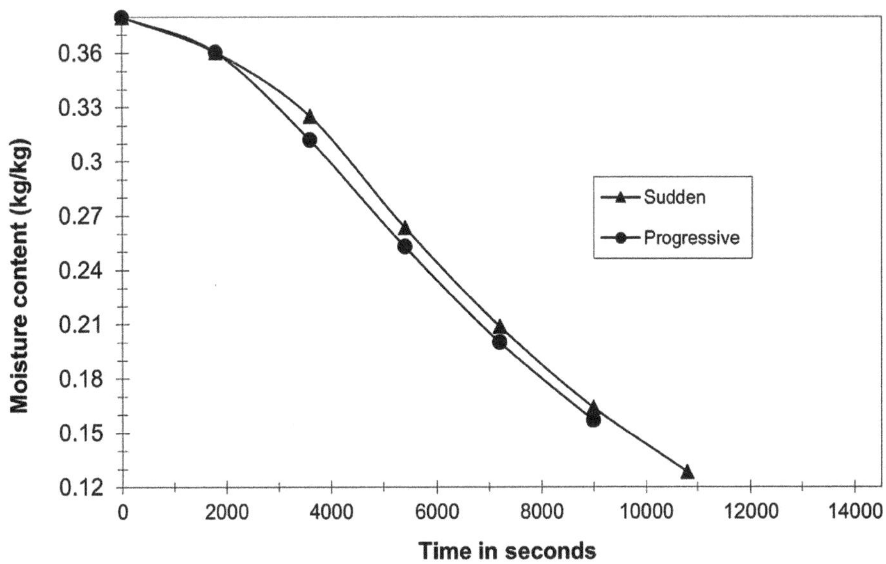

sudden change. It is important to note that during the progressive change, the temperature was increased 3.5 degrees for each 1000 second following the equation:

$$T_{ach} = \frac{14}{4000}\left(t - 2000\right) + 40\ (°C) \tag{27}$$

T_{ach} is the temperature of the heated air in °C and 't' represented in time in seconds.

The changes start at the time t=2000 seconds and consequently, the temperature of 54°C was reached at the time t=6000 seconds, which gives a reaction of about 4000 seconds.

PRESENTATION OF SOME OPTIMIZATION METHODS AND TOOLS USED IN DRYING

Kočî et al. (2012) presented an optimization method for determination of the coefficient of diffusion during drying of building materials. The computational optimization tool based on the 'GRADE' algorithm was developed at the Department of Mechanics, Faculty of Civil Engineering of the Czech Technical University, in Prague. This algorithm is using the standard genetic algorithm. Thus, 6 points of moisture diffusivity during desorption were introduced in the GRADE algorithm then the simulation tool called HEMOT, developed at the same university was used to calculate the moisture diffusivity of different building materials. Fealekari and Chayjan (2014) used Response Surface Method (RSM) to determine the parameters that should be studied for the drying of a local food. In fact, three parameters were selected, which are the temperature of the air, and its velocity and the thickness of the proposed samples. RSM gave three temperatures, three velocities and three thickness to test. The coefficient of diffusion for the test material was proposed to be a function of the temperature and the velocity of the air and thickness of the samples. Guimarães et al. (2017) used Luikov model that needs determination of the thermal and hydric diffusivity. The authors compared two simulation programs that allow determination of the diffusivities and in consequence the use of Luikov model with experimental results obtained by using gamma ray. Subsequently, the results obtained by the tool called 'TRHumidade' were compared to the WUFI Pro simulation tool. The authors pointed the importance of introducing the hydric diffusivity and comparison of the results showed that both simulation tools were in a good agreement with the experimental ones. However, 'TRHumidade' showed better results.

SOLUTIONS AND RECOMMENDATIONS

As drying is considered as an intensive process that has its influence on the price of the final product, it is important to choose the right operating conditions from the beginning. In this case, mathematical modeling and simulation of the drying process is an important step to consider. Instead of using complicated models for the prediction of the behavior of the material to be drier, the simple diffusion model can easily be used for the prediction of drying of the building materials, in particular bricks, subject of this study. In order to have accurate results using diffusion model, it is important to determine properly the coefficient of diffusion. Usually, drying is affected by the operating conditions which are the temperature of the drying air, its velocity and its humidity. Accordingly, we do recommend that coefficient of diffusion must not be just function of the temperature of the drying air but function of the temperature of the drying air, humidity of the air and velocity of the air as well. As it was proposed by different researchers such as Bennamoun and Belhamri (2006), Babalis and Belessiotis (2004) and Toğrul and Pehlivan (2003).

CONCLUSION

The operating conditions have a potential effect on the behavior of the tested material during application of drying process. This study confirms that the temperature of the drying air has more important effect on the drying kinetic than the air velocity. Generally, the change of the operating conditions, such as increasing the temperature of the drying air, will force the material to adapt itself to the new conditions, this adaptation is observed by a re-direction of the drying curve to the new conditions. However, even the change was instantly performed, a reaction time was observed. Two sorts of changes were tested: sudden and progressive changes, it was observed that the reaction time during the progressive change was lower than the sudden change and was reported that during the progressive change the material had more time to adapt itself to the new conditions.

Moreover, it is known that drying can be a complicated process, point of view mathematics. Accordingly, several non-practical models were developed. The diffusion model was proposed as a simple solution of mathematical modeling of drying process of building material (i.e. brick). However, in order to detect the changes in the drying kinetics or curves, it is important to define accurately the coefficient of diffusion and propose it as a function of the operating drying conditions, which are the temperature of the air, its velocity and its humidity.

FUTURE RESEARCH PERSPECTIVES

In addition to the fundamental knowledge and having simulation and modeling tools, it is important to have a good quality of the final product after processing it. Accordingly, it is important to perform mechanical and rheological tests on the dried samples and make the link between the operating conditions and the strength of the mechanical properties as well as the rheological behavior at the application of different conditions.

As it known, the behavior of the building materials during drying process is not the same, therefore it will be very useful to have a pre-determined database that gives the value of the coefficient of diffusion for different operating conditions and for different building materials. In this case, using diffusion equation can be really very easy and definitively practical for industrialist that perform drying of building materials.

REFERENCES

Ali Mohamed, A. (1992). *Validité des cinétiques de séchage sous des conditions d'air variables* (PhD thesis). Université de Poitiers.

Babalis, S. J., & Belessiotis, V. G. (2004). Influence of the drying conditions on the drying constants and moisture diffusivity during the thin-layer drying of figs. *Journal of Food Engineering*, *65*(3), 449–458. doi:10.1016/j.jfoodeng.2004.02.005

Belhamri, A. (1992). *Etude des transferts de chaleur et de masse à l'intérieur d'un milieu poreux au cours du séchage* (PhD thesis). Université de Poitiers.

Belhamri, A. (2003). Characterization of the first falling rate period during drying of a porous material. *Drying Technology*, *21*(7), 1235–1252. doi:10.1081/DRT-120023178

Belhamri, A., & Fohr, J. P. (1996). Heat and mass transfer along a wetted porous plate in an air stream. *AIChE Journal. American Institute of Chemical Engineers*, *42*(7), 1833–1843. doi:10.1002/aic.690420705

Bennamoun, L., Arlabosse, P., & Leonard, A. (2013a). Review on fundamental aspect of application of drying process to wastewater sludge. *Renewable & Sustainable Energy Reviews*, *28*, 29–43. doi:10.1016/j.rser.2013.07.043

Bennamoun, L., & Belhamri, A. (2003). Design and simulation of a solar dryer for agriculture products. *Journal of Food Engineering*, *59*(2-3), 259–266. doi:10.1016/S0260-8774(02)00466-1

Bennamoun, L., & Belhamri, A. (2006). Numerical simulation of drying under variable external conditions: Application to solar drying of seedless grapes. *Journal of Food Engineering*, *76*(2), 179–187. doi:10.1016/j.jfoodeng.2005.05.005

Bennamoun, L., Belhamri, A., & Ali Mohamed, A. (2009). Application of a diffusion model to predict drying changes under variable conditions: Experimental and simulation study. *Fluid Dynamics and Materials Processing*, *5*(2), 177–191.

Bennamoun, L., Chen, Z., & Afzal, M. T. (2016). Microwave drying of wastewater sludge: Experimental and modeling stydy. *Drying Technology*, *34*(2), 235–243. doi:10.1080/07373937.2015.1040885

Bennamoun, L., Crine, M., & Leonard, A. (2013b). Convective drying of wastewater sludge: Introduction of shrinkage effect in mathematical modeling. *Drying Technology*, *31*(6), 643–654. doi:10.1080/07373937.2012.752743

Chirife, J. (1983). Fundamentals of the drying mechanism during air dehydration of foods. In A.S. Mujumdar (Ed.), *Advances in drying II*. New York: Hemisphere Publication.

Crank, J. (1975). The mathematics of diffusion (2nd ed.). Oxford, UK: Clarendon.

Daguenet, M. (1985). *Les séchoirs solaires: théorie et pratique*. UNESCO.

Dong, W., Yuan, W., Zhou, X., & Wang, F. (2018). The fracture mechanism of circular/elliptical concrete rings under restrained shrinkage and drying from top and bottom surfaces. *Engineering Fracture Mechanics*, *189*, 148–163. doi:10.1016/j.engfracmech.2017.10.026

Fealekari, M., & Chayjan, R. A. (2014). Optimization of convective drying process for Persian shallot using response surface method (RSM). *CIGR Journal*, *16*(2), 157–166.

Fohr, J. P., Arnaud, G., Ali Mohamed, A., & Ben Moussa, H. (1990). Validity of drying kinetics. In A. S. Mujumdar & M. Roques (Eds.), *Drying 89* (pp. 269–275). New York: Hemisphere Publication.

Gerald, C. F., & Wheatly, P. O. (1989). *Applied numerical Analysis* (4th ed.). Addison Wisley.

Hammouda, I., & Mihoubi, D. (2017). Influence of stationary and non-stationary conditions on drying time and mechanical properties of a porcelain slab. *Heat and Mass Transfer*, *53*(12), 3571–3580. doi:10.100700231-017-2084-6

Jamayaha, S. E. G., Chou, S. K., & Wijeysundera, N. E. (1996). Drying of porous materials in a presence of solar radiation. *Drying Technology*, *14*(10), 2339–2369. doi:10.1080/07373939608917209

Khama, R., Aissani, F., & Alkama, R. (2016). Design and performance testing of an industrial-scale indirect solar dryer. *Journal of Engineering Science and Technology*, *11*(9), 1263–1281.

Kočí, J., Maděra, J., Jerman, M., & Černý, R. (2012). Optimization methods for determination of moisture diffusivity of building materials in the drying phase. *WIT Transactions on Ecology and The Environment*, *165*, 323–333. doi:10.2495/ARC120291

Korkida, M. K., Maroulis, Z. B., & Marinos-Kouris, D. (2002). Heat and mass transfer coefficients in drying: Compilation of Literature Data. *Drying Technology*, *20*(1), 1–18. doi:10.1081/DRT-120001363

Lampinen, M. J., & Ojala, K. T. (1993). Mathematical modeling of web drying. In R. A. Mashelkar (Ed.), *Advances transport process IX* (pp. 271–347). Mujumdar: Elsevier. doi:10.1016/B978-0-444-89737-4.50011-9

Liu, X. D., Wang, X. Z., Pan, Y. K., Cao, C. W., & Liu, D. Y. (2002). R & D of drying technology in China. In A. S. Mujumdar (Ed.), Drying 2002 (pp. 49–63). Academic Press.

Maake, W., Eckert, H. J., & Cauchepin, J. L. (1993). *Manuel technique du froid: bases-composant-calcul*. PYC.

Moyne C., & Roques M. (1986). Réalités et perspectives du séchage. *Revue Générale du Thermique*, 292.

Pel, L., Landman, K. A., & Kaasschieter, E. F. (2002). Analytic solution for the non-linear drying problem. *International Journal of Heat and Mass Transfer*, *45*(15), 3173–3180. doi:10.1016/S0017-9310(02)00025-X

Rabha, D. K., Muthukumar, P., & Somayaji, C. (2017). Experimental investigation of thin layer kinetics of ghost chilli pepper (Capsicum Chinense Jacq.) dried in a forced convection solar tunnel dryer. *Renewable Energy*, *105*, 583–589. doi:10.1016/j.renene.2016.12.091

Toğrul, I. T., & Pehlivan, D. (2003). Modelling of drying kinetics of single apricot. *Journal of Food Engineering*, *58*(1), 23–32. doi:10.1016/S0260-8774(02)00329-1

van Brakel, J. (1980). Mass transfer in convective drying. In Advances in drying I (pp. 217-265). New York: Hemisphere Publication.

Zhamalov, A. Z. (1989). Use of solar energy for drying construction materials. *Applied Solar Energy*, *25*(4), 68–70.

ADDITIONAL READING

Bennamoun, L., & Li, J. (2017). Drying of sewage sludge: Fundamental aspects, modeling and challenges. In E. Danso-Boateng (Ed.), *Sewage sludge. Assessment, treatment and environmental impact* (pp. 99–140). New York: Nova Science Publishers.

Boukadida, N., & Ben Nasrallah, S. (2002). Effect of the variability of heat and mass transfer coefficients on convective and convective radiative drying of porous media. *Drying Technology*, *20*(1), 67–91. doi:10.1081/DRT-120001367

Guimarães, A. S., Ribeiro, I. M., & de Freitas, V. P. (2017). Numerical models performance to predict drying liquid water in porous building materials: Comparison of experimental and simulated drying water content profiles. *Cogent Engineering*, *4*(1), 1365572. doi:10.1080/23311916.2017.1365572

Kaviany, M., & Mittal, M. (1987). Funicular state in drying of a porous slab. *International Journal of Heat and Mass Transfer*, *30*(7), 1407–1418. doi:10.1016/0017-9310(87)90172-4

Kowalski, S., & Mierzwa, D. (2013). Numerical analysis of drying kinetics for shrinkable products such as fruits and vegetables. *Journal of Food Engineering*, *114*(4), 522–529. doi:10.1016/j.jfoodeng.2012.08.037

Kowalski, S. J., Musielka, G., & Banaszak, J. (2010). Heat and mass transfer during microwave-convective drying. *AIChE Journal. American Institute of Chemical Engineers*, *56*(1), 24–35.

Ramos, I. N., Brandão, T. R. S., & Silva, C. L. M. (2015). Simulation of solar drying of grapes using an integrated heat and mass transfer model. *Renewable Energy*, *81*, 896–902. doi:10.1016/j.renene.2015.04.011

Suzuki, M., & Maeda, S. (1968). On the mechanism of drying of granular beds. *Journal of Chemical Engineering of Japan*, *1*(1), 26–31. doi:10.1252/jcej.1.26

Whitaker, S. (1980). Heat and Mass Transfer in Granular Porous Media. In A. S. Mujumdar (Ed.), *Advances in Drying* (Vol. 1, p. 146).

Zagrouba, F., Mihoubi, D., & Bellagi, A. (2002). Drying of clay. II Rheological modelisation and simulation of physical phenomena. *Drying Technology*, *20*(10), 1895–1917. doi:10.1081/DRT-120015575

KEY TERMS AND DEFINITIONS

Convective Drying: Is the application of drying process using convective transfer. Commonly, during convective drying, an ambient air is heated. This air will flow around the wet material. This contact between the heated air and the material conducts to an exchange of heat and mass between the two medias.

Diffusion Coefficient: Diffusion happens when there are two mediums with different concentrations. Diffusion coefficient is then a parameter to measure the molecular movement from one medium to another, usually from high concentration to low concentration. The common used unit in drying for this coefficient is m^2/s.

Diffusion Model: Is one of the simplest mathematical models that can well represent drying. This model is widely used during drying of foodstuff. It is represented by a partial differential equation of the second degree. In the equation the variables are time and space.

Drying: Is the process of removing moisture from a material by the application of a source of heat. This material can be exposed directly or indirectly to the source of heat.

Drying Kinetic: Is a graphical representation of the evolution of the moisture content (generally a decrease of the moisture content) inside the material. It is important to make the difference between the drying curve and the drying kinetic. The drying curve is usually represented by the variation of the moisture content versus drying time. However, the drying kinetic is the representation of the drying rate versus the drying time or the moisture content.

Drying Rate: Represents the speed that drying is taking. It can be calculated by knowing the moisture content in two different times. In the particular case the variation of the moisture is linear with time, the drying rate will be the slope of the line and the drying rate will be constant. This is known in drying by the constant drying rate phase.

Falling Drying Rate Period: Usually, drying passes by three phases. It starts by a short adaptation phase that most of the researchers ignore. The second phase is the constant drying rate phase, during this phase the drying rate is constant, then comes the third phase which is the falling drying rate phase. In this phase the surface of the material is not completely full of water, subsequently the drying rate starts decreasing.

Krischer's Curve: This graphical representation is the most adequate in drying field. It consists on representing the drying rate versus the moisture content. This representation allows determination of the different phases that a material can pass through during the application of the drying process.

Solar Drying: Is an application of a convective drying. During this process the ambient air is heated using solar radiation by the mean of a solar collector. In fact, the air passes through an air collector. This passage will increase the temperature of the air. This last will be directed to the material and flow around the material, which lead to the evaporation of the moisture contained in the material.

APPENDIX

It is important to mention that in this study all the parameters and dimensionless numbers were calculated based on the humid air. Accordingly, we are presenting in this appendix, the equations used for the humid air. Most of these equations were extracted from the reference Daguenet (1985).

$$Dv = 2.26\times 10^{-5}\frac{1}{p}\left(\frac{T_{ach}}{273}\right)^{1.81} \tag{28}$$

$$\rho_{as} = \frac{\rho_{ah}}{1+W} \tag{29}$$

$$W = 0.622\frac{\phi P_{vsat}}{P_{ah} - \phi P_{vsat}} \tag{30}$$

P and P_{ah} are commonly put equal to the atmospheric pressure.

$$\rho_{vap} = \rho_{ah} - \rho_{as} \tag{31}$$

$$P_{vsat} = 10^{17.433} - \frac{2795}{T_{ach}} - 3.868\log\left(T_{ach}\right) \tag{32}$$

Lampinen and Ojala (1993) wrote the viscosity of a humid air as follows:

$$\mu_{ah} = \frac{\mu_{as}\rho_{as} + \mu_{vap}\rho_{vap}}{\rho_{as} + \rho_{vap}} \tag{33}$$

With:

$$\mu_{as} = \frac{1.448\sqrt{T_{ach}}}{1+\dfrac{110.4}{T_{ach}}}10^{-6} \tag{34}$$

$$\mu_{vap} = \left(0.0361T_{ach} - 1.02\right)10^{-6} \tag{35}$$

Maake, Eckert and Chauchepin (1993) gave the density of the humid air:

$$\rho_{ah} = \frac{348.3}{T_{ach}} p_{ah} - \phi p_{vsat} \frac{131.6}{T_{ach}} \tag{36}$$

The pressure in this equation is in atmosphere.

Nomenclature

C: moisture content (kg/kg)
*C**: Dimensionless moisture
D: coefficient of diffusion (m^2/s)
Dv: diffusion of vapour in the air (m^2/s)
ep: overall product thickness (m)
F: flux ($kg/s.m^2$)
h: mass transfer coefficient (m/s)
P: pressure (Pa)
T: temperature (°C or K)
t: time (s)
*t**: dimensionless time
W: absolute humidity (kg/kg)
y: product thickness (m)
*y**: dimensionless thickness

Greek Symbols

μ: viscosity (kg/m.s)
ϕ: relative humidity
ρ: density (kg/m^3)

Subscripts

ach: heated air
ah: wet air
as: dry air
cr: critical
eq: equilibrium
s: product
vap: vapour
vsat: saturated vapour
wb: wet bulb

Dimensionless Number

Bi: Biot number (represented in this chapter by L)
Fo: Fourier number
Re: Reynolds number
Sc: Schmidt number
Sh: Sherwood number

Chapter 2
Improvement of Continuity Between Industrial Software and Research One by Object-Oriented Finite Element Formulation of Shell and Plate Element

Sabah Moussaoui
Setif 1 University, Algeria

Mourad Belgasmia
Setif 1 University, Algeria

ABSTRACT

This chapter shows, through the example of the addition of a plate and shell element to freeware FEM-object, an object-oriented (C++) finite element program, how object-oriented approaches, as opposed to procedural approaches, make finite element codes more compact, more modular, and versatile but mainly more easily expandable, in order to improve the continuity and the compatibility between software of research and industrial software. The fundamental traits of object-oriented programming are first briefly reviewed, and it is shown how such an approach simplifies the coding process. Then, the isotropic shell and orthotropic plate formulations used are given and the discretized equations developed. Finally, the necessary additions to the FEM-object code are reviewed. Numerical examples using the newly created plate membrane plate element are shown.

DOI: 10.4018/978-1-5225-7059-2.ch002

INTRODUCTION

The authors can divide all the software of modelization into 4 large families

- **General Search Software:** Which serves as a basis for the testing of innovative models or methods, must be flexible and capable of undergoing substantial changes. They should be conceived as constructs of "modules" independent and easily modifiable.
- **Specialized Research Software:** Models a particular type of phenomena and uses very specific methods. Like general cousins, specialized search software has a rough user interface and the data that the researcher manipulates is of very low level (matrix and vector in common); the commands are Fortran statements.
- **General Industrial Software:** Encompasses a maximum number of models and algorithms. They are equipped with a comfortable user interface and can communicate with many mesh and post-processors. The user here manipulates high-level data (physical structures, geometric data) and the commands are launched to the mouse. Because of their size these software are not very scalable.
- **Specialized industrial Software, or "Business" Software:** Are often ex-specialized research software that have been made more reliable, cleaned and documented. Laboratories strive to transform their novelties into a reliable product in order to gain fame. Similarly, industrialists naturally seek to quickly transfer the results of research to their code. One should therefore improve the continuity and the compatibility between software of research and industrial software that is to say to seek to professionalize the research software too often "artisanal".

In the last decades, the finite element method has gained a large acceptance as a general tool for modeling and simulating physical systems. It has become, in fact, the most widely used technique applicable in fields such as solid mechanics, fluid mechanics, electromagnetics etc.... In recent years, easy maintenance, extensibility and flexibility of the finite element code, encouraged many researchers to change from procedural to object oriented programming in order to make the finite element software more flexible.

The Object-Oriented programming approach appeared in the seventies and found its first convivial concretization with the SmallTalk language at the beginning of the eighties. Applications in the finite element field appeared around 1990. (Dubois, Zimmermann & Bomme, 1992 and Dubois & Zimmermann, 1993), and since then its use has spread. Object-Oriented approaches yield more modular programs than the standard procedural approach.

In this Chapter, an Object-Oriented approach of the finite element method for the study of bending of isotropic and orthotropic plates in dynamic and shell element in statics is presented.

The authors, shown how Object-Oriented programming features, through the concept of Object, Class, Inheritance and Polymorphism, make it possible to overcome those limits, and deals with plate and shell theory. Displacement and strain approximations are described, the equations of motion developed, the basics behind Mindlin-Reissner plate and shell theory are recalled, and stiffness matrices for this formulation are developed. A section is concerned with the addition of the aforementioned plate and shell element into the FEMObject code described elsewhere, (Commend & Zimmermann, 2001). Object-Oriented programming is shown to easily accomodate the addition of a plate and shell classes, thus demonstrating code extendibility. Finally, the authors present numerical results for isotropic and orthotropic thin plates and shells with conclusion at the end.

MODELING LANGUAGE

It is necessary to use a higher level of modeling language, to abstain from the computer details, and the ideal would be to code the formulations directly in mathematical language. It is necessary to define mathematical objects (vectors, matrices, tensors ...) as well as the operators that can be applied to them. We note that the c ++ language can be very useful in this field thanks to the overload of the operators.

CODING LANGUAGE

It must already help to implement the wishes formulated for the architecture and for the modeling language. It therefore requires a more powerful language with a built-in memory manager, free index pointer, a set of richer data structure types. It must be very modular and very readable. It must make the code reliable: to carry out the maximum checks at compilation and execution and finally it must be portable and fast.

We have put forward some guidelines that seem essential to the writing of modern software, and object-oriented programming seems to be a promising path. In the next subtitle, the authors will provide a reminder of object-oriented programming.

OBJECT ORIENTED PROGRAMMING

Procedural programming tends to show its limits especially for large finite-element programs, which contain hundreds of thousands of lines of code.

Such procedural programs are generally written in FORTRAN. This code contains a significant number of data structures which are accessible throughout the program, creating a loss of flexibility. It thus becomes difficult to modify the existing code and extend it to new uses. Loss of flexibility shows up in different manners:

- Detailed knowledge of the program is required to work on a small part of the code ;
- The re-use of the code is difficult ;
- A small change in the data structures can influence the whole system ;
- Many interdependences between the components of the design are dissimulated and difficult to establish ;
- The integrity of the data structures is not protected. (Zimmermann, Dubois & Bomme,1992).

RECALL ON OBJECT-ORIENTED PROGRAMMING

The main concept of object-oriented programming is the abstraction, which leads to the faithful representation of the entities of the real physical world through the concept of object. While Object-Oriented programming is intensively employed for graphical applications, where the concept of “object” appears naturally, its use for scientific computation is still limited and industrial software is still generally written in FORTRAN, using procedural programming. Object-Oriented programming is based on the concepts of Object, Class, Inheritance, Polymorphism and Encapsulation. (Delannoy, 2002), (Jamsa & Klander, 1999).The difference between an Object-Oriented and a procedural implementation is shown in Figure 1.

An object is characterized by its identity, state and behavior. The identity distinguishes an object from another. The state of an object is characterized by the current values of its attributes. The behavior shows the action of the object under its own control and how it behaves with the external requests (see Figure 2).

The behavior of the object is implemented through a set of operations working on its own attributes that are called methods. Methods and the data are encapsulated in the same enclosure (object), and they interact with the external world through an interface see Figure 3.

Figure 1. Object-Oriented structure

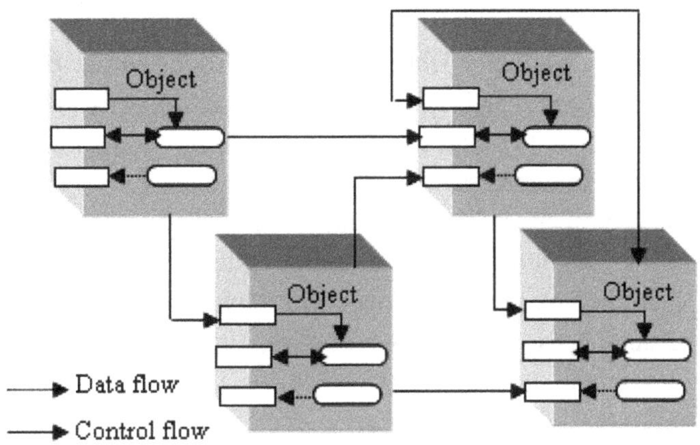

Figure 2. Behaviour and state of the object

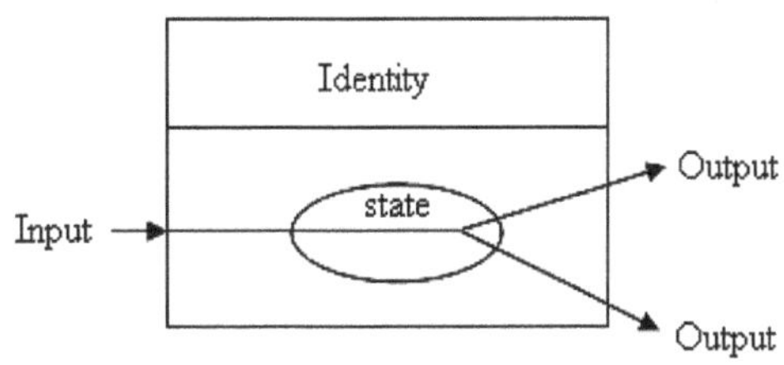

Figure 3. Data encapsulation

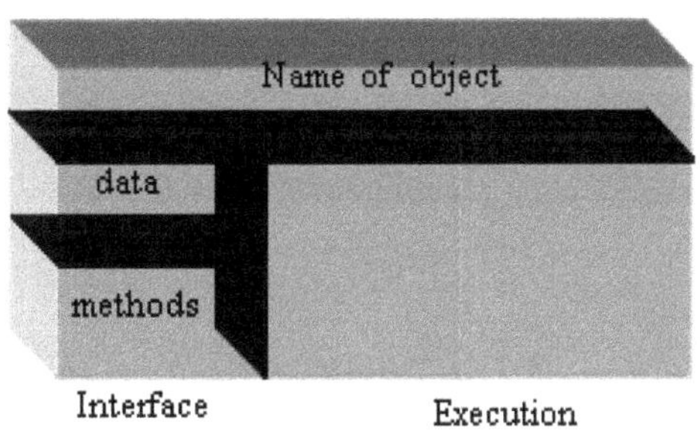

Communication by means of messages requires the simplification of:

- The destination object to which the request is addressed.
- Indication of the method to be carried out.

Key advantages related to the use of objects are:

- Ease of comprehension and re-use of the code.
- Improvement in reliability, through encapsulation.
- Increased abstraction and masking of information.

A class is an abstract data structure corresponding to objects sharing the same properties, attributes and methods. The class specifies its own attributes and the methods able to operate on these attributes. From classes, we instantiate objects which we activate by means of messages. The message activates the action, but the detail of the operation is left to the object. This « data hiding » is often referred to as encapsulation of the data, behavior, and state, (see Figure 4).

An abstract data type can be specified by the creation of a sub-type or of a subclass which inherits the structures of data and the methods from its ancestor often called superclass. The mechanism of inheritance can be total as in the Smalltalk language, for which all attributes and methods are inherited by the subclass, but some languages, such as C++ or Java, allow a mechanism of inheritance through the selection of inheritable variables and methods.

POLYMORPHISM

means the ability of objects of different classes to answer the same message in different manners, in other words, the corresponding method can have a different implementation in different classes. As Figure 5 indicates, the message « computeStiffnessMatrix » which can be addressed to different types of elements and the method will be different from one type of element to another.

Figure 4. Class concept

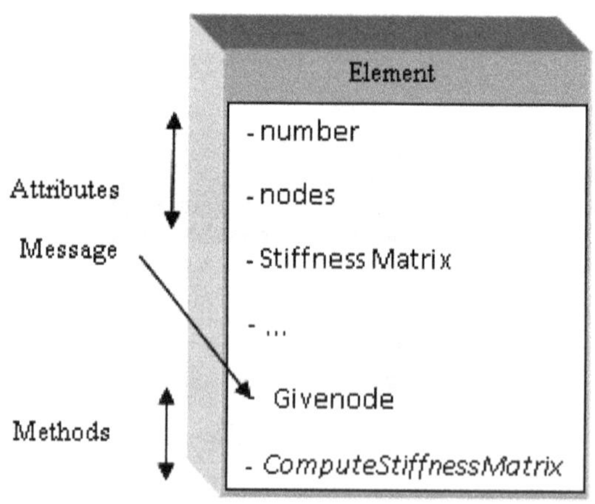

Figure 5. Polymorphism

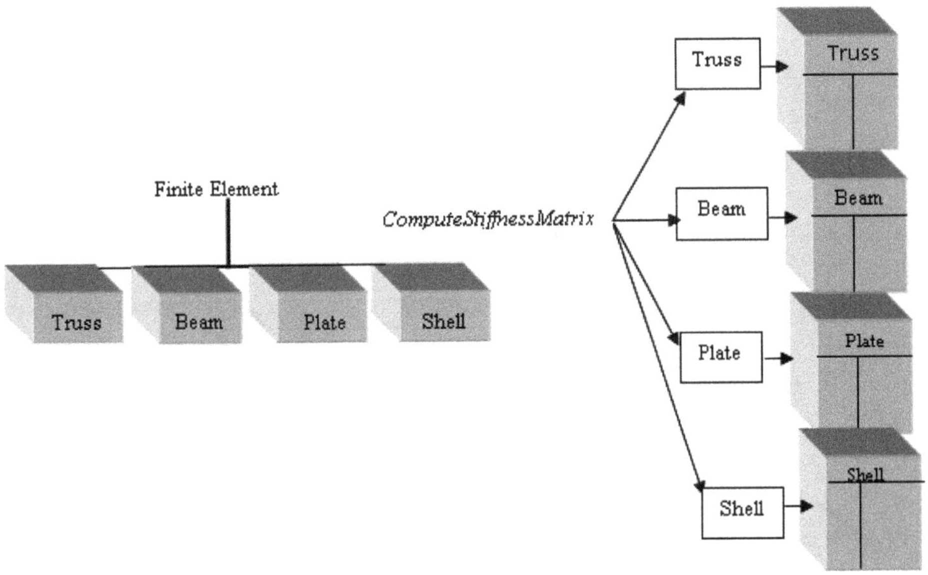

RECALL OF REISSNER-MINDLIN PLATE THEORY

The orthotropic plate formulation for finite elements in bending-shear is based on the theory of Reissner-Mindlin. Their conformity requires only C^0 continuity of w, β_x *and* β_y. Where w, β_x *and* β_y are transverse displacement, rotation of the normal on the middle surface in (x-z) and (y-z) plans respectively (Robert, David, Malkus & Plesha,1988), (see Figure 7).

Figure 6. Compute stiffness matrix method as an added class for shell element

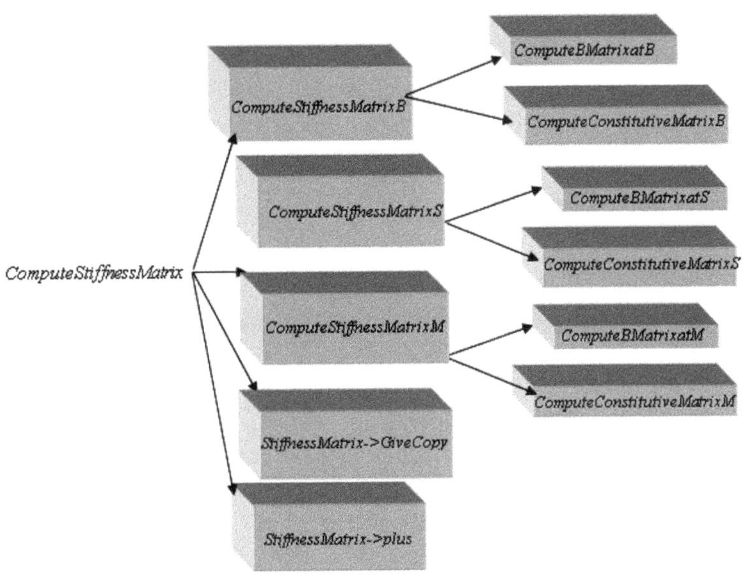

Figure 7. Sign conventions for displacements and rotations

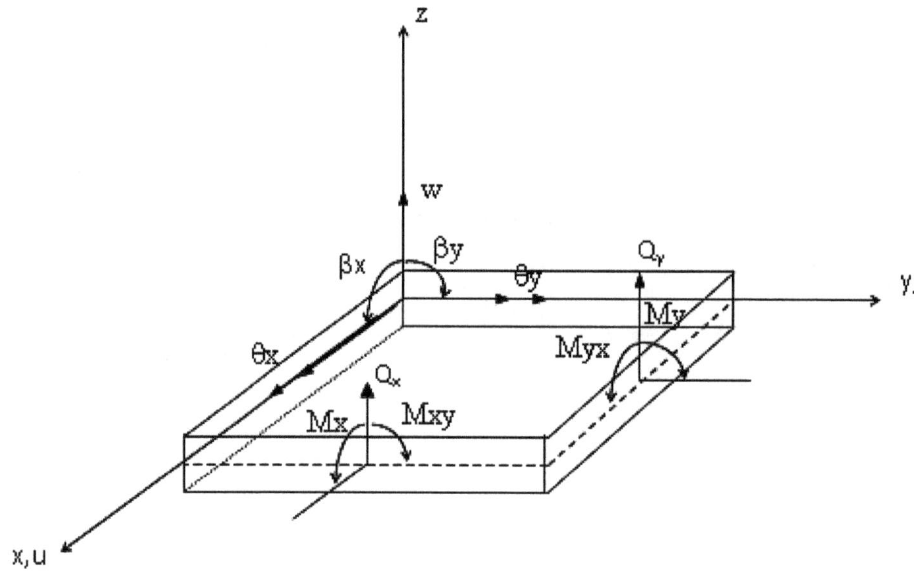

We consider isoparametric quadrilateral elements. (the approximations of w, β_x *and* β_y) (Batoz & Dhatt, 1990 and Dhatt & Touzot 1984).

$$w = N^T(\xi,\eta)W \ \beta_x = N^T(\xi,\eta)\widehat{\beta}_x \ \beta_y = N^T(\xi,\eta)\widehat{\beta}_y \tag{1}$$

use ^{T} for transposition.

The strain vector can be expressed:

$$\{\varepsilon\} = \left\{\{\varepsilon_f\}^t, \{\varepsilon_c\}^t\right\}^t = \left\{Z\{\chi\}^t, \{\gamma\}^t\right\}^t \tag{2}$$

where $\{\chi\}$ is curvature vector

1. The Contribution of bending effect is

$$\{\varepsilon_f\} = \begin{Bmatrix} \varepsilon_{xx} \\ \varepsilon_{yy} \\ \gamma_{xy} \end{Bmatrix} = z\{\chi\} = z\begin{Bmatrix} \dfrac{\partial \beta_x}{\partial x} \\ \dfrac{\partial \beta_y}{\partial y} \\ \dfrac{\partial \beta_x}{\partial y} + \dfrac{\partial \beta_y}{\partial x} \end{Bmatrix},$$

2. The Contribution of shear effect is

$$\{\varepsilon_c\} = \{\gamma\} = \begin{Bmatrix} \gamma_{xz} \\ \gamma_{yz} \end{Bmatrix} = \begin{Bmatrix} \beta_x + \dfrac{\partial w}{\partial x} \\ \beta_y + \dfrac{\partial w}{\partial y} \end{Bmatrix}. \tag{3}$$

The interpolation functions used are the usual interpolation functions of isoparametric quadrilaterals. In the case of the linear quadrilateral, one gets:

$$N^T = \begin{bmatrix} N_1 & N_2 & N_3 & N_4 \end{bmatrix} \text{with } N_i(\xi,\eta) = \frac{1}{4}(1+\xi\xi_i)(1+\eta\eta_i) \tag{4}$$

where ξ_i *or* η_i are taking values (+1) or (-1) according to the consideration of node .

$$W = \begin{Bmatrix} w_1 \\ w_2 \\ w_3 \\ w_4 \end{Bmatrix} \widehat{\beta}_x = \begin{Bmatrix} \beta_{x1} \\ \beta_{x2} \\ \beta_{x3} \\ \beta_{x4} \end{Bmatrix} \widehat{\beta}_y = \begin{Bmatrix} \beta_{y1} \\ \beta_{y2} \\ \beta_{y3} \\ \beta_{y4} \end{Bmatrix} \tag{5}$$

By substitution of (4), in the relations of deformations (3), we obtain the bending and shearing deformations matrices of interpolation as follows:

$$\{\chi\} = \{\overline{\varepsilon}_f\} = \begin{bmatrix} 0 & \dfrac{\partial N^T}{\partial x} & 0 \\ 0 & 0 & \dfrac{\partial N^T}{\partial y} \\ 0 & \dfrac{\partial N^T}{\partial y} & \dfrac{\partial N^T}{\partial x} \end{bmatrix} \begin{Bmatrix} W \\ \widehat{\beta}_x \\ \widehat{\beta}_y \end{Bmatrix} \tag{6}$$

$$\{\varepsilon_c\} = \{\gamma\} = \begin{bmatrix} \dfrac{\partial N^T}{\partial x} & N^T & 0 \\ \dfrac{\partial N^T}{\partial y} & 0 & N^T \end{bmatrix} \begin{Bmatrix} W \\ \hat{\beta}_x \\ \hat{\beta}_y \end{Bmatrix} \tag{7}$$

with

$$\{\chi\} = \{\overline{\varepsilon}_f\} = [\overline{\beta}_f]\{q\} \quad \{\gamma\} = [\beta_\gamma]\{q\} \tag{8}$$

The expression of the deformation energy makes it possible to calculate the stiffness matrix [3,4], we have: $U = U_F + U_C$

$$U = \frac{1}{2}\int_{S^e} \{\chi\}^T [C_b]\{\chi\} dxdy + \frac{1}{2}\int_{S^e} \{\gamma\}^T [C_s]\{\gamma\} dxdy = \frac{1}{2}\{q\}^T [K]\{q\} \tag{9}$$

$$\text{where } [C_b] = \frac{E}{(1-\nu^2)}\begin{bmatrix} 1 & \nu & 0 \\ \nu & 1 & 0 \\ 0 & 0 & \frac{1-\nu}{2} \end{bmatrix} [C_s] = \frac{E}{2(1+\nu)}\begin{bmatrix} \gamma & 0 \\ 0 & \gamma \end{bmatrix} \tag{10}$$

$[C]$ is the constitutive matrix, E_1, E_2 are the young moduli in two directions (x) (y), ν_{12}, ν_{21} are the Poisson's ratios, G_{12}, G_{13}, G_{23} are the shear moduli, k coefficient of correction of transverse shear.

Minimization of expression (9), leads to

$$[K] = [K_b] + [K_s] = \int_S [\bar{\beta}_b]^T [C_b][\bar{\beta}_b] dxdy + \int_S [\beta_\gamma]^T [C_s][\beta_\gamma] dxdy \tag{11}$$

$$[K] = \int_{-1}^{+1}\int_{-1}^{+1} [\bar{\beta}_b]^T [C_b][\bar{\beta}_b] \det[J] d\xi d\eta + \int_{-1}^{+1}\int_{-1}^{+1} [\beta_\gamma]^T [C_s][\beta_\gamma] \det[J] d\xi d\eta \tag{12}$$

where $[J]$ is: Jacobian matrix from the physical to the parent coordinate system. The stiffness matrix $[K]$ is evaluated numerically by selective numerical integration; ($[K_d]$) is integrated with (2x2) Gauss point formula $[K_s]$ is obtained by reduced integration).

The equivalent load vector is written as follows:

$$\{F\}^e = \int_{V^e} [N]^T \{f_V\} dV + \int_{S^e} [N]^T \{f_S\} dS \tag{13}$$

Lagrange equations allow the obtention of discrete system motion equations, using the expressions of kinetics, potential and dissipation energies, (Imbert, 1991). The Lagrangian defined by:

$$L = T - V \tag{14}$$

The kinetic and total potential energy can be written as below:

$$T = \frac{1}{2}\int_V \rho \dot{u}_i \dot{u}_i dV \tag{15}$$

and

$$V = U - W = \frac{1}{2}\int_V \sigma_{ij}\varepsilon_{ij} dV - \int_V f_i^V u_i dV - \int_S f_i^S u_i dS \tag{16}$$

U is the deformation energy and W the potential of the conservative surface and volume forces.

f_i^V volume forces and f_i^S surface forces

The nodal approximation for the displacement $\{u(t)\}$ of an arbitrary point in an element can be written as follows:

$$\{u(x,y,z,t)\}^e = [N(x,y,z)]^e \{q(t)\}^e \tag{17}$$

Similarly the expression of velocity components is:

$$\{\dot{u}(x,y,z,t)\}^e = [N(x,y,z)]^e \{\dot{q}(t)\}^e \tag{18}$$

By substitution of (18), in the relations of kinetic energy (15), we obtain

$$T = \frac{1}{2}\{\dot{q}\}^{eT} \int_{V^e} \rho [N]^T [N] dV \{\dot{q}\}^e \tag{19}$$

with $[M]^e = \int_{V^e} \rho [N]^T [N] dV$ (20)

The Lagrangian equation can be expressed using displacements at the nodes q_i and of their derivative:

$$L = T(\dot{q}) - V(q) \tag{21}$$

For a structure without damping the Euler-Lagrange equation is:

$$\frac{\partial}{\partial t}\left(\frac{\partial T}{\partial \dot{q}_i}\right) - \frac{\partial T}{\partial q_i} + \frac{\partial U}{\partial q_i} = F_i(t) \tag{22}$$

where $F_i(t)$ are the external forces. For small deformation of elastic systems, the kinetic and deformations energies are expressed as:

$$T = \frac{1}{2}\sum_{i=1}^{n}\sum_{j=1}^{n}\dot{q}_i M_{ij}\dot{q}_j = \frac{1}{2}\dot{q}^T M\dot{q}$$

$$U = \frac{1}{2}\sum_{i=1}^{n}\sum_{j=1}^{n}q_i K_{ij}q_j = \frac{1}{2}q^T Kq$$

The equations of Lagrange become:

$$\left[M\right]\left\{\ddot{q}\right\}+\left[K\right]\left\{q\right\}=\left\{F\left(t\right)\right\} \tag{23}$$

where $\left[M\right]$ is global mass matrix, $\left[K\right]$ is global stiffness mass matrix and $\left\{F(t)\right\}$ is total external vector.

RECALL OF SHELL THEORY

A shell is a solid whose volume V is generated by the average surface A, and its normal n extending by a distance t /2 on each side of the this latter. The thickness of the shell is small compared to the geometric quantities which characterize it.

Shell element (plate-membrane element), creating a faceted shell inscribed in the exact mean surface.

In most hull elements, the sixth degree of freedom, often called the rotation around the normal, is not supplied with stiffness, each node has only five degrees of local freedom, for example u, v, w, β_x and β_y see figure 7 ($u = u_0$ and $v = v_0$).

But the spatial assembly of the elements feeds the six degrees of rigidity Freedom of each node; Thus there are six unknowns in each node of a faceted shell (will be detailed later).

GENERAL CINEMATIC

Cartesian Description

Displacement Field

The behavior of the elemental portion of a shell, of Figure (8), is the superposition of a behavior in membrane and plate in flexion.

In a local orthonormal axes (s, t, n) the displacements are written:

Figure 8. Degree of freedom of Shell Element

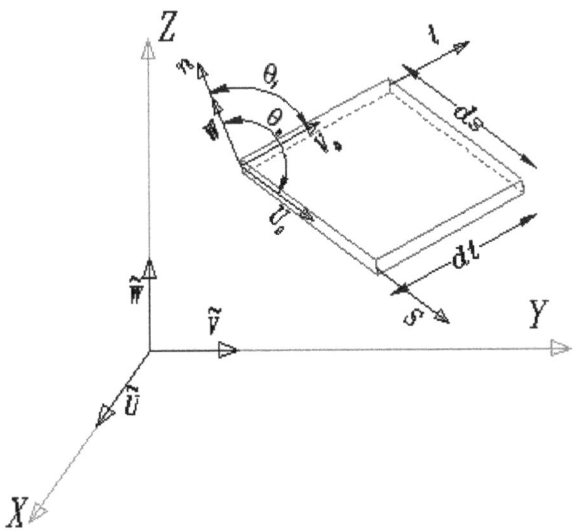

$$U(s,t,n) = U_0(s,t) + n\theta_s(s,t)$$
$$V(s,n,t) = V_0(s,t) + n\theta_t(s,t)$$
$$w(s,t,n) = w(s,t)$$

The displacement vector is defined by the following five degrees of freedom:

θ_s and θ_t Are the rotations in the planes n, s and n, respectively
U_0 or U and V_0 or V Are the displacements of the membrane in the plane s, t
W is the displacement transverse to the plane s, t

Deformation Field

The deformation tensor is written as below:

$$\begin{Bmatrix} \varepsilon_s \\ \varepsilon_t \\ \gamma_{st} \\ \gamma_{sn} \\ \gamma_{tn} \end{Bmatrix} = \begin{Bmatrix} \dfrac{\partial U}{\partial s} \\ \dfrac{\partial V}{\partial t} \\ \dfrac{\partial U}{\partial t} + \dfrac{\partial V}{\partial s} \\ \dfrac{\partial U}{\partial n} + \dfrac{\partial W}{\partial s} \\ \dfrac{\partial V}{\partial n} + \dfrac{\partial W}{\partial t} \end{Bmatrix}$$

It can be decomposed into

$$\{\varepsilon\} = \begin{Bmatrix} \{\varepsilon_m\} \\ \{0\} \end{Bmatrix} + \begin{Bmatrix} n\{\bar{\varepsilon}_b\} \\ \{\varepsilon_s\} \end{Bmatrix}$$

With

$$\{\varepsilon_m\} = \begin{Bmatrix} \dfrac{\partial U}{\partial s} \\ \dfrac{\partial V}{\partial t} \\ \dfrac{\partial U}{\partial t} + \dfrac{\partial V}{\partial s} \end{Bmatrix} \quad \{\bar{\varepsilon}_b\} = \begin{Bmatrix} \dfrac{\partial \theta_s}{\partial s} \\ \dfrac{\partial \theta_t}{\partial t} \\ \dfrac{\partial \theta_s}{\partial t} + \dfrac{\partial \theta_t}{\partial s} \end{Bmatrix} \quad \{\varepsilon_s\} = \begin{Bmatrix} \theta_s + \dfrac{\partial W}{\partial s} \\ \theta_t + \dfrac{\partial W}{\partial t} \end{Bmatrix}$$

$\{\varepsilon_m\}, \{\varepsilon_b\}, \{\varepsilon_s\}$ Are respectively the tensors of membrane, flexion and shear deformations. The tensor of generalized strains is defined by

$$\{\bar{\varepsilon}\} = \left\{ \{\varepsilon_m\}^T \{\bar{\varepsilon}_b\}^T, \{\varepsilon_s\}^T \right\}^T$$

Stress-Strain Relationship

In addition to the bending and shear stresses, we also have the contribution of membrane stresses in the structure

$$\{\sigma\} = \left\{ \{\sigma_m\}^T, \{\sigma_b\}^T, \{\sigma_s\}^T \right\}^T$$

The relationship of strain deformations, according to Hooke's law, is generalized

$$\{\sigma\} = [C]\{\varepsilon\}$$

For an isotropic material, they can be rewritten in the following form

$$\begin{Bmatrix} \{\sigma_m\} \\ \{\sigma_f\} \\ \{\sigma_c\} \end{Bmatrix} = \begin{bmatrix} [C_m] & [0] & [0] \\ [0] & [C_b] & [0] \\ [0] & [0] & [C_s] \end{bmatrix} \begin{Bmatrix} \{\varepsilon_m\} \\ \{\varepsilon_f\} \\ \{\varepsilon_c\} \end{Bmatrix}$$

With $\left[C_m\right] = \frac{E}{(1-\nu^2)}\begin{bmatrix} 1 & \nu & 0 \\ \nu & 1 & 0 \\ 0 & 0 & \frac{1-\nu}{2} \end{bmatrix}$

E: Young's modulus
ν : Poisson coefficient
γ : Shear Correction Coefficient

RELATION OF THE RESULTING FORCES-DEFORMATION

The bending and transverse shear forces shown in Figure 9, have the same expressions as those of the plates.

The resulting moments of bending are:

$$M_x = \int_{-t/2}^{t/2} Z\sigma_x dz$$
$$M_y = \int_{-t/2}^{t/2} Z\sigma_y dz$$
$$M_{xy} = \int_{-t/2}^{t/2} Z\sigma_{xy} dz$$

Figure 9. Bending and transverse shear forces

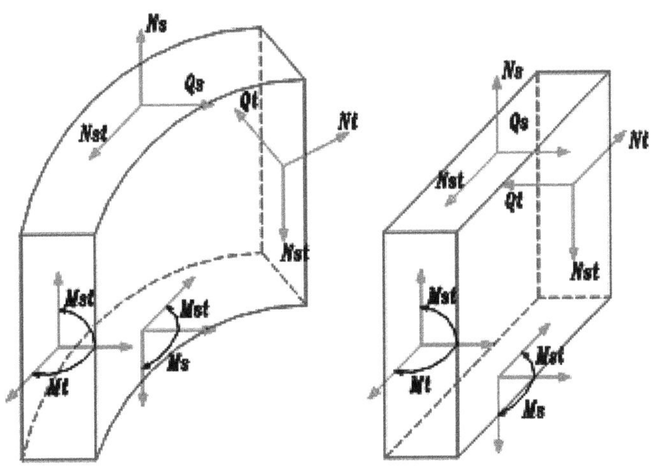

Shear efforts are:

$$Q_x = \int_{-t/2}^{t/2} \sigma_{xz} dz$$

$$Q_y = \int_{-t/2}^{t/2} \sigma_{yz} dz$$

With t is the thickness

The membrane behavior of the structure is characterized by the following forces see (Figure 10.a and Figure 10.b).

$$N_x = \int_{-t/2}^{t/2} \sigma_x dz$$

$$N_y = \int_{-t/2}^{t/2} \sigma_y dz$$

$$N_{xy} = \int_{-t/2}^{t/2} \sigma_{xy} dz$$

The vector of the resulting forces $\{E\}$ is given by:

$$\{E\} = \begin{bmatrix} [D_m] & [0] & [0] \\ [0] & [D_b] & 0 \\ [0] & [0] & [D_s] \end{bmatrix} \begin{bmatrix} \{\varepsilon_m\} \\ \{\overline{\varepsilon}_b\} \\ \{\varepsilon_s\} \end{bmatrix}$$

Figure 10a. Membrane state: Normal efforts Nx et Ny (Nx = t σ_x; Ny = t σ_y) Tangential efforts Nxy = Nyx (Nxy = t τ_{xy}). Shear forces Vx and Vy.

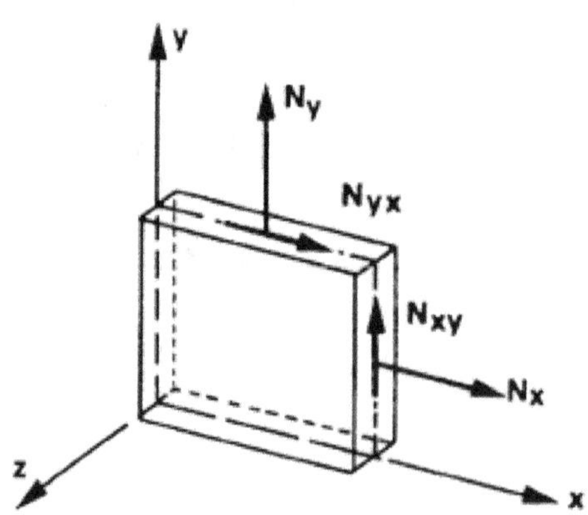

Figure 10b. Flexional state: Bending moments Mx et My , Moments of torsion Mxy = Myx , Shear forcex Vx and Vy.

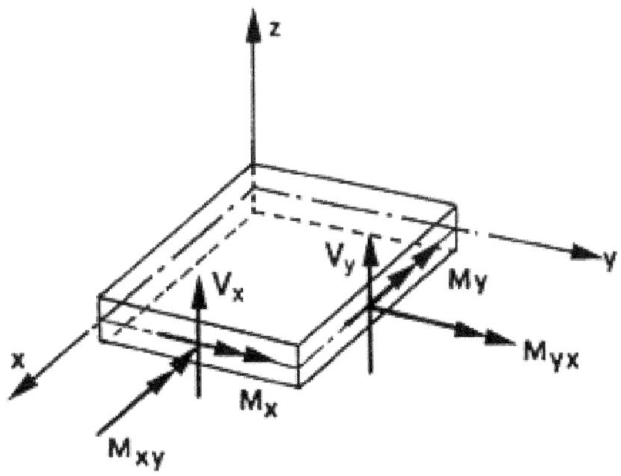

$$\left[D_m\right] = \frac{Et}{(1-\nu^2)}\begin{bmatrix} 1 & \nu & 0 \\ \nu & 1 & 0 \\ 0 & 0 & \frac{1-\nu}{2} \end{bmatrix}$$

FINITE ELEMENTS OF SHELL ELEMENT (MEMBRANE PLATE)

Discretization of the Displacement Field

In the case of the shell element, the displacement field has five components for a node, namely: u_i along the axis ox, v_i along the axis oy, w_i along the axis oz. Grouped in the nodal displacement vector $\left\{q_i\right\}$, (Geoffroy, 1983).

β_{xi} rotation of the normal around the x axis.
β_{yi} rotation of the normal around the y axis.

Dans chacun des éléments, les composantes u_i, v_i, w_i, β_{xi} et β_{yi} sont interpolées linéairement in each of the elements, the components u_i, v_i, w_i, β_{xi} and β_{yi} are interpolated linearly

The displacement field of a Q4 quadrilateral shell element (membrane plate) is as follows:

$$\left\{U\right\} = \sum_{l=1}^{n}\left[N_i\right]\left\{q_i\right\}$$

$$\begin{Bmatrix} u \\ v \\ w \\ \beta_x \\ \beta_y \end{Bmatrix} = \begin{bmatrix} N_1 & 0 & 0 & 0 & 0 & N_2 & 0 & 0 & 0 & 0 & \vdots & N_n & 0 & 0 & 0 & 0 \\ 0 & N_1 & 0 & 0 & 0 & 0 & N_2 & 0 & 0 & 0 & \vdots & 0 & N_n & 0 & 0 & 0 \\ 0 & 0 & N_1 & 0 & 0 & 0 & 0 & N_2 & 0 & 0 & \vdots & 0 & 0 & N_n & 0 & 0 \\ 0 & 0 & 0 & N_1 & 0 & 0 & 0 & 0 & N_2 & 0 & \vdots & 0 & 0 & 0 & N_n & 0 \\ 0 & 0 & 0 & 0 & N_1 & 0 & 0 & 0 & 0 & N_2 & \vdots & 0 & 0 & 0 & 0 & N_n \end{bmatrix} \begin{Bmatrix} u_1 \\ v_1 \\ w_1 \\ \beta_{x1} \\ \beta_{y1} \\ \vdots \\ \vdots \\ \vdots \\ \vdots \\ \vdots \\ u_n \\ v_n \\ w_n \\ \beta_{xn} \\ \beta_{yn} \end{Bmatrix}$$

Deformation matrices due to shear, flexion and membrane effect are:

$$\{\varepsilon\} = \begin{Bmatrix} \varepsilon_x \\ \varepsilon_y \\ \gamma_{xy} \\ \cdots \\ \gamma_{xz} \\ \gamma_{yz} \end{Bmatrix} = \begin{Bmatrix} \dfrac{\partial U}{\partial x} \\ \dfrac{\partial V}{\partial y} \\ \dfrac{\partial U}{\partial y} + \dfrac{\partial V}{\partial x} \\ \cdots \\ \dfrac{\partial U}{\partial z} + \dfrac{\partial w}{\partial x} \\ \dfrac{\partial V}{\partial z} + \dfrac{\partial w}{\partial y} \end{Bmatrix} = \begin{Bmatrix} \dfrac{\partial U_0}{\partial x} \\ \dfrac{\partial V_0}{\partial y} \\ \dfrac{\partial U_0}{\partial y} + \dfrac{\partial V_0}{\partial x} \\ \cdots \\ 0 \\ 0 \end{Bmatrix} + \begin{Bmatrix} z\dfrac{\partial \beta_x}{\partial x} \\ z\dfrac{\partial \beta_y}{\partial y} \\ z\left(\dfrac{\partial \beta_x}{\partial x} + \dfrac{\partial \beta_y}{\partial y}\right) \\ \cdots \\ \dfrac{\partial w}{\partial x} + \beta_x \\ \dfrac{\partial w}{\partial y} + \beta_y \end{Bmatrix}$$

The vector $\{\varepsilon\}$ is divided into two parts, one independent of z which represents membrane deformations.

$$\{\varepsilon_m\} = \begin{Bmatrix} \dfrac{\partial U_0}{\partial x} \\ \dfrac{\partial V_0}{\partial y} \\ \dfrac{\partial U_0}{\partial y} + \dfrac{\partial V_0}{\partial x} \end{Bmatrix}$$

The other dependent of z which represents the flexural deformations.

$$\left\{\varepsilon_f\right\} = \chi = \begin{Bmatrix} \dfrac{\partial \beta_x}{\partial x} \\ \dfrac{\partial \beta_y}{\partial y} \\ \dfrac{\partial \beta_x}{\partial x} + \dfrac{\partial \beta_y}{\partial y} \end{Bmatrix}$$

With $\varepsilon_f = z\chi$ and χ Is the curvature vector

Shear deformations are expressed by:

$$\left\{\varepsilon_c\right\} = \begin{Bmatrix} \dfrac{\partial w}{\partial x} + \beta_x \\ \dfrac{\partial w}{\partial y} + \beta_y \end{Bmatrix}$$

The authors can write:

$$\left\{\varepsilon\right\} = \begin{Bmatrix} \varepsilon_m \\ \cdots \\ \varepsilon_f \\ \cdots \\ \varepsilon_c \end{Bmatrix} + \begin{Bmatrix} \dfrac{\partial u}{\partial x} \\ \dfrac{\partial v}{\partial y} \\ \dfrac{\partial u}{\partial x} + \dfrac{\partial v}{\partial y} \\ \cdots \\ \dfrac{\partial \beta_x}{\partial x} \\ \dfrac{\partial \beta_y}{\partial y} \\ \dfrac{\partial \beta_x}{\partial y} + \dfrac{\partial \beta_y}{\partial x} \\ \cdots \\ \dfrac{\partial w}{\partial x} + \beta_x \\ \dfrac{\partial w}{\partial y} + \beta_y \end{Bmatrix} = \begin{Bmatrix} \dfrac{\partial N_i}{\partial x} u_i \\ \dfrac{\partial N_i}{\partial y} v_i \\ \dfrac{\partial N_i}{\partial y} u_i + \dfrac{\partial N_i}{\partial x} v_i \\ \cdots \\ \dfrac{\partial N_i}{\partial x} \beta_{xi} \\ \dfrac{\partial N_i}{\partial y} \beta_{yi} \\ \dfrac{\partial N_i}{\partial y} \beta_{xi} + \dfrac{\partial N_i}{\partial x} \beta_{yi} \\ \cdots \\ \dfrac{\partial N_i}{\partial x} w_i + N_i \beta_{xi} \\ \dfrac{\partial N_i}{\partial y} w_i + N_i \beta_{yi} \end{Bmatrix}$$

$$\{\varepsilon\} = \begin{bmatrix} \beta_1 & \beta_2 & \cdots & \beta_n \end{bmatrix} \begin{Bmatrix} u_1 \\ v_1 \\ w_1 \\ \beta_{x1} \\ \beta_{y1} \\ \vdots \\ u_n \\ v_n \\ w_n \\ \beta_{n1} \\ \beta_{n1} \end{Bmatrix}$$

With $\beta_1 = \begin{Bmatrix} B_{mi} \\ B_{fi} \\ B_{ci} \end{Bmatrix}$

$$\begin{bmatrix} B_m \end{bmatrix} = \begin{bmatrix} \frac{\partial N_i}{\partial x} & 0 & 0 & 0 & 0 \\ 0 & \frac{\partial N_i}{\partial y} & 0 & 0 & 0 \\ \frac{\partial N_i}{\partial y} & \frac{\partial N_i}{\partial x} & 0 & 0 & 0 \end{bmatrix}$$

$$\begin{bmatrix} B_f \end{bmatrix} = \begin{bmatrix} 0 & 0 & 0 & \frac{\partial N_i}{\partial x} & 0 \\ 0 & 0 & 0 & 0 & \frac{\partial N_i}{\partial y} \\ 0 & 0 & 0 & \frac{\partial N_i}{\partial y} & \frac{\partial N_i}{\partial x} \end{bmatrix} \quad (3.30)$$

$$\begin{bmatrix} B_\gamma \end{bmatrix} = \begin{bmatrix} 0 & 0 & \frac{\partial N_i}{\partial x} & N_i & 0 \\ 0 & 0 & \frac{\partial N_i}{\partial y} & 0 & N_i \end{bmatrix}$$

$$[K] = [K_f] + [K_c] + [K_m] = \int_{S^e} [B_f]^T [D_f][B_f] dS + \int_{S^e} [B_\gamma]^T [D_c][B_\gamma] dS + \int_{S^e} [B_m]^T [D_m][B_m] dS$$

SIXTH DEGREE OF FREEDOM

In the local axes of the element, there are five unknowns per node: one degree of freedom is not used, which is the rotation of the normal to the plane of the element around itself. In the stiffness matrix K of such an element, the matrix block of the node is represented in the figure 11. (Frey&Jirousek 2001)

To be able to work in 3D where there is 6 degrees of freedom nodal in the general case. It is necessary to make an expansion of each matrix block by the addition of a sixth line and a sixth column of zeros (figure 10. After rotating the local axes to the global axes, the zeros disappear and each matrix block seems full, but it is obvious that one of the rotation components remains linearly dependent on the two other (components). It is the problem of the sixth degree of freedom which is practically presented as follows.

Let's examine the mesh of a pleated structure (Figure 12) two cases can occur.

- All elements surrounding a neoud are in the same plane (A nodes for example) or have a common tangent plane: then in this node there is no rigidity around the normal to this plane and there is only five unknowns;
- The elements joining a node are not all in the same plane (nodes B); in this case, there are three independent rotations in such node, and six degrees of freedom

Figure 11. Appearance of a matrix block of a finite element membrane plate

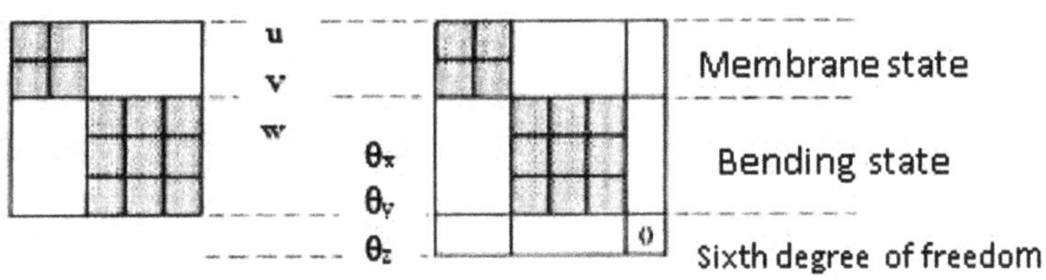

Figure 12. Nodes situated or not in the same plane; A nodes with five degree of freedom; B nodes with six degree of freedom

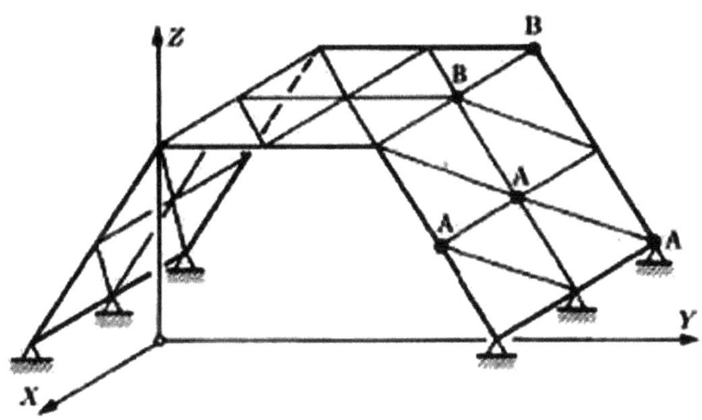

IMPLEMENTATION INTO FEMOBJECT CODE

The enrichment of FEMObject code was carried out so that the concept of object was preserved by leaving the existing methods unchanged and creating a new class (Plate) inheriting the methods of the former base « Element » class, and re-implementing (polymorphism) some of them.

Starting from the existing class hierarchy a new subclass of class « Element » is introduced in table1.

One of the principal classes in FEMOBJ is the Element class which allows to calculate all the characteristics of an element. By adding the plate class, it was found meaningful to split the stiffness matrix into two, one due to shearing and the other due to bending. Once these two matrices are calculated the stiffness matrix of the element is the assembled from the two contributions (bending and shearing).In order to create a new class of shell and plate type element we proceed as follows (see table4 and table 2) respectively.

In the class Element of FEMOBJ, a method *ComputeStiffnessMatrix* calls three other methods: *ComputeBmatrix* (for the calculation the shape functions derivative) ; *Computeconstitutivematrix* (for

Table 1. The C++ finite element class

Dictionary		
Dof		
Domain		
FemComponent		
	Element	
		PlaneStrain
		Truss2D
		Plate
		Orplate
		SHELLS
	Load	
		BodyLoad
	DeadWeight	
		BoundryCondition
		NodalLoad
	LoadTimeFunction	
		ConstantFunction
		PeakFunction
	Material	
	Node	
	TimeIntegrationScheme	
		Newmark
		Static
	Timestep	
FileReader		
FloatArray		
	Column	
GaussPoint		
IntArray		
LinearSystem		
	BandSystem	
	SkylineSystem	
List		
Matrix		
	FloatMatrix	
		BandMatrix
	PolynomialMtrix	
Polynomial		
	PolynomialXY	
Skyline		
String		

Table 2. Creation of a new class: Plate type element

```
Element*  Element :: ofType (char* aClass)

  {
   Element* newElement ;

    if (! strncmp(aClass,"Pl",2))
    newElement = new Plate (number,domain) ;
     else if (! strncmp(aClass,"Pl",2))
   newElement = new Orplate (number,domain) ;

   else {
   printf ("%s : unknown element type \n",aClass);
   exit(0) ;
  }

return newElement ;
        }
```

Table 3. implementation of the computeStiffnessMatrix method for plate element

```
FloatMatrix* Plate :: computeStiffnessMatrix ()


 {

this->computeStiffnessMatrixB();
this->computeStiffnessMatrixS();
stiffnessMatrix = stiffnessMatrix ->GiveCopy() ;
stiffnessMatrix->plus(stiffnessMatrixS);

 return stiffnessMatrix ;
 }
```

the calculation of the constitutive matrix), and *Computevolumearound* (to calculate the Jacobian). This last method remains unchanged because the jacobian is identical for all types of 4-node quadrilateral isoparametric elements. On the other hand, the other existing methods must be modified, because of the need to compute the shearing and bending contributions for plate element add to this (shear and bending) the membrane contribution to modelize the shell element .

The unchanged method *ComputeVolumeAround* is inherited from the base class Element, whose implementation will be required for the computation of the Jacobian for the plate element.

On the other hand, the *ComputeStiffnessMatrix* method implemented in Element is overridden by another implementation in class Plate and SHELLS.

The plate class has to implement a *ComputeStiffnessMatrix* and *ComputeMassMatrix ;* the first new method now makes use of the new methods *ComputeStiffnessMatrixB* and *ComputeStiffnessMatrixS* and sums the results to obtain the correct stiffness matrix (see table3) and

Table 4. creation of a new class: Shell type element

```
Element* Element :: ofType (char* aClass)

  {
   Element* newElement ;

    if (! strncmp(aClass,"SH",2))
    newElement = new SHELLS (number,domain) ;
     else if (! strncmp(aClass,"SH",2))
   newElement = new Orplate (number,domain) ;

   else {
   printf ("%s : unknown element type \n",aClass);
   exit(0) ;
  }

return newElement ;
        }
```

Table 5. Implementation of the computeStiffnessMatrix method for shell element

```
FloatMatrix* SHELLS :: computeStiffnessMatrix ()

 {

this->computeStiffnessMatrixB();
this->computeStiffnessMatrixS();
this->computeStiffnessMatrixM();
stiffnessMatrix = stiffnessMatrix ->GiveCopy() ;
stiffnessMatrix->plus(stiffnessMatrixS);
stiffnessMatrix->plus(stiffnessMatrixM);
 return stiffnessMatrix ;
```

On the level of these two methods one can create the elementary matrices. The *ComputeStiffnessMatrixB* method uses *ComputeBmatrixB* (which calculates the bending matrix form the discretized symmetric gradient operator) and *ComputeconstitutivematrixB* (which calculates the bending constitutive matrix); similarly, method *ComputeStiffnessMatrixS* computes the shearing contribution to the stiffness matrix.

For shell element the authors have implemented the membrane contribution leads by *ComputeStiffnessMatrixM* method which uses *ComputeBmatrixM* (which calculates the membrane matrix form the discretized symmetric gradient operator) and *ComputeconstitutivematrixM* (which calculates the membrane constitutive matrix) (see table 8).

The implementation of strainvector and mass matrix method of plate type element is detailed in the appendix.

Table 6. Implementation of Bending stiffness matrix of shell element

```
FloatMatrix* SHELLS :: computeStiffnessMatrixB ()
{
  int      i ;
  double    dV ;
  FloatMatrix *b,*db,*d ;
  GaussPoint *gp ;

  stiffnessMatrix = new FloatMatrix() ;
  for (i=0; i<4 ; i++) {
    gp = gaussPointArray[i] ;
    b  = this -> ComputeBmatrixAtB(gp) ;
```

Table 7. Implementation of Shear stiffness matrix of shell element

```
FloatMatrix* SHELLS :: computeStiffnessMatrixS ()
{
  int      i ;
  double    dV ;
  FloatMatrix *b,*db,*d ;
  GaussPoint *gp ;

  stiffnessMatrixS=new FloatMatrix();
  for (i=0 ; i<4 ; i++) {
           gp = gaussPointArray[i];
    b  = this -> ComputeBmatrixAtS(gp);
```

APPLICATION

This chapter aims at showing the effectiveness of Object-Oriented programming, illustrated by the code FEMObject, when adding a new class, which here treats dynamic isotropic and orthotropic plates.

The numerical validation examples are an isotropic square plate of constant thickness, simply supported and clamped on its four sides, subjected to a constant load of intensity F. The geometrical and mechanical data of the structure are given in Figures 13, as well as the boundary conditions, which are imposed on the boundary of the plate.

Table 8. Implementation of membrane stiffness matrix of shell element

```
FloatMatrix* SHELLS :: computeStiffnessMatrixM ()
 {
 int      i;
 double   dV;
 FloatMatrix *b,*db,*d;
 GaussPoint *gp;
 stiffnessMatrixM=new FloatMatrix();
 for (i=0; i<4; i++) {
   gp = gaussPointArray[i];
   b = this -> ComputeBmatrixAtM(gp);
   d = this -> computeConstitutiveMatrixM();
```

In order to study convergence, Figure 14 shows evolution in time of isotropic plate response, which is clamped on its sides. Our numerical results are compared with SAP2000 software. As we can see on Figure 14 the two results are very close.

The boundary condition has a big influence in the response of the structure as shown in Figure 15 with finite element Object oriented programming, same influence in figure 16 with classical manner of programming using Fortran.

Figure 13. Isotropic square plate simply supported on its four sides under concentrated load F

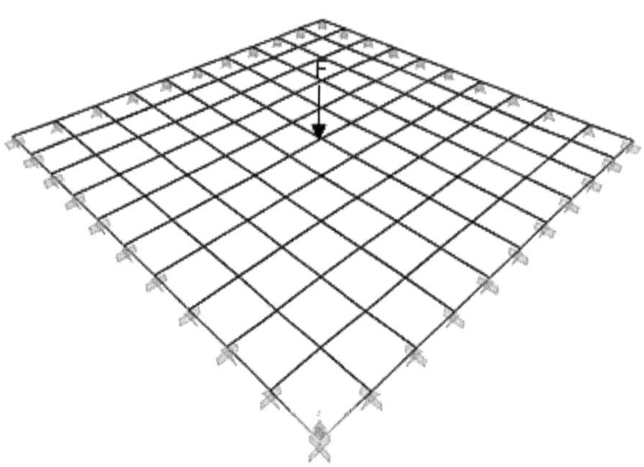

Data

$E=210 \times 10^9$ N/m²

a =2m

h = 0.03m

ν =0.3

ρ=7800 kg/m³

γ = 0.5 β=0.25 (newmark)

DT=0.0025 s

P = 800 N

Table 9. Implementation of rotation matrix of shell element

```
FloatMatrix* SHELLS :: giveRotationMatrix ()
{
  double sine,cosine ;
  if (! rotationMatrix) {
    sine      = sin (this->givePitch()) ;
    cosine    = cos (pitch) ;
    rotationMatrix = new FloatMatrix(24,24) ;
    rotationMatrix ->at(1,1) = sine ;
    rotationMatrix ->at(1,3) = -cosine ;
    rotationMatrix ->at(2,2) = 1 ;
    rotationMatrix ->at(3,1) = cosine ;
    rotationMatrix ->at(3,3) = sine ;
    rotationMatrix ->at(4,1) = 1 ;
    rotationMatrix ->at(4,4) = 1 ;
    rotationMatrix ->at(5,5) = sine ;
    rotationMatrix ->at(5,6) = cosine ;
    rotationMatrix ->at(6,5) = -cosine ;
    rotationMatrix ->at(6,6) = sine ;
    rotationMatrix ->at(7,7) = sine ;
    rotationMatrix ->at(7,9) = -cosine ;
    rotationMatrix ->at(8,8) = 1 ;
    rotationMatrix ->at(9,7) = cosine ;
    rotationMatrix ->at(9,9) = sine ;
    rotationMatrix ->at(10,7) = 1 ;
    rotationMatrix ->at(10,10) = 1 ;
    rotationMatrix ->at(11,11) = sine ;
    rotationMatrix ->at(11,12) = cosine
```

Figure 17 shows the influence of the mesh in the response of structure using classical manner of programming.

The geometrical and mechanical data of a cantilever plate are given in figure 18 subjected to concentrated load in the center of the free side of the structure of intensity F. In order to study convergence, Figure 19 shows evolution in time of plate response, which is clamped on one side and free on the other sides. Our numerical results are compared with those from (Geoffroy,P. 1983), his maximum value is 0.163m compared to 0.159m given by the FEMOBJ program.

Figure 14. Dynamic response of finite element solution for an isotropic plate simply supported on its four opposite sides

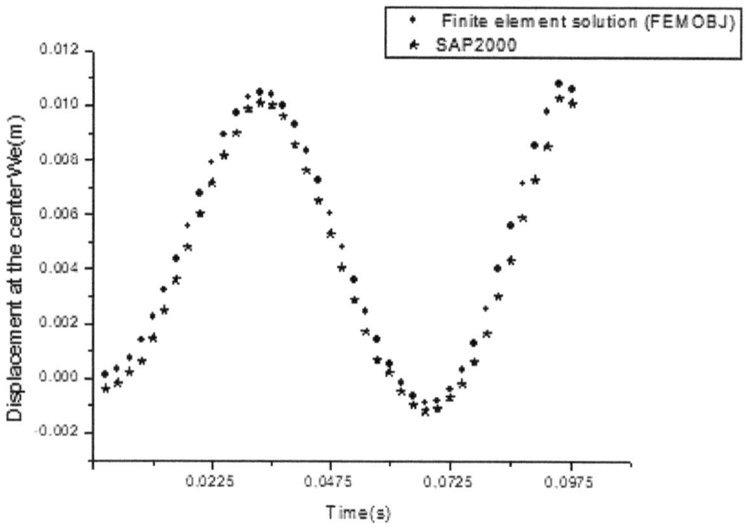

Figure 15. Influence of the supports in dynamic response of isotropic square plate with FEMObj

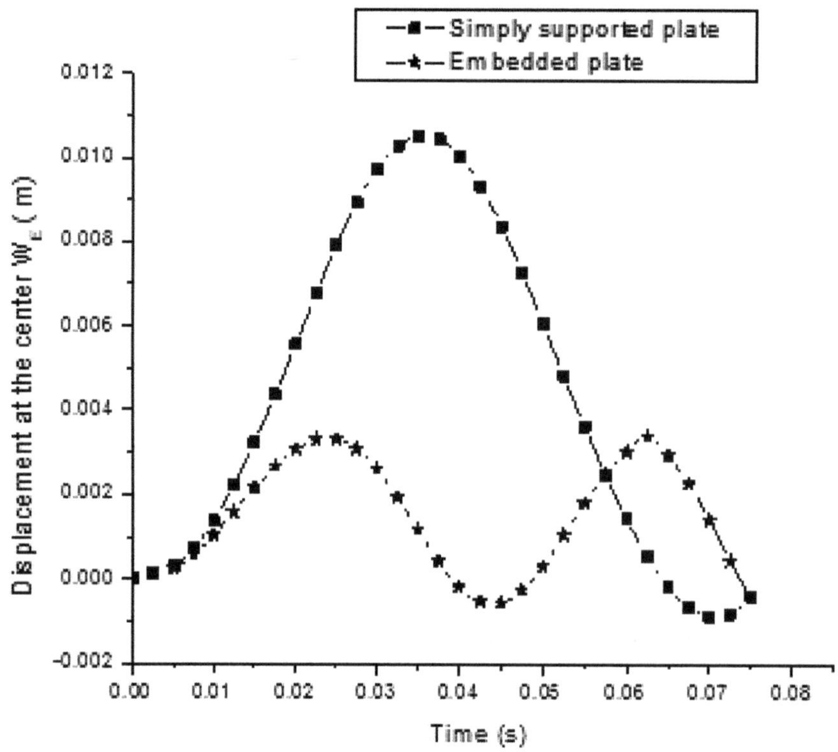

Figure 16. Supports influence response of isotropic square plate using classical manner of programming

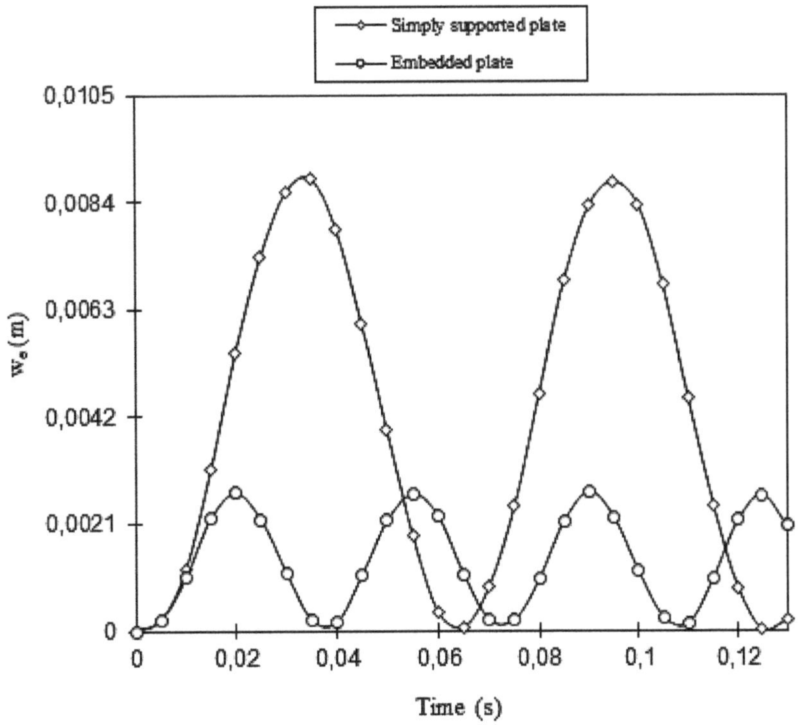

Figure 17. Influence of the mesh in the response of structure using classical manner of programming

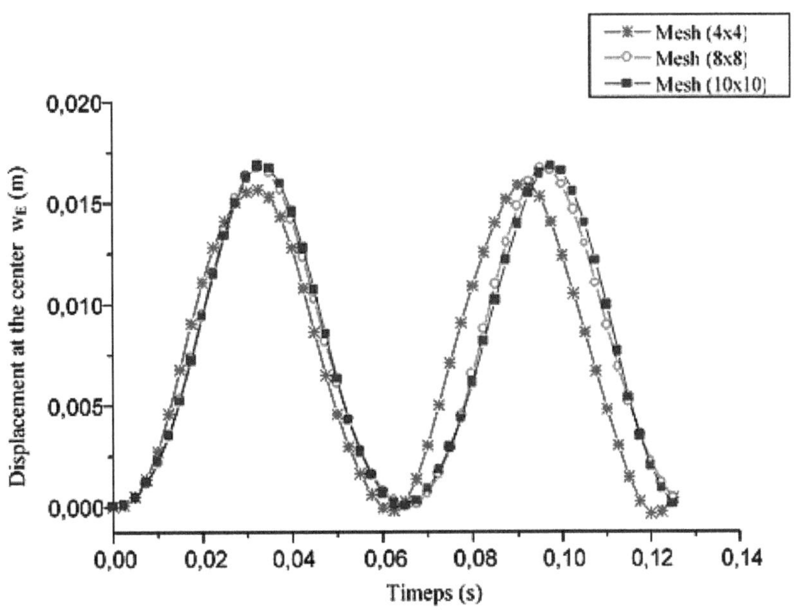

Figure 18. Concentrate load F on the free side of isotropic cantilaver plate

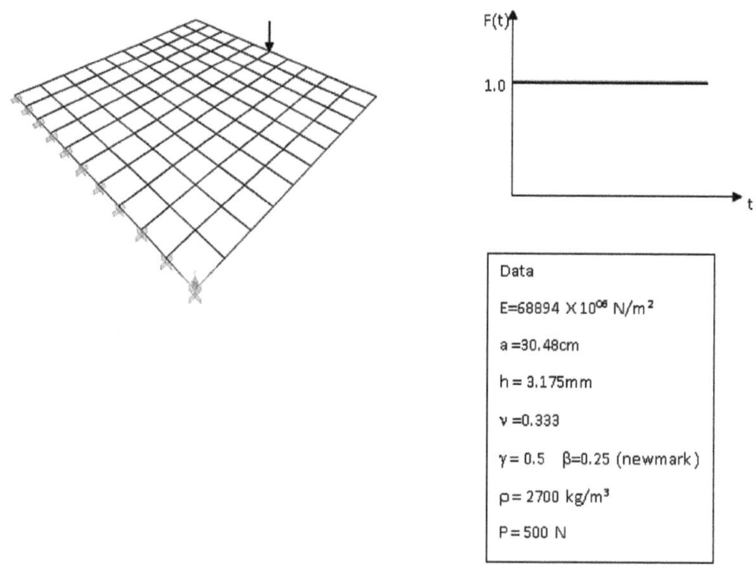

Figure 19. Dynamic response of finite element solution for an isotropic cantilaver plate

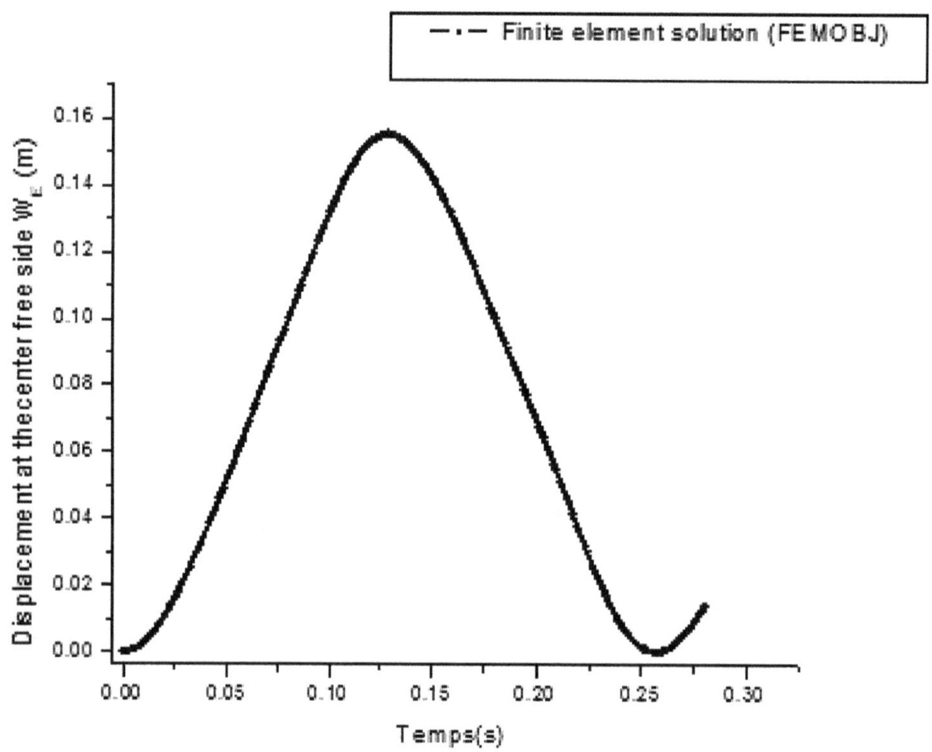

CONCLUSION

Through the example of the addition of a plate element in existing C++ code, FEMObject, this paper shows how Object-Oriented programming makes it possible to easily develop new features in finite element codes, showing such Object-Oriented programs to be more modular, reusable and more easily expandable.

The Object-Oriented approach is shown to offer undeniable advantages compared to earlier programming structures. The encapsulation of the data largely improves the modularity, and thus the reliability and legibility of the code are increased. Inheritance allows an automatic reusability of the already developed methods, and polymorphism is a powerful means to raise the level of abstraction. The Object-Oriented approach offers more generic structures, making it possible to overcome a number of difficulties of the traditional implementations of the Finite Element Method.

Object-Oriented Programming also supports the upgrading needs of the finite element method implementation by enhancing modularity and flexibility of evolution. Thanks to the derivation of classes and a high level of abstraction, Object-Oriented Programming can be an effective tool to keep control of the codes of tomorrow. The late binding capability (polymorphism) facilitates adaptability.

REFERENCES

Batoz, J. L., & Dhatt, G. (1990). *Modélisation des structures par élément finis* (Vol. 2). Hermès Paris.

Commend, S., & Zimmermann, T. (2001). Object-oriented nonlinear finite element programming: A primer. *Advances in Engineering Software*, *32*(8), 611–628. doi:10.1016/S0965-9978(01)00011-4

Delannoy, C. (2002). *Programmer en langage c* (3rd ed.). Eyrolles Paris.

Dhatt, G., & Touzot, G. (1984). Une présentation de la méthode des éléments finis (2nd ed.). Edition Maloine S.A.

Dubois-pèlerin, Y., & Zimmermann, T. (1993). Object-oriented finite element programming III. An efficient implementation in C++. *Computer Methods in Applied Mechanics and Engineering*, *108*(1-2), 165–183. doi:10.1016/0045-7825(93)90159-U

Dubois-pèlerin, Y., Zimmermann, T., & Bomme, P. (1992). Object-oriented finite element programming II. A prototype programming in smaltalk. *Computer Methods in Applied Mechanics and Engineering*, *98*(3), 361–397. doi:10.1016/0045-7825(92)90004-4

Frey, F., & Jirousek, J. (2001). *Analyse des structures et milieu continus méthode des élément finis*. Suiss Federal institute of Technology.

Geoffroy, P. (1983). *Development and evaluation of a finite element for the static, dynamics nonlinear analysis of thin shells* (PhD thesis). University of Technology of Compiègne.

Imbert, J. F. (1991). *Analyse des Structures par Eléments finis* (3rd ed.). Cépaduès Toulouse.

Kris, J., & Lars, K. (1999). *La bible du programmeur c/c*. Editions Reynauld Goulet.

Robert, D. C., David, S., & Malkus, M. E. P. (1988). *Concepts and Applications of Finite Element analysis*. University of Wisconsin Madison.

Zimmermann, T., Dubois-pèlerin, Y., & Bomme, P. (1992). Object-oriented finite element programming I. Governing principles. *Computer Methods in Applied Mechanics and Engineering*, *98*, 291–303.

ADDITIONAL READING

Forde, B. W. R., Foschi, R. O., & Stiemer, S. F. (1990). Object-oriented finite element analysis. *Comp.&Struct*, *34*(3), 355–374. doi:10.1016/0045-7949(90)90261-Y

Yu, G., & Adeli, H. (1993). Object-oriented finite element analysis EER models. *Journal of Structural Engineering*, *119*(9), 2763–2781. doi:10.1061/(ASCE)0733-9445(1993)119:9(2763)

KEY TERMS AND DEFINITIONS

Abstract Data Types: Abstract data types play a central role in object-oriented programming. They make it possible to distinguish the concept (what is being done) from the actual implementation (how it is done).

Attribute: The attributes are the data used by the method of the object.

Constructor: The constructor of a class is a special function which is used when an object of that class is created, either automatically or dynamically.

Data Encapsulation: This is the property of objects to possess attributes and to hide them from the other objects.

Destructor: The destructor of a class is a special function which is used when an object of that class is destroyed, either automatically or dynamically.

Message: Objects communicate through message. These messages are requested to activate a method of the receiver of the message.

APPENDIX

Implementation of the computeStrainVector and Compute Mass Matrix Method of Plate Type Element

```
FloatArray*  Plate:: computeStrainVector (GaussPoint* gp, TimeStep* stepN)
   // Computes the vector containing the strains at the Gauss point gp of
   // the receiver, at time step stepN. The nature of these strains depends
   // on the element's type.
{
        this->computeStrainVectorS (gp,  stepN);
        this->computeStrainVectorB (gp,  stepN);
        return 0;
}
FloatArray*  Plate:: computeStrainVectorS (GaussPoint* gp, TimeStep* stepN)
{
   FloatMatrix *b ;
   FloatArray  *u,*Epsilon ;
   b       = this -> ComputeBmatrixAtS(gp) ;
   u       = this -> ComputeVectorOf('d',stepN) ;
   Epsilon = b -> Times(u) ;
   gp -> letStrainVectorBe(Epsilon) ;
   delete b ;
   delete u ;
   return Epsilon ;
}
FloatArray*  Plate:: computeStrainVectorB (GaussPoint* gp, TimeStep* stepN)
{
        FloatMatrix *b ;
    FloatArray  *u,*Epsilon ;
   b       = this -> ComputeBmatrixAtB(gp) ;
   u       = this -> ComputeVectorOf('d',stepN) ;
   Epsilon = b -> Times(u) ;
   gp -> letStrainVectorBe(Epsilon) ;    // gp stores Epsilon, not a copy
   delete b ;
   delete u ;
   return Epsilon ;
}
FloatMatrix*  Plate:: computeMassMatrix ()
{
   FloatMatrix* consistentMatrix ;
```

```
   consistentMatrix = this ->ComputeConsistentMassMatrix() ;
   massMatrix       = consistentMatrix -> Lumped() ;
   delete consistentMatrix ;
   return massMatrix ;
}
FloatMatrix*  Plate:: ComputeConsistentMassMatrix ()
{
   int   i ;
   double      density,dV,thickness ;
   FloatMatrix *n,*answer ;
   GaussPoint  *gp ;
   thickness   = this -> giveMaterial() -> give('t') ;
   answer  = new FloatMatrix() ;
   density = this -> giveMaterial() -> give('d') ;
   for (i=0 ; i<numberOfGaussPoints ; i++) {
      gp      = gaussPointArray[i] ;
      n       = this -> ComputeNmatrixAt(gp) ;
      dV      = this -> computeVolumeAround(gp) ;
      answer -> plusProduct(n,n,density*dV*thickness) ;
      delete n ;}
return  answer->symmetrized() ;
}
```

Chapter 3
Optimization of Single Row Layout in Construction Site Planning:
A Comparative Study of Heuristics Algorithms

Amalia Utamima
Institut Teknologi Sepuluh Nopember, Indonesia

Arif Djunaidy
Institut Teknologi Sepuluh Nopember, Indonesia

Angelia Melani Adrian
Universitas Katolik De La Salle Manado, Indonesia

ABSTRACT

Several heuristics algorithms can be employed to solve single row layout in construction site planning. Firstly, this chapter builds Tabu Search to deal with the problem. Other heuristics methods which are genetic algorithm (GA) and estimation distribution algorithm (EDA) are also developed against Tabu Search. A comparative study is performed to test the effectiveness and efficiency of the algorithms. The statistical test, ANOVA followed by the t-test, compares the results of the three algorithms. Then, the pros and cons of using the algorithms are stated.

INTRODUCTION

The arrangement of machines in construction site layout can improve in cost reduction and time savings during the construction process. In construction site planning, there is a case when the facilities need to be placed in a row. The case is also known as a single row facility layout problem. The objective is to minimize the sum of the distance between facilities for the given traffic loads of the facilities.

DOI: 10.4018/978-1-5225-7059-2.ch003

This problem is categorized as an NP-Complete problem and heuristics method is needed to provide a near-optimal solution (Samarghandi & Eshghi, 2010). Several heuristics methods can be used to deal with this problem. Some research use metaheuristics method like Genetic Algorithm (Datta, Amaral, & Figueira, 2011), Particle Swarm Optimization (Samarghandi, Taabayan, & Jahantigh, 2010), and Ant Colony Optimization (Solimanpur, Vrat, & Shankar, 2005), while others employed heuristics methods like Estimation Distribution Algorithm (Ou-Yang & Utamima, 2013) and Variable Neighborhood Search (Guan & Lin, 2016).

An early work on the benefits of heuristics for facility layout problem (FLP) was introduced by Kumar et al. (Kumar, Hadjinicola, & Lin, 1995). The method used is greedy heuristic to minimize the cost of material for a single-row layout. The main contribution is the two premises can be included at any step of the proposed method when generating an optimal solution. Another evolutionary-based approach, namely psychoclonal algorithm, was demonstrated to resolve the layout design in a manufacturing system (Khilwani, Shankar, & Tiwari, 2012). The main objective functions considered are activity relationship and distance from one department to other departments.

Uno and Hirabayashi (2009) suggested the correlated mutation technique as a mutation operator for evolution strategies (ES) in solving facility layout. ES is a method that similar to genetic algorithm uses a self-adaptive strategy. The experiments indicated that the correlated mutation is better than the simple mutation approach. To deal with the multi-objective case, Ripon et al. (Ripon et al., 2009) propose a genetic algorithm with a support from Pareto-optimum to get trade-off solutions from various objectives. There are two parameters considered for two objective functions, i.e. material handling cost, which based on quantitative model and closeness rating score that represents qualitative approach. Ripon et al. (2013) improve their work by employing an adaptive local search based on a variable of neighborhood search and genetic algorithm for solving a multi-objective FLP.

Genetic Algorithm (GA) simulates natural evolution process which generates a group candidate solution with selection to the best candidates, crossover, and mutation. GA become a useful optimization technique to deal with many kinds of problems since it is flexible and only need pieces of knowledge (Goldberg, 1989). Several applications of facility layout problem used many types of GA as the optimization tool. Datta et al. (2011) use permutation-based GA for several instances of single row facility layout problem and they develop customized crossover and mutation operators. Therefore, SRFLP is modeled as an unconstrained optimization case. El-Baz (2004) use GA for facility layout problem in the different manufacturing environment (El-Baz, 2004) and the author considers various aspects such as materials flow and handling costs between manufacturing machines. The machines in the manufacturing department are modeled as the grids, and these grids then become chromosomes in the proposed GA (El-Baz, 2004). Said and El Rayes (2013) build and compare GA and dynamic programming for dynamic site layout planning which deals with nonlinear constraints. The authors reported that GA is outperformed by Approximate Dynamic Programming method, but GA still has the advantage to support multi-objective problem. Said & El-Rayes (2013) consider not only construction space, but also the schedule to build the construction site.

Estimation Distribution Algorithm (EDA) is an algorithm that mimics the statistical process of the sampling technique. EDA also uses randomization technique like GA in the beginning but choose the candidate solution based on the probabilistic model that defined before (Hauschild & Pelikan, 2011). Numerous applications use EDA to deal with their problem, i.e., flow shop problem (Zhang & Li, 2011), single machine scheduling (Chen, et al., 2010), and stochastic permutation flow shop (Wang, Choi, & Lu, 2015). Regarding construction site planning, EDA is employed to solve facility layout problem (Ou-

Yang & Utamima, 2013). EDA is combined with Particle Swarm Optimization, and Tabu Search to keep the diversity loss low when iterating through the possible solutions.

Tabu Search is a heuristic algorithm that searches in the neighborhood of the candidate solution. Similar to EDA, Tabu Search (TS) basically use random solution only once in the beginning. Then TS checks whether the improvement of the current solution can be found by exploring the neighborhood (Jason, 2011). Tabu Search employed in several optimization problems like scheduling road vehicle in sugarcane transport (Higgins, 2006) and vehicle routing problem in agriculture (Seyyedhasani & Dvorak, 2017). In the context of construction layout, TS has successfully tackled single row facility layout case (Samarghandi & Eshghi, 2010). TS is able to consider various dimension or size of manufacturing machine, facility location, and can run faster to find the optimal solution (Samarghandi & Eshghi, 2010).

Despite the variation of algorithms used to deal with the problem, however, the statistical analysis to compare the performance of the algorithms was not done yet. This comparative study examines Genetic Algorithm, Estimation Distribution Algorithm, and Tabu Search. The effectiveness and efficiency of the three algorithms are tested. Then, this study also performs statistical tests to analyze the results of three algorithms. The ANOVA test and t-test are chosen to test the effectiveness of the three algorithms to deal with single row layout in the construction site. Then, the pros and cons of utilizing Tabu Search, GA, and EDA will be stated.

HEURISTICS ALGORITHM FOR SINGLE ROW LAYOUT IN CONSTRUCTION SITE

The single row layout problem is related to find the best layout to optimize the flow's distance. Suppose there are several facilities in the single row construction layout and the material are needing to be processed between these facilities. The required flow between these facilities is known as well as the distance between facility. The objective function of the single row layout optimization problem is to minimize the total distance of the flow between the facilities by planning the facilities' placement in the site. Equation (1) listed the objective function to compute the distance. The d_{ij} is the distance of the facilities i and j, c_{ij} is the flow between facilities i and j, while f is the total number of facilities.

$$z = \min \sum_{i=1}^{f-1} \sum_{j=i+1}^{f} c_{ij} d_{ij} \tag{1}$$

This research focuses on three heuristics algorithm that was used to deal with single row layout problem; those are Genetic Algorithm, Estimation Distribution Algorithm, and Tabu Search. Table 1 shows the general strength and weakness of GA, EDA, and TS in single row layout optimization problem. The similarity of the weakness of these three algorithms is they are often trapped in the local optima (Haupt & Haupt, 2004). Hence several studies tried to combine the algorithms (Glover, Kelly, & Laguna, 1995). As shown in Table 10, both GA and EDA have weak local search ability because mainly they produce solutions by exploring several previous candidate solutions. This weakness is not found in TS since the strong local search technique is the pros of Tabu Search. This kind of search which focus on the neighborhood was not found in GA and EDA.

Table 1 also listed that Genetic Algorithm has an active crossover technique that usually produces the candidate solution from two best solutions found before. However, it is hard to encode a problem to fit with crossover and mutation technique in GA. On the other hand, the pros of EDA is the use of the probabilistic model by evaluating the distribution of the previous generation to produce candidate solution.

Figure 1 presents the general flow process of GA, EDA, and TS. The similarity of EDA and GA are they generate a group of the random solution in the start of the iteration. Therefore the objective function calculation is performed at the beginning of the iteration. On the other hand, Tabu Search only generate a single solution at the beginning of the iteration then continue to search in the current solution's neighborhood. Tabu Search constructs swaplist that contains the list of random swaps between the element in the current solution and the value of that move. The 'move' that has the best value is chosen if that

Table 1. The strength and weakness of GA, EDA, and TS

Algorithm	Strength	Weakness
Genetic Algorithm	Has a crossover technique that combines two best solutions	Often trapped in the local optima Hard to encode a problem Weak local search
Estimation Distribution Algorithm	Use the probabilistic model to produce the solution	Often trapped in the local optima Weak local search
Tabu Search	Strong local-search ability	Often caught in the local optima

Figure 1. The general process of GA (a), EDA (b), and TS (c)

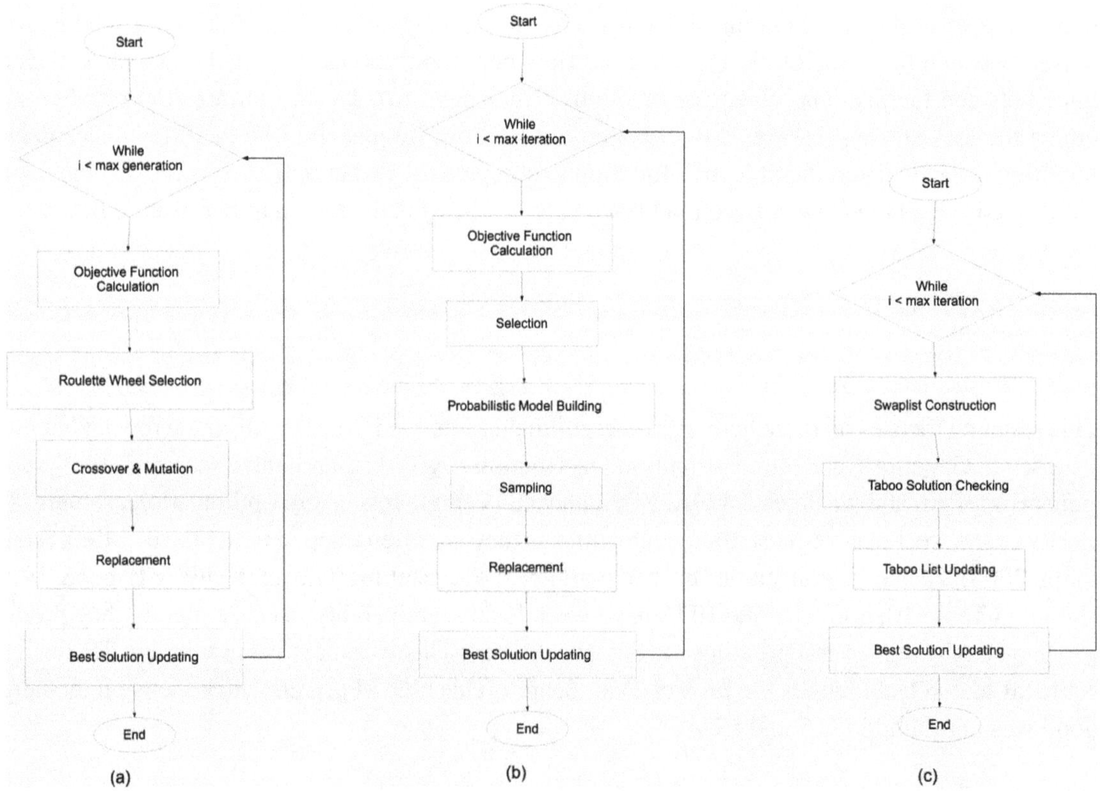

'move' is not taboo. The 'move' is labeled as taboo if it has chosen in the previous iteration. Tabu tenure parameters define how long the 'move' will stay in the Taboo List (Jason, 2011).

At the end of every iteration of the three algorithms, the best solution is updated until the maximum iteration reached. The different between GA and EDA is the method that use to produce a solution. GA uses roulette wheel solution then do crossover and mutation with defined rate (Haupt & Haupt, 2004), while EDA uses truncation selection, probabilistic model, and sampling strategy (Hauschild & Pelikan, 2011). The replacement process of GA and EDA is similar since the old solution is updated

In this work, some parameters were defined to test and compare some algorithms, specifically GA, EDA, and TS as these three algorithms can provide an optimum solution as indicated in the previous works. Table 2 presents the value assigned to the parameters in the GA, EDA, and TS for solving the single row layout in construction site planning. The values were obtained by tuning the parameters via a series of experiments to the algorithm. This research use n in the parameters to make the algorithms can run flexible with the increase of the size of the problem. The number of generation in Tabu Search is different with GA and EDA because it was determined by some experiments to test how many generation the TS can achieve the optimum solution in a small sized problem. In other words, Tabu Search needs more generations to produce a good solution compared to GA and EDA.

COMPARISON RESULTS

In this comparative study, GA, EDA, and TS algorithms are run ten times to solve ten datasets of single row facility layout problem. The ten datasets are listed in the Table 3 column 'Problem Code' (Datta, Amaral, & Figueira, 2011). The number in the problem code indicates the number of facilities while the alphabet indicates the first alphabet of the founders' name of this dataset. This research records the minimum solution, mean of the solution, and percentage error of the three algorithms.

Table 3 listed the minimum value found in 10 replications of the algorithms. The first and second column is the code of the problem and the size of them. The third column listed the optimum solution to the problem found in the literature (Datta et al., 2011; Ou-Yang & Utamima, 2013). Then, the following column provided the minimum value achieved by GA, EDA, and TS in solving this single row layout in construction site planning. Both EDA and TS can meet the optimum solution of all the ten datasets. However, GA only can achieve the optimum value for 8 out of 10 problems.

Table 2. The settings of parameters in the three algorithms

Algorithm	Parameters Settings
Genetic Algorithm	Crossover rate: 0.7 Mutation rate: 0.3 Number of Population: 5*n* Number of Generation: 10*n*
Estimation Distribution Algorithm	Number of Population: 5*n* Number of Generation: 10*n*
Tabu Search	Tabu Tenure: *n*/4 Number of Generation: $25n^2$

**n: the size of the problem*

Table 3. The minimum objective function of GA, EDA, and TS

Problem Code	Size	Optimum Solution	GA	EDA	TS
S4	4	638.00	638.00	638.00	638.00
LW5	5	151.00	151.00	151.00	151.00
N6	6	1.99	1.99	1.99	1.99
S8	8	801.00	801.00	801.00	801.00
S8H	8	2324.50	2324.50	2324.50	2324.50
S9	9	2469.50	2469.50	2469.50	2469.50
S9H	9	4695.50	4695.50	4695.50	4695.50
S10	10	2781.50	2781.50	2781.50	2781.50
S11	11	6933.50	6953.50	6933.50	6933.50
LW11	11	6933.50	6943.50	6933.50	6933.50

Table 4 shows the average of the solution reached in ten replications and the percentage of error for GA, EDA, and TS. The error rate is calculated by comparing the mean of each algorithm and the optimum value. Overall, EDA is outperformed GA and TS in the lowest mean and lowest percentage error. The smallest error found in EDA that just gets the 0.488%, followed by GA that gets 1.290%. GA and Tabu's error rate is nearly similar with only 0.135% difference. GA and EDA consistently get the same mean with the optimum value in S4 until N6 resulting in the 0% error. However, TS cannot achieve the 0% error in S4 and N6.

On the other hand, Table 5 shows the average running time that is attained by GA, EDA, and TS. GA gets the fastest runtime compare to EDA and TS. There is a slight difference between the average runtime of GA and EDA with 0.015 seconds, but the gap with TS's runtime is higher with more than 0.07 seconds difference.

Table 4. The mean and error percentage in 10 runs of the algorithms

Problem Code	Size	Optimum Value	GA		EDA		TS	
			Mean	Error	Mean	Error	Mean	Error
S4	4	638	638.00	0.000%	638.00	0.000%	661.20	3.636%
LW5	5	151	151.00	0.000%	151.00	0.000%	151.00	0.000%
N6	6	1.99	1.99	0.000%	1.99	0.000%	2.05	2.764%
S8	8	801	807.10	0.762%	801.40	0.050%	805.90	0.612%
S8H	8	2324.5	2331.00	0.280%	2324.50	0.000%	2336.00	0.495%
S9	9	2469.5	2498.80	1.186%	2481.00	0.466%	2472.50	0.121%
S9H	9	4695.5	4703.80	0.177%	4697.50	0.043%	4702.80	0.155%
S10	10	2781.5	2877.90	3.466%	2839.90	2.100%	2828.30	1.683%
S11	11	6933.5	7174.30	3.473%	7013.80	1.158%	7103.30	2.449%
LW11	11	6933.5	7180.20	3.558%	7007.30	1.064%	7095.50	2.336%
Average				1.290%		0.488%		1.425%

Table 5. The average running time of the algorithms

Problem	GA	EDA	TS
S4	0.020	0.013	0.008
LW5	0.038	0.024	0.025
N6	0.039	0.038	0.041
S8	0.094	0.098	0.112
S8H	0.095	0.101	0.117
S9	0.133	0.134	0.190
S9H	0.130	0.148	0.191
S10	0.160	0.199	0.293
S11	0.216	0.265	0.472
LW11	0.227	0.281	0.473
Average:	0.115	0.130	0.192

STATISTICAL TEST

The statistical test of GA, EDA, and TS were done to problem LW11 with 11 facilities. Firstly, this research records the minimum value achieved from 10 runs of the three algorithms that are listed in Table 6. The test begins with single factor ANOVA to test whether there is a significant difference between the mean in 10 runs of the three algorithms. Table 7 lists the ANOVA results. Since F > F crit (15.196 > 3.443), the null hypothesis is rejected and there is unequal variance found. This result indicates that the means of the three populations are not all equal or there is one or more algorithm that has a significantly different means compare to others. Thus, at least one of the means is different. Since the ANOVA does not describes where the difference lies, a post-hoc test is needed to test each pair of the algorithm. The t-test is chosen in this research as a post-hoc test to know which algorithm has significantly different means (Urdan, 2011).

Table 6. The minimum value attained in 10 replications of the algorithms

Run	GA	EDA	TS
1st	7080.5	6933.5	6953.5
2nd	7151.5	6933.5	7171.5
3rd	7399.5	6995.5	6933.5
4th	7274.5	7101.5	6953.5
5th	7389.5	7021.5	7324.5
6th	7185.5	6970.5	7062.5
7th	7100.5	6953.5	7274.5
8th	6943.5	7095.5	7274.5
9th	7083.5	7062.5	7073.5
10th	7193.5	7005.5	6933.5

The t-tests between each pair of GA, EDA, and TS are conducted, and the results are described in Table 8 – 10. These t-test use the assumption of unequal variances between each pair of the algorithms. The first test is for GA and EDA's pair and the result is provided in Table 7. As shown in Table 7, the t Stat is greater than the t Critical two-tail (3.507>2.178) which concludes that the average solutions of GA and EDA differ significantly.

On the other hand, the t-test results in Table 9 concludes that GA and TS's average solutions are not significantly different, while the t-test for EDA and TS's pair (listed in Table 10) is also received

Table 7. The ANOVA results

Source of Variation	SS	df	MS	F	P-value	F crit
Between Groups	149492.5	2	74746.23	4.635546	0.018593	3.354131
Within Groups	435363.7	27	16124.58			
Total	584856.2	29				

Table 8. The t-test results between GA and EDA

	GA	EDA
Mean	7180.2	7007.3
Variance	20387.7889	3909.511111
Observations	10	10
Hypothesized Mean Difference	0	
Df	12	
t Stat	3.50764784	
P(T<=t) one-tail	0.00216029	
t Critical one-tail	1.78228756	
P(T<=t) two-tail	0.00432058	
t Critical two-tail	2.17881283	

Table 9. The t-test results between GA and TS

	GA	TS
Mean	7180.2	7095.5
Variance	20387.78889	24076.44444
Observations	10	10
Hypothesized Mean Difference	0	
df	18	
t Stat	1.270217249	
P(T<=t) one-tail	0.110090845	
t Critical one-tail	1.734063607	
P(T<=t) two-tail	0.22018169	
t Critical two-tail	2.10092204	

similar conclusion. Since t Stat lays in between -t Critical two-tail and t Critical two-tail, then the null hypotheses are accepted. The case in Table 8 is -2.10 < 1.27 < 2 (-t Critical two-tail t Stat < t Critical two-tail), while the case in Table 10 is -2.179 < 1.667 < 2.179. Hence, the observed difference between the sample means is not convincing enough to say that the average solutions of the pairs differ significantly. Thus, between three pairs of the tested algorithms, only GA and EDA's pair that significantly different in the average solutions.

DISCUSSION AND CONCLUSION

This research focused on the application of the basic heuristics algorithm to solve the single row facility layout in the construction site. GA, EDA, and TS were chosen, and their behaviors are reflected from the results. Based on the experiments and results in the previous section, this research listed the strength and weakness of GA, EDA, and TS.

Table 11 shows the findings of this research by comparing GA, EDA, and TS in single row layout optimization problem. The focus of neighborhood search technique in TS makes the running time longer than the other algorithms. In other words, TS usually needs longer running time than GA and EDA to produce a good solution. On the other hand, Genetic Algorithm is usually faster than EDA and Tabu Search. However, the solution quality in achieving the optimum value is not as good as EDA and TS. Concerning the effectiveness, EDA produces the lowest percentage of error compared to GA and TS.

Although EDA gets the lowest error and Tabu Search get the highest one, the statistical test concludes that there is no significant difference between the average solutions of EDA and TS. However, the t-test

Table 10. The t-test results between EDA and TS

	EDA	TS
Mean	7007.3	7095.5
Variance	3909.511111	24076.44444
Observations	10	10
Hypothesized Mean Difference	0	
df	12	
t Stat	-1.667241512	
P(T<=t) one-tail	0.060665584	
t Critical one-tail	1.782287556	
P(T<=t) two-tail	0.121331168	
t Critical two-tail	2.17881283	

Table 11. The findings based on the comparative study

Algorithm	Findings
Genetic Algorithm	Faster than EDA and TS
Estimation Distribution Algorithm	Gets the lowest error compared to GA and TS
Tabu Search	Less efficient than GA and EDA regarding the running time

concludes that EDA and GA's solutions are significantly different. Furthermore, EDA's running time also faster than TS. Hence, this research finds that EDA outperforms GA and TS when solving single row layout in construction planning in solution quality.

The future research can focus to improve the performance of the algorithm or to add more algorithms to be compared. In case of enhancing the effectiveness, there are two kinds of technique that can be considered. The first one is to hybridize the basic algorithm with another heuristic. The second one is by adding a part in the algorithm to avoid local optima. This part can be another research topic again since rigorous experiments are needed to be done.

REFERENCES

Chen, S.-H., Chen, M.-C., Chang, P.-C., Zhang, Q., & Chen, Y.-M. (2010). Guidelines for developing effective Estimation of Distribution Algorithms in solving single machine scheduling problems. *Expert Systems with Applications*, *37*(9), 6441–6451. doi:10.1016/j.eswa.2010.02.073

Datta, D., Amaral, A. R. S., & Figueira, J. R. (2011). Single row facility layout problem using a permutation-based genetic algorithm. *European Journal of Operational Research*, *213*(2), 388–394. doi:10.1016/j.ejor.2011.03.034

El-Baz, M. A. (2004). A genetic algorithm for facility layout problems of different manufacturing environments. *Computers & Industrial Engineering*, *47*(2–3), 233–246. doi:10.1016/j.cie.2004.07.001

Glover, F., Kelly, J. P., & Laguna, M. (1995). Genetic algorithms and tabu search: Hybrids for optimization. *Computers & Operations Research*, *22*(1), 111–134. doi:10.1016/0305-0548(93)E0023-M

Goldberg, D. E. (1989). *Genetic Algorithms in Search, Optimization, and Machine Learning. Addison-Wesley Publishing Company*, Inc.

Guan, J., & Lin, G. (2016). Hybridizing variable neighborhood search with ant colony optimization for solving the single row facility layout problem. *European Journal of Operational Research*, *248*(3), 899–909. doi:10.1016/j.ejor.2015.08.014

Haupt, R. L., & Haupt, S. E. (2004). *Practical Genetic Algorithms* (2nd ed.). Wiley Interscience.

Hauschild, M., & Pelikan, M. (2011). An introduction and survey of estimation of distribution algorithms. *Swarm and Evolutionary Computation*, *1*(3), 111–128. doi:10.1016/j.swevo.2011.08.003

Higgins, A. (2006). Scheduling of road vehicles in sugarcane transport: A case study at an Australian sugar mill. *European Journal of Operational Research*, *170*(3), 987–1000. doi:10.1016/j.ejor.2004.07.055

Jason, B. (2011). *Clever Algorithm* (1st ed.). Melbourne, Australia: LuLu.

Khilwani, N., Shankar, R., & Tiwari, M. K. (2012). *Facility layout problem: an approach based on a group decision-making system and psychoclonal algorithm*. Academic Press.

Kumar, K. R., Hadjinicola, G. C., & Lin, T. (1995). A heuristic procedure for the single-row facility layout problem. *European Journal of Operational Research*, 87.

Ou-Yang, C., & Utamima, A. (2013). Hybrid Estimation of Distribution Algorithm for Solving Single Row Facility Layout Problem. *Computers & Industrial Engineering*, *66*(1), 95–103. doi:10.1016/j.cie.2013.05.018

Ripon, K. S. N., Glette, K., Khan, K. N., Hovin, M., & Torresen, J. (2013). Adaptive variable neighborhood search for solving multi-objective facility layout problems with unequal area facilities. *Swarm and Evolutionary Computation*, *8*, 1–12. doi:10.1016/j.swevo.2012.07.003

Ripon, S. K. N., Glette, K., Mirmotahari, O., Høvin, M., & Tørresen, J. (2009). *Pareto Optimal Based Evolutionary Approach for Solving Multi-Objective Facility Layout Problem*. Academic Press.

Said, H., & El-Rayes, K. (2013). Performance of global optimization models for dynamic site layout planning of construction projects. *Automation in Construction*, *36*, 71–78. doi:10.1016/j.autcon.2013.08.008

Samarghandi, H., & Eshghi, K. (2010). An efficient tabu algorithm for the single row facility layout problem. *European Journal of Operational Research*, *205*(1), 98–105. doi:10.1016/j.ejor.2009.11.034

Samarghandi, H., Taabayan, P., & Jahantigh, F. F. (2010). A particle swarm optimization for the single row facility layout problem. *Computers & Industrial Engineering*, *58*(4), 529–534. doi:10.1016/j.cie.2009.11.015

Seyyedhasani, H., & Dvorak, J. S. (2017). Using the Vehicle Routing Problem to reduce field completion times with multiple machines. *Computers and Electronics in Agriculture*, *134*, 142–150. doi:10.1016/j.compag.2016.11.010

Solimanpur, M., Vrat, P., & Shankar, R. (2005). An ant algorithm for the single row layout problem in flexible manufacturing systems. *Computers & Operations Research*, *32*(3), 583–598. doi:10.1016/j.cor.2003.08.005

Uno, Y., & Hirabayashi, N. (2009). *Facility Layout Method Using Evolution Strategies with Correlated Mutations*. Academic Press.

Urdan, T. C. (2011). *Statistics in Plain English. Routledge: Taylor & Francis Group* (3rd ed.). New York: Routledge.

Wang, K., Choi, S. H., & Lu, H. (2015). A hybrid estimation of distribution algorithm for simulation-based scheduling in a stochastic permutation flowshop. *Computers & Industrial Engineering*, *90*, 186–196. doi:10.1016/j.cie.2015.09.007

Zhang, Y., & Li, X. (2011). Estimation of distribution algorithm for permutation flow shops with total flowtime minimization. *Computers & Industrial Engineering*, *60*(4), 706–718. doi:10.1016/j.cie.2011.01.005

ADDITIONAL READING

Adrian, A. M., Utamima, A., & Wang, K.-J. (2015). A comparative study of GA, PSO and ACO for solving construction site layout optimization. *KSCE Journal of Civil Engineering*, *19*(3), 520–527. doi:10.100712205-013-1467-6

Amaral, A. R. S. (2009). A new lower bound for the single row facility layout problem. *Discrete Applied Mathematics*, *157*(1), 183–190. doi:10.1016/j.dam.2008.06.002

Ceberio, J., Irurozki, E., Mendiburu, A., & Lozano, J. A. (2012). A review on estimation of distribution algorithms in permutation-based combinatorial optimization problems. *Progress in Artificial Intelligence*, *1*(1), 103–117. doi:10.100713748-011-0005-3

Drira, A., Pierreval, H., & Hajri-Gabouj, S. (2007). Facility layout problems: A survey. *Annual Reviews in Control*, *31*(2), 255–267. doi:10.1016/j.arcontrol.2007.04.001

Heragu, S. S., & Alfa, A. S. (1992). Experimental analysis of simulated annealing based algorithms for the layout problem. *European Journal of Operational Research*, *57*(2), 190–202. doi:10.1016/0377-2217(92)90042-8

Heragu, S. S., & Kusiak, A. (1991). Efficient Models for the facility layout problem. *European Journal of Operational Research*, *53*(1), 1–13. doi:10.1016/0377-2217(91)90088-D

Hungerländer, P., & Rendl, F. (2013). A computational study and survey of methods for the single-row facility layout problem. *Computational Optimization and Applications*, *55*(1), 1–20. doi:10.100710589-012-9505-8

Sule, R. D. (2009). Manufacturing Facilities Location, Planning, and Design (3rd ed.). CRC Press, Taylor & Francis Group.

KEY TERMS AND DEFINITIONS

Algorithm: A method that conducts a series of specified functions to solve a problem.

Dataset: The problem data that commonly used and solved in literature.

Heuristics: A method that is designed to solve a problem faster to find an approximate solution when classic methods cannot find an exact solution.

Iteration: The whole looping inside an algorithm that need to be done to produce a good solution.

Local Optima: A best solution that is relative within a set of neighbor solution.

Probabilistic Model: A function that is built with probability concept.

Single Row Layout: A layout arranged in single line.

Chapter 4
Offshore Structures:
Fire-Based Structural Design Criteria

Mavis Sika Okyere
Ghana National Gas Company, Ghana

ABSTRACT

Fire will always be a major threat to the offshore structure as oil and gas always passes through the installation. The design against accidental fire situation should be included in the structural design of offshore structures in collaboration with safety engineers. The design of offshore structures for fire safety involves considering fire as a load condition, assessment of fire resistance, use of fire protection materials, and so on. This chapter presents a methodology that will enable an engineer to design an offshore structure to resist fire. It aims to highlight the major requirements of design and to establish a common approach in carrying out the design.

INTRODUCTION

During a major fire incident, it is our aim to be able to sustain the integrity of the offshore structure for an ample period to perform critical task such as platform shutdown, fire-fighting response, disembarkation, etc, whiles permitting evacuation of operation and maintenance personnel.

The offshore structure must be able to maintain its safety functions for a required performance time, if subjected to fire and explosion damage. Therefore, there is a need to design the offshore structure to have redundant elements appropriately placed, be able to withstand increased loadings, and have the ability to redistribute loadings.

Offshore structural engineers concerning structural fire endurance use the slenderness ratio and the wall thickness modulus. Protecting people from injuries is therefore linked to designing structures. Structures should be designed to withstand loads without creating dangerous fragments or falling down.

According to Design Buildings Ltd (2018), the key design options to ensure fire safety are:

- **Prevention:** Controlling ignition and fuel sources so that fires do not start.
- **Communications:** If ignition occurs, ensuring occupants are informed and any active fire systems are triggered.

DOI: 10.4018/978-1-5225-7059-2.ch004

- **Escape:** Ensuring that occupants and surrounding areas are able to move to places of safety.
- **Containment:** Fire should be contained to the smallest possible area, limiting the amount of property likely to be damaged and the threat to life safety.
- **Extinguishment:** Ensuring that fire can be extinguished quickly and with minimum consequential damage.

This chapter describes the major items that need to be incorporated in the structural design of an offshore structure for fire resistance. Structural engineers in collaboration with safety engineers should perform the structural fire design of offshore structures.

Objective

- The main objective is to reduce to within standard limits the possibility for death or injury to the occupants of the offshore structure and others who may become involved, such as maintenance and operation personnel, fire and rescue team, as well as to protect contents and ensure that the structure can continue to function after a fire and that it can be repaired.
- To characterize the offshore fire problem and develop an engineering methodology to improve the offshore platform fire resistance by extending structural fire endurance.

DESIGN LOAD

The offshore structure design must account for a variety of loads demands including, dead and live loads, environmental loads (ie. Wind, wave, ice and current loads), seismic loads, operational loads (such as drilling) and transportation and construction loads, accidental loads such as collision, fire, hydrocarbon explosion should also be accounted for during the design process (Bea, Williamson & Gale, 1991).

The structure must be adequately robust in terms of capacity, redundancy, and ductility, to transfer the environmental and deck loads to the pile foundation without loss of serviceability over a wide range of conditions.

Fire as a Structural Load

Fire is a combustible vapour or gas that combines with an oxidiser in a combustion process that is manifested by the evolution of light, heat, and flame (Paik, 2011). An ignition source, fuel and a supply of oxygen are the three main components required for a fire to start. Since it is difficult to eliminate oxygen from a building, fire prevention tends to concentrate on the other two components.

The British standard BS 5950 Part 8 (BSI 1990b), and Eurocode 3 part 1.2 (CEN 2000b, hereafter referred to as EC3) are the main source of information for fire resistant design of steel structures. For structural fire resistant design, reduced partial safety factors for structural loads should be used. Reduced load factors for the fire limit sate design can be obtained from BS 5950 Part 8 (BSI 1990b), and Eurocode 3 part 1.2 (CEN 2000a).

Determine relevant fire sizes and loads to be applied for structural integrity analyses. This can be used as input to the structural calculation programs (e.g. FAHTS/USFOS) for advanced structural fire design (Lloyds, 2014).

The British Standard adopts load ratio whiles in Eurocode load level is used. Hence, the load level at exposure time in fire condition can be expressed in equation as given in Eurocode 3 and by Lawson and Newman as

$$\eta_{fl,t} = \frac{E_{f,d,t}}{R_d} = k_{y,\theta}\left[\frac{\gamma_M}{\gamma_{M,fl}}\right] \leq \frac{R_{f,d,t}}{R_d}. \tag{1}$$

where

$\eta_{fl,t}$ = load level at exposure time in fire condition
$E_{f,d,t}$ = actions effect at the fire limit state (N, Nm)
R_d = design value of resistance (N, Nm)
$R_{f,d,t}$ = reduced memb resistance (design resistance in fire condition) (N, Nm)
γ_M = partial safety factor on materials
$\gamma_{M,fl}$ = partial safety factor on materials in fire condition
$k_{y,\theta}$ = reduction factor for yield strength with temperature, θ

The Equation above is considered effective at the fire limit state.

Fire Limit State

In the limit state definition, the safety of structures is assured when

$$\Sigma \gamma_f E \leq \frac{R}{\gamma_M} \tag{2}$$

where

γ_f = the partial safety factors on actions
γ_M = partial safety factors on materials
R = member resistance (N, Nm)
E = Young's modulus (N/m^2); actions effect (N, Nm)

At the fire limit state, the partial safety factors for permanent and variable actions are put to unity to account for the likelihood of reduced loading in the event of a fire thereby leading to the design value of the actions adjusted as follows

$$E_{fi,d,t} = G_d + \varphi_1 Q_{k,1} + \varphi_2 Q_{k,2} \tag{3}$$

where

$E_{fi,d,t}$ = actions effect at the fire limit state (N, Nm)
φ_1, φ_2 = load combination factors depending on the limit state under consideration
G_d = permanent actions (N)
$Q_{k,i}$ = variable actions (N, i = 1, 2,)

Explosion Load

Explosion is a sudden and violent release of energy the violence of which depends on the rate at which the energy is released. The energy stored in a car tyre for example can cause an explosive burst but it can be dissipated by gradual release.

Three basic types of energy released in an explosion are

1. **Physical Energy:** Includes energy in gases, strain energy in metals etc.
2. **Chemical Energy:** Derived from a chemical reaction usually combustion
3. **Nuclear Energy:** Is not considered here. It is chemical explosions and particularly gas phase explosions that are of interest.

Explosions may ether be categorized as *deflagrations* or *detonations*. In deflagrations, the flammable mixture burns relatively slowly.

Blast or explosion loading may also be categorized as *confined explosion* and *unconfined explosion*. Unconfined explosion occurs in the free air, may occur near the ground. Confined explosion occurs in fully confined, partially confined and fully ventilated area. Confined explosion occurs inside the structure, and the combined effect of high temperature, accumulation of gaseous products of chemical reaction in the blast may lead to the collapse of the structure if not designed to withstand internal pressure.

Often accidental release of gas will occur in some form of containment provided by a building or section of industrial plant. This will lead to a *confined gas explosion*. Under conditions of complete confinement, most fuel gases can produce a maximum pressure rise of about 8bar. Most buildings and heating plant are incapable of withstanding such pressures. However, internal gas explosions rarely cause complete destruction because either by design or fortuitously the pressure is relieved at an early stage of the explosion by failure of weak components (Harris, 1983). Such explosions are termed *vented confined explosions*.

Hydrocarbon Explosion Load

A hydrocarbon explosion is a process in which combustion of a premixed hydrocarbon gas-air cloud causes a rapid increase of pressure waves that generate blast loading (Farid et al, 2003). Common types of explosions include accidental explosions resulting from natural gas leaks or other chemical/explosive materials. The three elements necessary for explosion to occur are:

1. Oxidiser
2. Fuel

3. Ignition source

Gas explosions can occur inside process equipment or pipes, in buildings or offshore modules, in open process areas or in unconfined areas. The worst-case that is possible to occur on the platform is a hydrocarbon explosion that generates blast loading. A topside structure, which forms a percentage of the superstructure where most facilities of the process plant are sited, must be protected against hydrocarbon explosion loading. Usually, safety aspects and prevention measures on offshore installations are very tough; hence, the likely incident of an explosion is reasonably low.

Hydrocarbons can explode through ignition when combined with an oxidiser (usually air). Thus, when the temperature rises to the point at which hydrocarbon molecules react spontaneously to an oxidiser, combustion takes place. This hydrocarbon explosion causes a blast and a rapid increase in overpressure (Paik, 2011).

In an unconfined or free area, the pressure waves are released/ discharged in all possible directions within the duration of few milliseconds as a pressure impulse. The magnitude of overpressure for the unconfined area is lower than a confined/ restricted area.

Most facilities are in-placed at the topside, and it consists of structural members, piping, equipment, cables and other appurtenances that can hinder the free movement of these waves. Therefore, introducing congestion and confinement significantly increases the magnitude of overpressure loads (Farid et al, 2003).

In the Piper Alpha incident (1988), where one hundred and sixty seven (167) men died, a major part of the installation was burnt down. A series of explosions destroyed the Piper Alpha oil platform in the North Sea. An inquiry blamed the operator, Occidental, for poor maintenance and safety procedures. The incident started with a rather small gas explosion in a compressor module that caused a fire, which subsequently resulted in the rupture of a riser. The explosion-ruptured firewalls that were designed to withstand oil fires, but not gas explosions, and were not retrofitted when the rig was modified to accept gas.

It is therefore, a challenge for offshore structural engineers, especially for a new development project that the effects of blast loadings should be taken into serious consideration from the onset of the design stage (Ngo, 2007).

Overpressure Load

The impact of overpressure from explosions and that of elevated temperature from fire are the primary concern in terms of the actions that result from hazards within the risk assessment and management framework (Paik, 2011).

Quantitative Risk Assessment (QRA) study, provide engineers with the nominal overpressure values or alternatively the overpressure exceedance (probability of any particular level of overpressure being exceeded) for design purposes. A compilation of design overpressure is given in Table 1.

As shown in Figure 1, low overpressure values are dominated by a higher frequency of exceedance values and vice versa. The maximum peak pressure (at low frequency of exceedance) of 3.0 bar up to 4.0 bar is recommended for the design of primary supporting trusses, while nominal load for open deck flooring is recommended between 0.5 bar to 1.0 bar (at high frequency of exceedance), (Farid et al, 2003).

In a process module on an offshore platform, a large separator can block off venting across it, effectively behaving like a wall resulting in large over-pressure load. This effect can be significant even when confinement is minimal (Ali, 2007).

Table 1. Nominal blast overpressure

Blast Area	Pressure (bar)
Wellhead deck	2.0 – 2.5
Gas separation facilities	1.5 – 2.0
Gas treatment compression facilities	1.0 – 1.5
Process area, large or congested	0.5
Process area, small and not congested	0.2
Open drill floor	0.2
Totally enclosed compartment	4.0

(Ali, 2007).

Figure 1. The frequency of exceedance: Overpressure
(Ali, 2007)

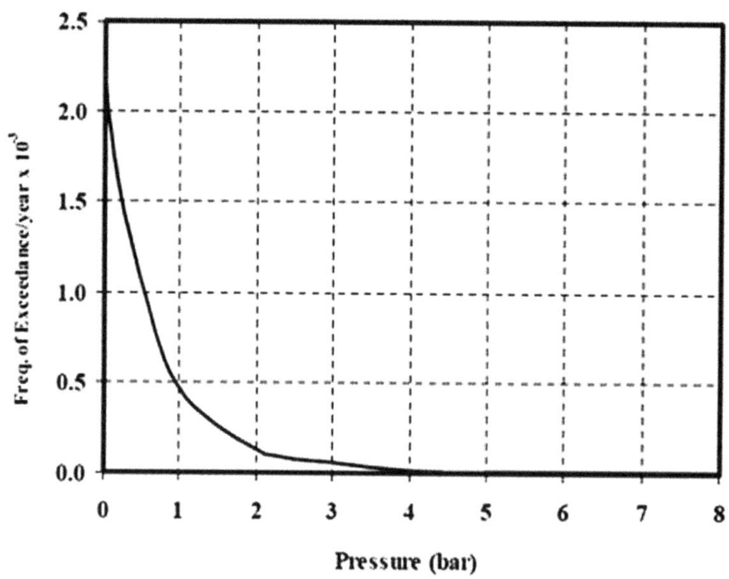

The criteria for failure of steel members, firewalls, and risers under hydrocarbon fire impact are shown below. They are derived from various sources (Technica) for the piper Alpha investigation. Under explosion overpressures, the following criteria was established:

- Decks are blown out at 9psi
- Firewalls and steel walls are blown out at 2psi
- Process equipment within modules is ruptured at 5psi
- External module walls are punctured at 1.5psi
- Structural failure of a platform could occur due to an explosion of a vapour cloud located below the module support frame, ie. Surrounding the jacket. A vapour cloud explosion occurring above or beside the upper deck level would only be expected to severely impact the topsides.

- No structural impact is anticipated from a "flash-fire".

Drag Load

Gas explosion on an offshore platform will generate drag loads in the offshore installations because of their high velocity affecting pipe work and vessels possibly leading to damage to safety critical systems, loss of containment of hydrocarbon and escalation of the event.

Structural engineers have become aware of overpressure effects through working on the design of blast walls, steel work and consideration of escalation effects.

The Steel Construction Institute's "Interim guidance notes" (SCI IGN, 1992), are used extensively for reference and can be used to evaluate the structural performance of structural steel work under blast load (Ali, 2007).

The wind of the blast load causes drag loads (drag force), and the impact of these drag forces is significant on small objects such as electrical cable trays, piping. Drag force is calculated as

$$F_D = C_D \frac{1}{2} \rho A u_{gas}^2 \tag{4}$$

where

F_D = drag force or drag load from the blast per unit length
ρ = density of gas
u_{gas} = flow velocity of gas
A = reference area or cross-sectional area (projected area of the object normal to the flow direction)
C_D = drag coefficient for blast (it is dimensionless),

Equation above can be expressed in terms of gas temperature as

$$F_D = \frac{1}{2} C_d \rho_{amb} \left(283 / T\right) A u_{gas}^2 \tag{5}$$

where

ρ_{amb} = the density of the air at 283 K
T = the temperature of the burnt gas (K)

All other parameters are as defined above.

The density of gas and velocity of gas flow are unpredictable, therefore Equation (4) is modified into a simplified expression as follows (Eknes & Moan, 1994)

$$q_D = C_D p\left(t\right) D \tag{6}$$

where

q_D = line load on pipe
$p(t)$ = maximum overpressure time-history
D = outer pipe diameter
C_D = drag coefficient (it is dimensionless)

In many types of topside or deck, grating have been installed to improve ventilation and allow venting of possible explosion gases. Grating allows partial venting of the explosions allowing some of the gases to pass through. The resultant loading on the grating is a combination of the overpressure differential across the grating and drag load created by the gases passing through the voids in the grating (Ali, 2007).

Therefore, loading from a gas explosion on grating is mainly from drag. The drag loading on platform grating can be calculated assuming the following:

- Differential pressure across constituent elements of the grating (i.e. plates and cylinders) is negligible.
- The total load on the grating plate is the sum of drag load acting on individual constituent elements.
- Drag coefficients are similar to those measured under steady state situations.
- Flow is turbulent.
- The aspect ratio of the constituent element: length/*D* is large, where *D* is the diameter in the case of cylinders, and the width in the case of plates.

The appropriate drag coefficient (CD) for all the elements comprising the grating is 2.0. Drag loading on a grated deck is calculated as

$$Drag_{grating} = \frac{1}{2} C_d \rho A u_{gas}^2 \tag{7}$$

where, A is the cross-sectional area presented by the grating of the wind.

The drag loading for a square metre of grating is

$$Drag_{grating} = 0.4 \rho u_{gas}^2 \tag{8}$$

Missile Load

The consequent effect of the drag loads are missile loads. The rupturing of equipment causes all kinds of flying objects as well as fragments; these are considered missiles (Ali, 2007). The equation for calculating the missile load is generated as

$$F_{missile} = \frac{1}{2} C_d \rho A \left(u_{gas} - u_{missile}\right)^2 \tag{9}$$

where

$F_{missile}$ = drag force on the missile
$u_{missile}$ = velocity of the missile

Tan and Simmonds (1991) proposed an equation for calculating the peak velocity using correlated maximum gas velocity ($u_{g\max}$), duration of the gas flow ($t_{g\max}$) and mass of missile to empirical parameter 'α' as

$$\alpha^2 = \frac{1}{2} C_d \rho A u_{g\max} \frac{t_{g\max}}{M} \quad (10)$$

where

$u_{g\max}$ = maximum gas velocity
M = mass of the missile
$t_{g\max}$ = duration of the gas flow

The peak velocity of the missile is calculated as

$$u_{missile} = u_{g\max} \left[1 - \frac{\sqrt{2}}{\alpha} \tan^{-1} \left(\tanh \frac{\alpha}{\sqrt{2}} \right) \right] \quad (11)$$

The Tam and Simmond's (1991) model was based on the following assumptions:

1. That the gas was burnt, and
2. That the velocity increased linearly to a peak and then decreased linearly with equal rise and fall times

External Explosion Load From High Explosives

An explosion is a rapid release of stored energy characterized by a bright flash and an audible blast. Part of the energy is released as thermal radiation (flash); and part is coupled into the air as air blast and into the ground as ground shock, both as radially expanding shock waves.

The explosion load is one of the most important parameters to determine whether the structural integrity is satisfactory. Explosion load is an uncertain parameter because it depends on variable factors, such as maximum pressure, rise time and pulse duration.

Compared with gases, solids possess thermal conductivities which are typically ten times as great and diffusivities which are many powers of ten lower. This coupled with their very high energy density produces very energetic combustion; either detonation or deflagration producing very high-pressures (Harris, 1983).

Detonating explosives are normally termed high explosives and may produce peak pressures of up to 300,000 atmospheres. Deflagrating explosives are termed low-explosives in the USA and produce peak pressures below 4000 atmospheres.

Of the high explosives, some are detonated by all normal ignition sources, eg heat, spark, mechanical impact, and are termed primary explosives. Secondary explosives on the other hand will only detonate under influence of an externally applied shock or detonation wave and merely deflagrate when ignited by a flame. Small quantities of primary explosives are used in detonators to induce explosion in bulk secondary explosives.

High explosives produce high amplitude, short duration pressure profiles compared with gas explosions and this is a weakness in the application of correlations based on high explosives to other explosive situations. Nevertheless, the TNT model is widely used for estimating the effects of explosions on plant and buildings (Harris, 1983).

Assuming the blast occurs at a distance outside the platform and the entire structure is to be affected by the blast waves. The maximum overpressure is calculated to determine loads that can be exerted on the platform due to blasting. The proposed equations by Kinney and Graham, and TM5-1300 diagram (see Figure 2) can be used to accurately estimate the overpressure due to blast (TM5, 1990).

Using Kinney and Graham suggestion for calculating maximum overpressure (P_s),

$$\frac{P_s}{P_0} = \frac{808\left[1+\left(\frac{z}{4.5}\right)\right]^2}{\left[1+\left(\frac{Z}{0.048}\right)^2\right]^{\frac{1}{2}} \cdot \left[1+\left(\frac{Z}{0.32}\right)^2\right]^{\frac{1}{2}} \left[1+\left(\frac{Z}{1.35}\right)^2\right]^{\frac{1}{2}}} \tag{12}$$

where

P_0 = ambient air pressure (atmospheric pressure), bar
P_s = peak static wave front overpressure, bar
Z = scaled distance in m/kg$^{1/3}$, $Z = \frac{R}{W^{\frac{1}{3}}}$
R = the distance from detonation point to the point of registered pressure
W =the weight equivalent TNT (Trinitrotoluene) charge, expressed in kg (see Table 2)

An equivalent TNT weight (W) is computed according using an equation that links the weight of the chosen design explosive to the equivalent weight of TNT by utilizing the ratio of the heat produced during detonation (Vasilis &George, 2013):

$$W = W_{\exp}\frac{H^d_{\exp}}{H^d_{TNT}} \tag{13}$$

where

W = the TNT equivalent weight (kg)
$W_{\exp}$ = the weight of the actual explosive (kg)
$H^{d}_{\exp}$ = the heat of detonation of the actual explosive (MJ/kg)
H^{d}_{TNT} = the heat of detonation of the TNT (MJ/kg)

Selection of the blast charge size 'W' is based on the perceived risk to the design structure and any structures nearby. Various factors play a role here, such as the social and economic significance of the structure, security measures that deter terrorists, and data from previous attacks on similar facilities. The minimum standoff distance 'R' is determined from the layout of a structure's surroundings and reflects the expectation of how close to the structure the design charge could explode. 'W' and 'R' are two important inputs for the scaled distance parameter Z (Stolz, Klomfass, & Millon, 2016).

The velocity of the blast wave propagation in the air (U_s) may be calculated as

$$U_s = \sqrt{\frac{6P_s + 7P_0}{7P_0}}a_0 \tag{14}$$

where, a_0 is the sound velocity in air (m/s)

The blast wave propagation in the air produces dynamic pressure, which is calculated as

Table 2. Indicative values of heat of detonation of common explosives

Explosives	Heat of Detonation (MJ/kg)	Characteristics	Uses
TNT (trinitrotoluene)	4.10 – 4.55	Safe to handle. Contains insufficient oxygen for complete combustion	Popular as military and commercial explosive. used in conjunction with ammonium nitrate - AMATOLS
C4	5.86	-	-
RDX (Cyclotrimethyl-enetrinitramine)	5.31 – 6.19	Thermally stable and powerful. Requires some densitization	Powerful military explosive. may be used with TNT
PETN (pentaerythritol tetranitrate)	6.69	Very powerful but too sensitive. Expensive to manufacture	Used as military explosive in combination with other explosives, eg with TNT-PENTOLITES
PENTOLITE 50/50	5.86	-	-
NITROGLYCERIN (NG)	6.30	Liquid. Very sensitive explosive when solid	Used in dynamites
NITROMETHANE	6.40	-	-
NITROCELLULOSE (NC)	10.60	Sensitive and dangerous explosive	Used in mixtures with nitro-glycerine - GELIGNITE. commercial and military explosives
AMON. /NIT. (AN)	1.59	-	-

(Vasilis &George, 2013)

$$q_s = \frac{5P_s^2}{2\left(P_s + P_0\right)} \tag{15}$$

As the blast waves collide with a perpendicular surface, reflection pressure is generated which can be evaluated using the following equation.

$$P_T = 2P_s\left[\frac{7P_0 + 4P_s}{7P_0 + P_s}\right] \tag{16}$$

The duration of the blast loading on a structure, (t_s) can be calculated by using the diagram presented by TM5-1300 or the equation proposed by Izadifard and Maheri (2010) as

$$\log_{10}\left(t_s / W^{\frac{1}{3}}\right) = \begin{cases} 2.5\log_{10}\left(Z\right) + 0.28, & For\, Z \le 1 \\ 0.31\log_{10}\left(Z\right) + 0.28, & For\, Z \ge 1 \end{cases} \tag{17}$$

Blast waves caused by gas or dust explosions differ from those of TNT or other high explosives as they generate smaller overpressures and larger impulses in the near field.

Due to their highly ductile features, structural steel frames provide additional ultimate resistance for a blast event exceeding in severity the design blast.

Bolted connections, such as those using top and bottom flange angles, can sustain significant inelastic deformations and sometimes are preferred in blast-resistant design (Stolz, Klomfass, & Millon, 2016).

The book by J. M. Biggs, which is a revision of an earlier book written by several authors including J. M. Biggs (1964), contains excellent simple methods for the design of structures subjected to blast loads produced by blast from nuclear weapons.

Brode (1955), proposed the following equation for calculating the peak static overpressure for near (when the p_s is greater than 10 bar) and for medium to far away (when the p_s is between 0.1 and 10 bar):

$$P_s = \begin{cases} \dfrac{6.7}{Z^3} + 1, & For\, P_s > 10bar \\ \dfrac{0.975}{Z} + \dfrac{1.455}{Z^2} + \dfrac{5.85}{Z^3} - 0.019, & For\, 0.1 < P_s < 10bar \end{cases} \tag{18}$$

where

Z = scaled distance
R = distance from centre of a spherical charge, m
W = charge mass expressed in kilograms of TNT
P_s = peak static wave front overpressure,

Figure 2. Comparisons between Izadifard and Maheri (2010) equation with TM5-1300 diagram

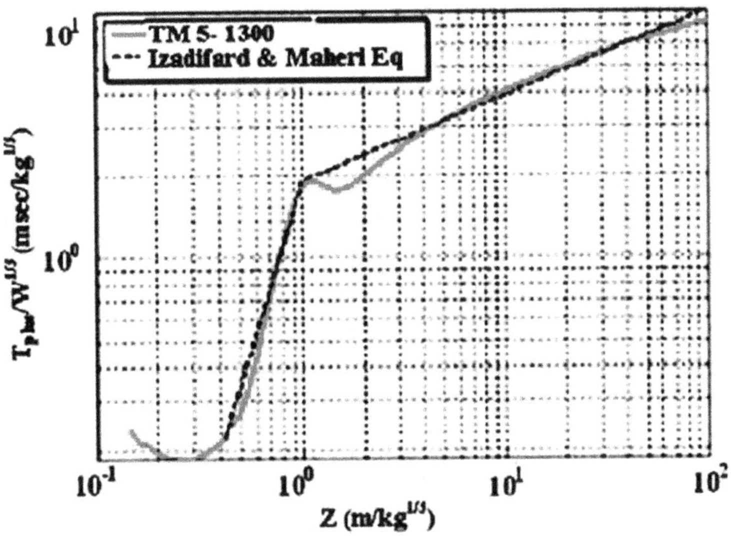

To calculate the maximum value of negative pressure (pressure below ambient pressure) in the negative phase of the blast, Brode (1955), proposed the following value for $\bar{P}$ as

For Z>1.6,

$$\bar{P} = -\frac{0.35}{Z} bar \tag{19}$$

where, $\bar{P}$ is the maximum value of negative pressure in the negative phase of the blast.

And the corresponding specific wave impulse at this stage, $\bar{i}_s$, is given by:

$$\bar{i}_s \approx i_s \left(1 - \frac{1}{2Z}\right) \tag{20}$$

Mills (1987), proposed the following

$$P_s = \frac{1772}{Z^3} + \frac{114}{Z^2} + \frac{108}{Z} - 0.019 kPa \tag{21}$$

where, Z = scaled distance (m/kg$^{1/3}$)

Newmark also proposed an equation for calculating peak overpressure for ground surface blast as

$$P_{so} = 6784 \frac{W}{R^3} + 93 \sqrt{\frac{W}{R^3}} \tag{22}$$

where

P_{so} = peak overpressure, (bars)
W = the charge mass in metric tonnes (=1000kg) of TNT
R = the distance of the surface from the centre of a spherical explosion (m)

Load Combination for Accidental Fire Situation

The applied load is obtained by considering the accidental combination of the mechanical actions such as: dead load, live load, wind (only for bracing), snow.

Due to the low probability that both fire and extreme severity of external actions occur at the same time, only the following accidental combinations are considered.

$$1.0G_k + y_1 Q_{k,1} + S y_{2,i} Q_{k,i} \quad (23)$$

where

G_K = the characteristic value of permanent actions (permanent / dead load)
$Q_{K,1}$ = the characteristic value of the main variable actions
$Q_{K,i}$ = the characteristic value of other variable actions
y_1 = the frequent value of the main variable actions
$y_{2,i}$ = the average of the other variable actions

Generally, in fire: $y_1 = 0.5$ and $y_{2,i} = 0$

Apart from bracings, $Q_{K,1}$ and $Q_{K,2}$ generally correspond to imposed loads and snow loads (ESDEP, 2017).

DESIGN OF TUBULAR MEMBERS

A member may be required to permanently, temporarily or only once carry a load. Structures can be designed to 'allowable stresses' or 'limit state' methods.

1. **Allowable Stress:** Permanent and temporary structures
2. **Ultimate Limit State:** Accidental damage

For the design of the tubular members of the offshore structures for accidental fire condition, there is a need to use the 'Ultimate limit state methods' for design. The stress calculations can be based on the requirements of chapter thirteen (13) of ISO 19902:2007.

Slenderness Ratio

The slenderness ratio is calculated as

$$slenderness\,ratio\left(\lambda\right)=\frac{effective\,length}{least\,radius\,of\,gyration} \tag{24}$$

$$\lambda=\frac{kL}{r} \tag{25}$$

where

kL = effective member length depending on end restraint conditions
L = actual member length
k = the effective length factor or effective length coefficient
r = cross-sectional radius of gyration

$$r=\left(\frac{I}{A}\right)^{1/2} \tag{26}$$

As slenderness ratio increases, permissible stress or critical stress reduces. Consequently, load carrying capacity also reduces.

Depending on the structure end conditions or boundary conditions, the value of 'k' may vary. The effective length factor, k, can be obtained from Table 13.5-1 of ISO 19902.

In the seismic zone, the slenderness ratio ($\frac{KL}{r}$) of primary bracing in vertical frames shall be limited to no more than 80 (with the corresponding column slenderness parameter λ, not exceeding $\frac{80}{\pi}\times\sqrt{\frac{fyc}{E}}$, and $fy\frac{D}{E}\times t\leq 0.069$

For values of, 30 <kL/r>100, the slenderness ratios are considered to be within the middle range.

Most designers aim to maintain slenderness ratios between 60 and 90; within this range, the member strength depends on the tangent modulus of the material and on end restraint design (Profire, 2014).

In a fire situation, ductile collapse of the structure may be accelerated in designs with high slenderness ratios, even though such designs may employ a higher degree of redundancy. Lower slenderness ratio encourages high D/t ratios for tubular members that may compound local buckling problems. The fire resistance of unprotected steel members and the endurance of the overall system capacity may be improved by limiting D/t and kL/r ratios of critical above-water braces to a maximum of 30 and 60 respectively, and optimizing structural configurations of the framing system using X-frame configurations to maximize residual strength (Profire, 2014).

For a brace, it is advisable to limit the slenderness ratio to 90, in order to avoid Euler and local buckling (James, 1988). A brace with a $\frac{kL}{r}>100$ is subject to Euler elastic buckling, which is independent of the material's yield strength. Designers seek to avoid both Euler and local buckling by limiting the upper end of the ratio less than 90 in order to take advantage of high strength steels (Bea, Williamson

& Gale, 1991). For optimum deck plate design, plate slenderness in the range of 70 to 100 is recommended for design. At this range, the change of response to blast is more predictable.

In seismic zones, the slenderness ratio of primary diagonal bracing in vertical frames is limited to a maximum of 80, and D/t ratio restricted to 1900/Fy (ksi). To reduce the wave induced lateral loads on the jacket structure, it is important to keep the diameter of the structural members subject to hydrodynamic forces as small as possible, thereby reducing drag. Use high slenderness ratio for the design to improve reserve strength. The residual strength is a measure of the offshore structures ability to sustain damage without failure.

Wall Thickness Modulus

The wall thickness modulus is used to classify a tubular section as thin or thick wall members, and is a measure of buckling resistance.

$$\frac{D}{t} \tag{27}$$

Where

D = diameter of tubular member
t = wall thickness of the tube

It is practical to keep the D/t ratios between 30 and 60. Tubular members with D/t ratios less than or equal to 60 are normally not subjected to local buckling from axial compression and can be designed based on material failure. D/t greater than 60 can present local buckling problems. By using compact sections for offshore designs, local buckling of tubular compression members can be avoided. Sizes of braces may be reduced to lower the drag force and loadings on the structure (Bea, Williamson & Gale, 1991).

Hence, thinner-wall tubular members are practically sensitive to failure from local buckling, either due to fabrication defects or due to thermal impact.

Members with D/t ratios below 25 are considered thick-walled and will not float. Therefore, their use offshore has been limited to date. On the other hand, Members with low D/t ratios have much greater inherent thermal mass and fire endurance than thinner wall members, and may find greater use in the future for above-waterline applications for reasons of their increased thermal robustness (Bea, Williamson & Gale, 1991).

FIRE RESISTANCE

The offshore structure must be designed and constructed in such a way that in the event of an outbreak of fire within the structure, the load-bearing capacity of the platform will continue to function until all occupants (operation and maintenance personnel) have escaped, or been assisted to escape, from the platform and any fire containment measures have been initiated.

The four methods used to assess and define the structural fire endurance or fire resistance of steel structural members are listed below.

1. Empirically derived correlations
2. Heat transfer analyses
3. Structural performance evaluation
4. Structural (mechanical fire) analysis

Using these approaches, the fire endurance of offshore structural components such as platform jackets and module support trusses (frames) can be analyzed and predicted (Bea, Williamson & Gale, 1991).

For fire resistant design of offshore structures, the standard hydrocarbon fire curve should be used (see Figure 3). This fire has a much faster rate of initial increase in temperature.

Fire resistance of a steel member is a function of its mass, its geometry, and the actions to which it is subjected, its structural support condition, fire protection measures adopted and the fire to which it is exposed. Design provisions to resist fire are briefly discussed below.

Empirically Derived Correlations

W/D Calculation Method

The W/D ratio is defined as the weight per unit length (W) of the steel member tested divided by the heated perimeter (D) of the member. Steel members with lower W/D ratio than the referenced or tested steel member require an increased fireproofing thickness. On the other hand, steel members with a higher W/D ratio than the referenced steel member may have the Sprayed fire-resistive materials (SFRM) thickness reduced.

W/D, A/P and M/D ratios are used to size structural steel members with the objective of fire protection. Usually, as the W/D, A/P and M/D ratio increases, the fire resistance increases and/or the required thickness of directly applied fire protection material decreases for a given rating (Profire, 2017).

$$\frac{W}{D} = \frac{M}{D} \times 0.017 \quad (28)$$

Figure 3. Standard fire curve

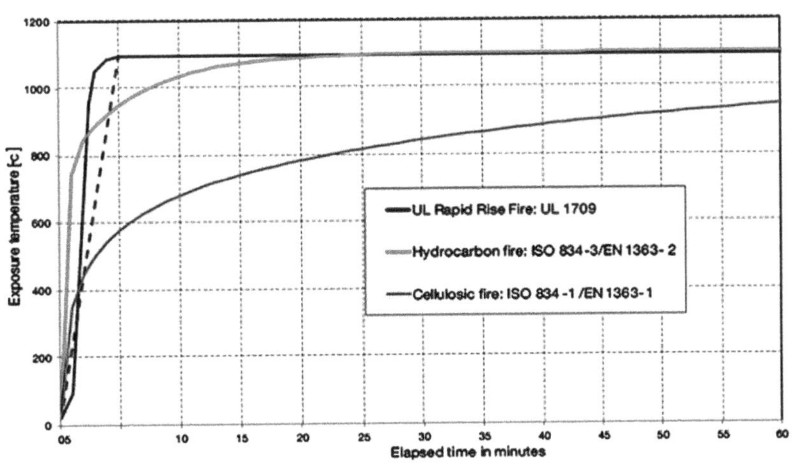

$$\frac{M}{D} = \frac{(W / D)}{0.017} \tag{29}$$

$$\frac{A}{P} = \frac{W}{D} \times \frac{144}{490} \tag{30}$$

where

D = Heated perimeter of steel section in inches (in.)
A = Cross sectional area of steel section (in^2)
P = Heated perimeter of steel section in inches (in.)
M = Mass of steel section (kg/m)
D = Heated perimeter of steel section in meters (m)

Structural steel columns should be insulated to avoid failure at temperatures of approximately 1000°F when exposed to fire. Laboratory experiments have revealed that for slenderness ratio from 42 to 112, the temperature at which the steel structure will fail can be approximately calculated by the equation below.

$$\left[1040 + 1.8(1 / r)\right] \pm 50F \tag{31}$$

The ensuing empirical relationship developed by Stanzak and Lie can be used to determine the fire endurance of steel columns (James, 1988).

For $\frac{W}{D} < 10$

$$R = 10.3\left(\frac{W}{D}\right)^{0.7} \tag{32}$$

And for $\frac{W}{D} > 10$

$$R = 8.3\left(\frac{W}{D}\right)^{0.8} \tag{33}$$

where

R = fire resistance in minutes or time in minutes for the column to reach 1000°F
W = linear density or weight of the steel section in lb/ft
D = heated perimeter of the steel section in inches

$$R = \left(C_1 W / D + C_2\right) h \tag{34}$$

where

R = fire endurance or fire resistance (hr)
W= steel weight per lineal foot (lb/ft)
D= heated perimeter of the steel at the insulation interface (in.)
h = thickness of insulation (in.)

The constants $C1$ and $C2$ need to be determined for each protection material. The constants take into account the thermal conductivity and heat capacity of the insulation material (James, 1988).

The W/D ratio is used to express characteristic fire resistance of steel members. W/D ratio is used by laboratories for normalizing structural steel fire resistive ratings for fireproofed members. Steel members having a higher W/D ratio than the rated member size (for a given thickness of fireproofing) are considered larger than the specified minimum size required for achieving the preferred degree of fire resistance (AkzoNobel, 2017).

W/D ratio has a disadvantage of not accounting for geometry of the structural member or how it is used. W/D ratio is used for wide-flange sections and A/P for hollow sections.

The thickness for the protection of steel beams is determined based on the following equation

$$h_1 = \left[\frac{\left(W_2 / D_2\right) + 0.6}{\left(W_1 / D_1\right) + 0.6}\right] h_2 \tag{35}$$

where

h_1 = thickness of spray-applied fire protection (in)
W = weight of steel beam (lb/ft)
D = heated perimeter of the steel beam (in.)

And where the subscripts

1=substitute beam and required protection thickness
h_2=the beam and protection thickness specified in the referenced tested design or tested assembly

Limitations of this equation are noted as follows

1. $W/D \geq 0.37$
2. $h \geq 3/8$ in. (9.5 mm)
3. The unrestrained beam rating in the referenced tested design or tested assembly is at least one (1) hour

The above equation only applies to the calculation of the protection thickness for a beam in a floor or roof assembly. All other features of the assembly, including the protection thickness for the deck, must remain unaltered (James, 1988).

A study by Aramco shows that unprotected self-supporting columns legs with 1.50" thick wall sections can withstand fire offshore for approximately one hour.Hp/A Calculation Method

Figure 4. Heated perimeter of steel beams
(James, 1988)

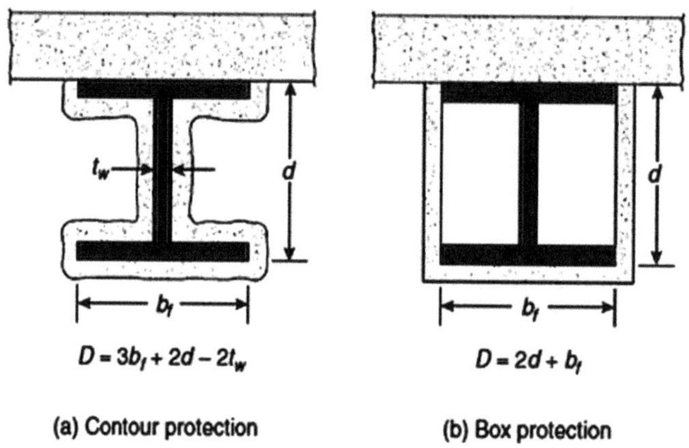

Table 3. Empirical equations for steel columns

Member/Protection	Solution	Symbol
Column/unprotected	$R = 10.3\left(\frac{W}{D}\right)^{0.7} \; for \frac{W}{D} < 10$ $R = 8.3\left(\frac{W}{D}\right)^{0.8} \; for \frac{W}{D} \geq 10$ For critical temperatures of 1000ºF	R = fire resistance time or time in minutes for the column to reach 1000ºF W = linear density or weight of the steel section in lb/ft D = heated perimeter of the steel section in inches
Column/gypsum wallboard	$R = 130\left(\frac{hW'/D}{2}\right)^{0.75}$ Where $W' = W + \left(\frac{50hD}{144}\right)$	h = thickness of protection (in.) W'= weight of steel section and gypsum wallboard (lb/ft)
Column/spray-applied materials and board products-wide flange shapes	$R = \left[C_1(W/D) + C_2\right]h$	C_1 & C_2 = constants for specific protection material

continued on following page

Table 3. Continued

Member/Protection	Solution	Symbol
Column/spray-applied materials and board products-hollow sections	$R = C_1\left(\frac{A}{P}\right)h + C_2$	C_1 & C_2 = constants for specific protection material The A/P ratio of a circular pipe is determined by $A / Ppipe = \frac{t\left(d - t\right)}{d}$ Where d = outer diameter of the pipe (in.) t = wall thickness of the pipe (in.) The A/P ratio of a rectangular or square tube is determined by $A / Ptube = \frac{t\left(a + b - 2t\right)}{a + b}$ Where a = outer width of the tube (in.) b = outer length of the tube (in.) t = wall thickness of the tube (in.)
Column/concrete cover	$R = R_0\left(1 + 0.03m\right)$ Where $R_0 = 10\left(W / D\right)^{0.7} + 17\left(\frac{h^{1.6}}{K_c^{\ 0.2}}\right)$ $\times\left\{1 + 26\left[\frac{H}{PC_c h\left(L + h\right)}\right]^{0.8}\right\}$ $D = 2\left(b_f + d\right)$	R_0 = fire resistance at zero moisture content of concrete (min.) m = equilibrium moisture content of concrete (% by volume) b_f = width of flange (in.) d = depth of section (in.) k_c = thermal conductivity of concrete at ambient temperature (Btu/hr.ft.°F. °F) h = thickness of concrete cover (in.)
Column/concrete encased	For concrete-encased columns use $H = 0.11W + \frac{PC_c}{144}\left(b_f d - A_s\right)$ $D = 2\left(b_f + d\right)$ $L = \left(b_f + d\right) / 2$	H = thermal capacity of steel section at ambient temperature (= 0.11W Btu/ft. °F) c_c = specific heat of concrete at ambient temperature (Btu/lb. °F) L = inside dimension of one side of square concrete box protection (in.) A_s = cross-sectional area of steel column (in^2)

(James, 1988)

The section factor (Hp/A) is the ratio of the fire exposed perimeter to the cross sectional area of the steel member. Essentially, it is a measure of how quickly the steel section will heat in a fire, and therefore how much fire protection is required. The advantages of using Hp/A methodology is that the British have developed values for this modulus based on extensive fire tests for most standard structural shapes, including tubular members. In recent years, this method has been approved for hydrocarbon fuelled fire scenarios and is now accepted by offshore operators and regulators in the North Sea.

Fire resistance test in accordance with the requirements of BS476, Part B, proves that for a totally stressed unprotected steel section, columns exposed on four sides that have a section factor, Hp/A, of up to 50m^{-1} can achieve a ½-hour fire rating (AkzoNobel, 2017).

For a tubular member,

$$Section\,Factor = \frac{Heat\,Perimeter\left(H_p\right)}{Cross-sectional\,Area\left(A\right)} = \frac{H_p}{A} \tag{36}$$

$$\frac{H_p}{A} = \frac{12.56\left(D\right)}{\left(D^2 - ID^2\right)} \tag{37}$$

where, D is the outer diameter of the member, and ID, is the internal diameter of the member.

Fire Resistance Test

Fire testing involves live fire exposures upwards of 1100 °C, depending on the fire-resistance rating and duration one is after. More items than just fire exposures are typically required to be tested to ensure the survivability of the system under realistic conditions.

Fire resistance test should be performed in accordance with the requirements of ASTM E119, BS476, Part B, API RP 6F, ISO 5660-1, ISO 5657, API Std 607, BS 6755-2, DIN 53436, FTP Code, HSE (UK) Offshore safety reports, IEC 60331, IEC 60332-3, ISO 834, ISO 1182, ISO 1716, IMO res. A.653(16), IMO res. A.754(18), and client requirement.

Basic test standards for walls and floors, such as BS 476: Part 22: 1987, BS EN 1364-1: 1999 & BS EN 1364-2: 1999 or ASTM E119.

Walls, floors and electrical circuits are required to have a fire-resistance rating.

Heat Transfer Equation

The general approach to studying the increase of temperature in structural elements exposed to fire is based on the integration of the Fourier heat transfer equation for non-steady heat conduction inside the member (ESDEP, 2017).

$$\frac{dQ_{cond}}{dx} = \frac{d}{dx}\left(\lambda_i \frac{dT}{dx}\right) = \rho c_p \frac{dT}{dt} \tag{39}$$

For a simplified calculation use $c_p = 600J / kgK$, $\lambda_i = 45W / mK$, $\rho = 7850kg / m^3$ for all grades of steel

Temperature Development of Steel Sections During Fire

$$\frac{dT}{dt} = \alpha \frac{d^2T}{dx^2} \tag{40}$$

$$\alpha = \lambda_i / \rho c_p$$

Figure 5. Thermal properties of steel at elevated temperature

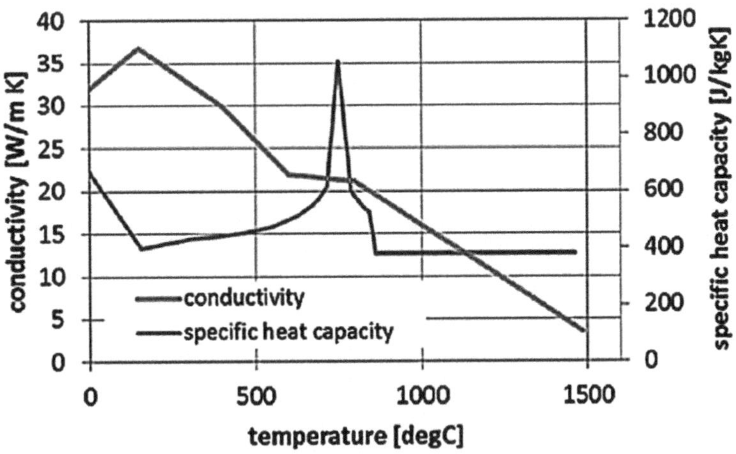

The quantity α is known as thermal diffusivity and varies with the temperature.

The quantity of heat transferred per unit length in the time interval Δt is

$$\Delta Q = \lambda_i A_m \left(T_f - T_s\right) \cdot \Delta t \tag{41}$$

where

λ_i = the total heat transfer coefficient (W/m$^{2°}$C)
A_m = the perimeter surface area per unit length exposed to fire (m^2/m)
T_f = the temperature of hot gases (°C)
T_s = the temperature of steel during the time interval Δt (°C)

If no loss of heat is considered, the internal energy of the unit length of a steel element increases by the same quantity ΔT_s

$$\Delta Q = C_p \times \rho \times A \times \Delta T_s \tag{42}$$

where

A = the cross-sectional area of the member (m^2).

The temperature rise of the steel is given by combining the two equations above as follows

$$\Delta T_s = \left[\lambda_i / \left(C_p / \rho\right)\right] \times \left[A_m / A\right] \times \left(T_f - T_s\right) \Delta t \tag{43}$$

Solving the above incremental equation systematically gives the temperature development of the steel element during the fire. In Eurocode 3 Part 1.2 (2005), it is suggested that

$$\Delta t \leq \frac{2.5 \times 10^4}{\left(\frac{A_m}{A}\right)} \tag{44}$$

Where

Δt is in seconds
A_m /A is in m^{-1}

There is a need to calculate the critical temperature of structural steel members, although 1000°F is usually taken as the critical temperature by AISC (2011). The critical temperature (θ_{cr}) which leads to the failure is calculated for a steel structure assuming a uniform temperature distribution along and across the members.

For unprotected elements the equation is

$$\theta_{cr} = 1.85t\left(A_m / A\right)^{0.6} + 50 \tag{45}$$

$$A_m / A = 0.36\left[\frac{\left(\theta_{cr} - 50\right)}{t}\right]^{1.67} \tag{46}$$

$$t = 0.54\left(\theta_{cr} - 50\right)\left(A_m / A\right)^{-0.6} \tag{47}$$

These three equations above are valid within following ranges: t = 10 to 80 min, θ_{cr} = 400 to 600°C.

In the same way, for sections protected by a light insulation material, the equations are

$$t = 40\left(\theta_{cr} - 140\right) \times \left[dA / \lambda_i A_m\right]^{0.77} \tag{48}$$

$$d = 0.0083\left[t / \left(\theta_{cr} - 140\right)\right]^{1.3}\left[A_m / A\right]\lambda_i \tag{49}$$

Where

d = the protection thickness (in meters)
λ_i = the thermal conductivity of the material (in W/m°C).

All other parameters are as defined above.

These equations can be expressed also in a nomogram that is very practical for the purpose of design (see figure below).

Heat Transfer Analysis

The purpose of the heat transfer analysis is to determine the time required for the structural member to attain a predetermined critical temperature or to provide input to a structural analysis. The temperature endpoint criteria cited by ASTM E119 are often accepted as the critical temperatures (James, 1988).

Heat transfer analysis should be performed to define the steel temperature taking into account the effects of radiation, convection, and conduction. The offshore platform should be able to resist accidental and catastrophic fires.

A simple FEA simulation may be used to evaluate the temperature distribution and relevant stresses (TMR, 2009). The modelling of a structure involves three stages.

1. The first stage is to model the fire scenario to determine the heat energy released from the fire and the resulting atmospheric temperatures within the platform.
2. The second stage is to model the heat transfer between the atmosphere and the structure. Heat transfer involves three phenomena (conduction, convection and radiation) all contribute to the rise in temperature of the structural materials during the fire event.
3. The third stage is the determination of the response of the structure – basic simple checks, engineering advanced models and sophisticated discrete models based on all data available.

Heat conduction, heat accumulation and exchange of radiation is calculated based on the refined FEM model. The transient behaviour is based on the following equation.

Figure 6. Relation between utilization factor, section factor and fire resistance (ESDEP, 2017)

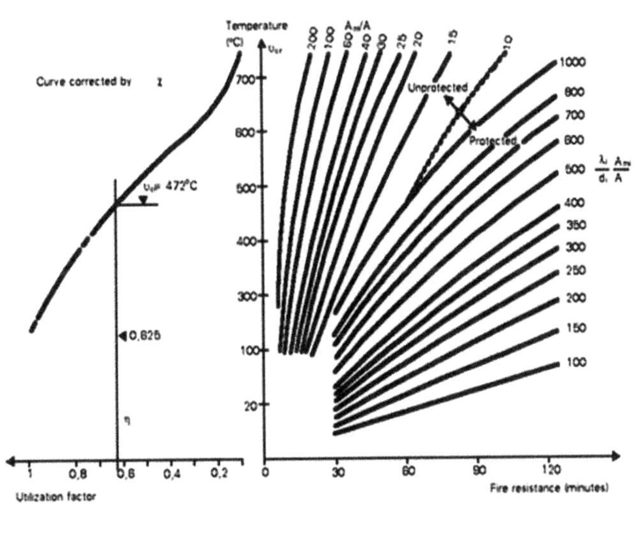

$$KT + C\dot{T} = Q \tag{38}$$

where

K = the conductivity matrix of the structure
C = the heat capacity matrix
T and T are the vector of nodal point temperatures and the corresponding rates respectively
Q = the heat load vector.

The above equation is solved in the time domain using a standard integration procedure. The temperatures in each structural member are transferred to the space frame model.

Structural (Mechanical Fire) Analysis

During fire, the load bearing resistance of the steel decreases, this leads to decrease of mechanical properties such as yield stress, young modulus, and ultimate compressive strength.

Applied Load

The applied load is calculated as

$$P = P_u\left(\theta_{cr}\right) \tag{50}$$

where

P = the applied load in fire conditions
θ_{cr} = the critical temperature
P_u = the load bearing resistance at room temperature

Load Combination for Accidental Fire Situation

The applied load is obtained by considering the accidental combination of the mechanical actions such as: dead load, live load, wind (only for bracing), snow.

Due to the low probability that both fire and extreme severity of external actions occur at the same time, only the following accidental combinations are considered.

$$1.0G_k + y_1 Q_{k,1} + Sy_{2,i} Q_{k,i} \tag{51}$$

where

G_K = the characteristic value of permanent actions (permanent / dead load)
$Q_{K,1}$ = the characteristic value of the main variable actions
$Q_{K,i}$ = the characteristic value of other variable actions

Figure 7. Examples of decrease of load-bearing resistance of a structure during exposure to fire (ESDEP, 2017)

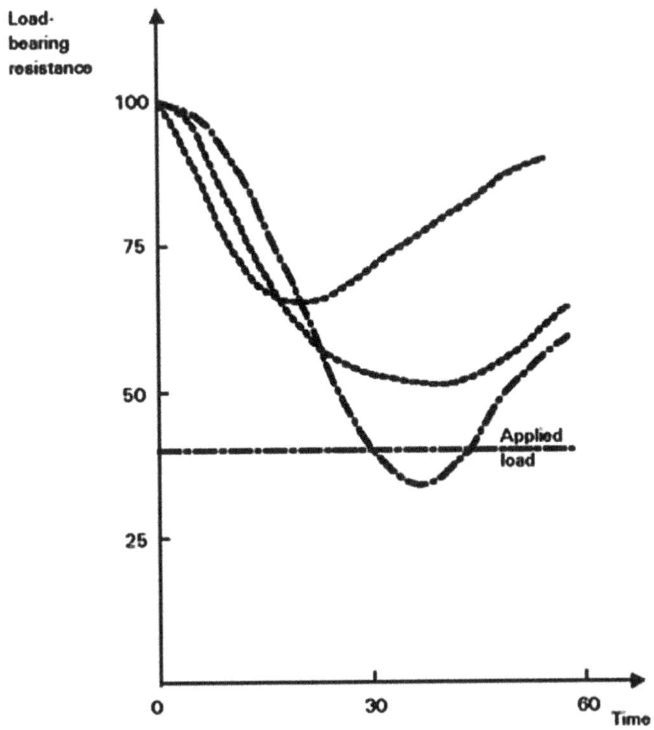

y_1 = the frequent value of the main variable actions
$y_{2,i}$ = the average of the other variable actions

Generally, in fire: $y_1 = 0.5$ and $y_{2,i} = 0$

Apart from bracings, $Q_{K,1}$ and $Q_{K,2}$ generally correspond to imposed loads and snow loads (ESDEP, 2017).

Calculating the Load Bearing Resistance of Steel Members

The critical temperature (θ_{cr}) leads to the failure and should be calculated for a steel structure assuming a uniform temperature distribution along and across the members.

Tension Member

At room temperature, the ultimate thermal resistance is calculated as

$$N_p = A \times f_y \tag{52}$$

where

N_p = ultimate thermal resistance
A = cross-section area of the member,
f_y = yield stress

At a given uniform temperature q, through the member, the ultimate tensile resistance is

$$N_p(q) = A \times y(q) \times f_y \tag{53}$$

where, y (q) is the strength reduction of steel at q,
The collapse of the member will occur at the temperature θ_{cr}. when

$$N_p(\theta_{cr}) = N \tag{54}$$

where

N = the applied load in fire conditions

The above Equation for 'N' can also be written as

$$A \times y(\theta_{cr}) \times f_y = A \times s \tag{55}$$

where

s = applied stress in fire conditions

Thus

$$y(\theta_{cr}) = \frac{s}{f_y} \tag{56}$$

or

$$y(\theta_{cr}) = \frac{A \times s}{A \times f_y} = \frac{N}{N_p} = \frac{P}{P_u} \tag{57}$$

Therefore, knowing P/P_u it is possible to determine the value of the steel critical temperature, θ_{cr} for which $y(\theta_{cr})$ is equal to $P/P_{u,}$ using Figure 8.

Figure 8. Parameters of structural steel at elevated temperature
(ESDEP, 2017)

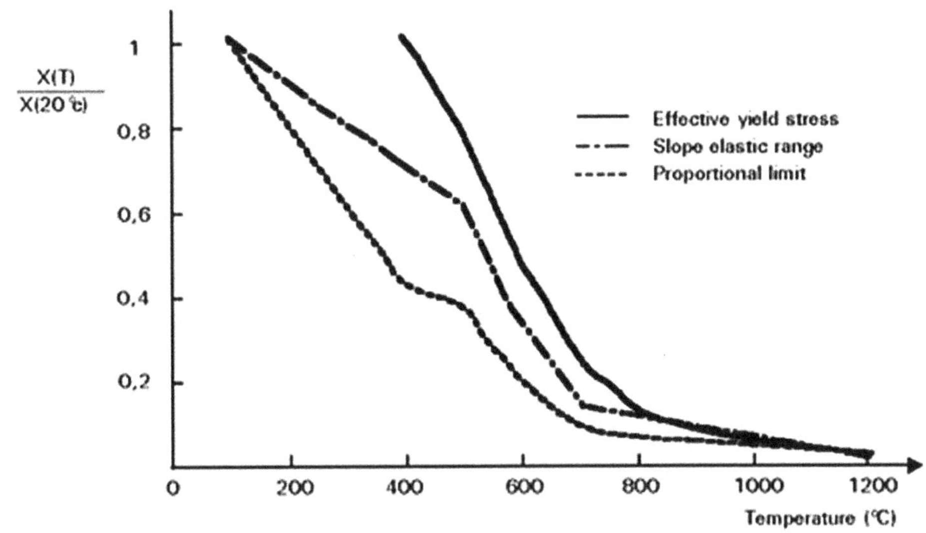

Columns

In the analysis of the columns the effects of column buckling is considered by modifying the ultimate bearing resistance by buckling coefficient.

$$y\left(\theta_{cr}\right) = kP / P_{u} \tag{58}$$

where, k is the correction factor, k = 1.2

The correction factor is used to compensate the choice of, f_y, which is related to the effective yield stress (the stress level at which the stress-strain relationship of steel tends to a yield plateau for a certain temperature) and not to the yield stress at 0.2% strain. Both P and Pu should be evaluated using the appropriate bucking coefficient.

$$\gamma\left(\theta\right) = \frac{1}{\varnothing\left(\theta\right)} + \left[\varnothing\left(\theta\right)^{2} - \lambda\left(\theta\right)^{2}\right]^{\frac{1}{2}} \leq 1 \tag{59}$$

where

$$\varnothing\left(\theta\right) = 0.5\left(1 + \alpha\left(\lambda\left(\theta\right) - 0.2\right) + \lambda\left(\theta\right)^{2}\right) \tag{60}$$

and

$$\lambda(\theta) = \frac{\lambda}{\pi}\sqrt{\frac{f_y(\theta)}{E(\theta)}} \approx \sqrt{\frac{f_y}{E}} \tag{61}$$

In the case of a braced frame in which each storey comprises a separate fire compartment with sufficient fire resistance, the Eurocode 3 part 1-2 (2005), recommends to use 0.5 as the effective buckling length coefficient (γ) of a steel column at high temperatures for columns when there are members of ordinary temperatures adjacent to both the upper and lower ends and to use 0.7 for columns on the top story.

However, the Recommendation for Fire Resistant Design of Steel Structures by AIJ advises us to use 1.0 regardless of the boundary conditions of the column at both ends since elongation of adjacent beams may cause local buckling of the column at the upper and lower ends and loosen the rotational restriction of the column.

Simply Supported Beams

For a simply supported beam which is uniformly loaded, the maximum bending moment is

$$M = \frac{PL}{8} \tag{62}$$

And the corresponding maximum stress is

$$s = \frac{M}{S_e} \tag{63}$$

where, S_e is the minimum elastic modulus of the section

Failure will occur when the total load on the beam is

$$P_u = \frac{8M_u}{L} \tag{64}$$

where

M_u = plastic bending moment resistance given by

$$M_u = z \times F_y \tag{65}$$

and z is the plastic modulus of the section

When the temperature is equal to θ, this plastic bending moment resistance is equal to

$$M_u(\theta) = z \times y(\theta) \times F_y \tag{66}$$

For a beam subject to a load of P, the collapse will occur at θ_{cr} when

$$P_u(\theta_{cr}) = P \; or \; M_u(\theta_{cr}) = M \tag{67}$$

That is when

$$y(\theta_{cr}) = S_e \times \frac{s}{z} \times F_y = \frac{s}{\left(\left(\frac{z}{s_e}\right) \times F_y\right)} = \frac{P}{P_u} \tag{68}$$

where, f=Z/S_e is the shape factor of the steel section (~ 1.10 to 1.3).

Continuous Beam

For a continuous beam, the maximum bending moment is

$$M = \frac{PL}{8} \tag{69}$$

In a fire situation, a plastic hinge will form at the middle support as the temperature increases when

$$M_u(\theta_1) = M \tag{70}$$

The load bearing resistance of this continuous beam is

$$P_u(\theta_{cr}) = \frac{12 M_u(\theta_{cr})}{L} \tag{71}$$

$$y(\theta_{cr}) = 8 \times S_e \times \frac{s}{12} \times z \times F_y = \frac{P}{P_u} \tag{72}$$

The ratio 12/8 = 1.5 = c is the statically indeterminate coefficient, or plastic redistribution coefficient.

Beam Column

When axial force and bending moment act together on the same structural element, its critical temperature can be obtained from the following formula

$$y\left(\theta_{cr}\right)=\frac{N}{c_{min}N_{p}}+\frac{k_{y}M_{y}}{M_{p}c_{y}}+\frac{k_{z}M_{z}}{M_{p}c_{z}} \tag{73}$$

where, c_{min} is the lesser of the buckling coefficients, c_y and c_z about the yy or zz axis and k_y and k_z are the reduction factors for the yy and zz axes respectively.

Steel Elements With Non-Uniform Temperature Distribution (ESDEP, 2007)

For non-uniform temperature distribution in the structure, global coefficient called the Kappa-factor accounts for the beneficial influence of thermal gradient for beams.

For a beam the general formula becomes

$$y\left(\theta_{cr}\right)=k\times\frac{s}{c}\times f\times F_{y} \tag{74}$$

where

k= 1 for simply supported beams exposed to fire on all sides
k= 0.7 for simply supported beams exposed on 3 sides
k= 0.85 for continuous beams exposed on all sides
k= 0.60 for continuous beams exposed on 3 sides

Load Bearing Resistance of Composite Components

Structural components, members, and structural steel frame systems should be designed to maintain their load-bearing function and to satisfy other performance requirements specified for the client.

Figure 9. Effect of non-uniform temperatures on bending capacity (ESDEP, 2017)

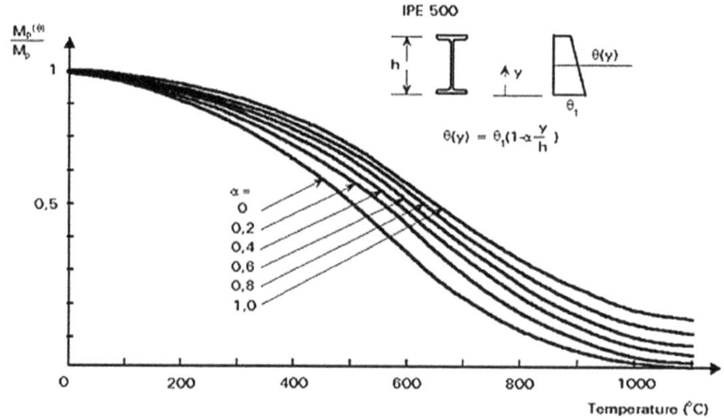

There is a need to calculate the load bearing resistance of composite columns, composite slabs and composite members as a whole. Also, determine the fire resistance of connections between columns and beams.

Composite Beam

The tensile force summed over the three parts of the steel section is

$$T = \sum_{i=1}^{3} A_i y\left(\theta_i\right) F_y \tag{75}$$

where

A_i = the area of lower flange, web and upper flange of the steel profile
θ_i = the respective temperature

The point of application of this force is the plastic neutral axis at elevated temperature of the three (3) parts of the steel action.

In order to balance this tensile force, a layer of the concrete slab is compressed such that

$$T = c = btf_{ck} \tag{76}$$

where

b = the effective width of the slab
t = the thickness of the compressive zone
f_{ck} = the ultimate strength of concrete

The ultimate bending resistance of the composite section is

$$M_{u(\theta)}^{+} = T \times z \tag{77}$$

where, z is the distance between the points of application of the tensile and the compressive forces.

For continuous beams, the determination of the full load bearing resistance also requires the calculation of the negative plastic bending moment ($M_{u(\theta)}^{-}$) assumed that stability in fire is maintained if the isostatic bending moment (M) of the applied load in a span is lower or equal to the sum of the positive and negative moment resistances of the composite section,

$$M_{u(\theta)}^{+} + M_{u(\theta)}^{-} \geq M \tag{78}$$

Composite Slabs

The calculation of the fire behaviour of composite concrete slabs with profiled steel sheet is made using the same theories as for composite beams.

In this state after 30 minutes of ISO fire, the steel sheet is not taken into account when calculating the mechanical behaviour of the element (ESDEP, 2017).

Structural Performance Evaluation

Design the offshore structure to have sufficient residual strength and structural robustness. Design the offshore structure for damage tolerance. The structural frame and foundation should be able to provide the strength and deformation capacity to withstand, as a system, the structural actions developed during the fire within the prescribed limits of deformation. The structural system shall be designed to sustain local damage with the structural system as a whole remaining stable.

Deflection Criteria

Calculate the total deflection and rate of deflection for loaded and heated steel beams by considering elastic and plastic strain due to the applied load, thermal strain due to thermal expansion, and creep strain (James, 1988).

Compare the calculated deflection and rate of deflection with the Robertson-Ryan criteria. Deflection of unheated beams requires computer solutions.

Robertson-Ryan Criteria

Based on the examination of fifty (50) fire test reports, Robertson and Ryan (1959), found that structural failure of floors and beams could be satisfactorily defined as the point at which

$$y_c \geq \frac{1}{800}\frac{L^2}{h} \tag{79}$$

$$\frac{dy_c}{dt} \geq \frac{1}{150}\frac{L^2}{h} \tag{80}$$

The temperature criterion of failure for truss, joist, during fire exposure is defined as

$$T_k = \frac{\frac{\Delta H}{R}}{\ln\left[37.5\left(\frac{L_k}{L}\right)Z_k\right]} \tag{81}$$

The temperature criterion of structural failure for beam during fire exposure is defined as

$$T_{ac} = \frac{\frac{\Delta H}{R}}{\ln\left[\left(\frac{150}{\pi^2}\right) Z_{ac}\right]} \quad (82)$$

where

ΔH = activation energy of the creep
y_c = midspan deflection resulting from fire exposure (inches or centimetres)
$\frac{dy_c}{dt}$ = deflection rate (in/hr or cm/hr)
L = span length (inches or centimetres)
h = distance between upper and lower extreme fibres of structural components (inches or centimetres)

Bresler and Iding prefers the use of Robertson and Ryan performance criteria based on midspan deflection and the rate of change. The following limit state is usually used to predict failure of an end-supported beam or floor/roof assembly subject to a standard fire test (Bea & Williamson, 1991).

Failure occurs if

$$D \geq (L / 800)(L / d)\, and\, R \geq (L / 150)(L / d).$$

where D is midspan deflection in inches, R is the hourly deflection rate (in/Hr), L is the span length in inches, and d is the distance between the upper and lower extreme fibers of the member in inches.

Critical Temperature

The critical temperature is defined as the temperature at which the material properties have decreased to the extent that the steel structural member is no longer capable of carrying a specified load or stress level (James, 1988). Eurocode allows the use of the critical temperature method for tension members and restrained beams but not columns or unrestrained beams

$$\theta_{cr} = 39.19 \ln\left[\frac{1}{0.9674 \mu_0^{3.833}} - 1\right] + 482 \quad (83)$$

where

θ_{cr} = critical temperature
μ_0 = is the degree of utilization which is the ratio of the design loading in fire to the design loading resistance at ambient temperature

$$\mu_0 = \frac{E_{fi,d}}{R_{fi,d,o}} \tag{84}$$

$E_{fi,d}$ = the design effect of actions for the fire design situation as outlined in EC1-1-2.

$R_{fi,d,o}$ = the design resistance of the isolated member as described in EC3-1-2 clause 4.2.3 but at a time t=0, i.e. ambient temperature

Default values for critical temperatures can be obtained from BS EN 1993-12:2005

The critical steel temperature which must be protected against should be defined. For example, structural steel between 200 and 750 C, vessels between 200 and 350 C, or a 140 C temperature rise for divisions where the critical temperature requirement is much lower to protect personnel on the other side of the division or in a safety refuge.

Critical Stress Equations

Elastic design

$$\sigma_{yT} = \frac{1}{F_e}\frac{Z_e}{Z_p} \tag{85}$$

Plastic design

$$\frac{\sigma_{yT}}{\sigma_y} = \frac{1}{F_p} \tag{86}$$

where

σ_{yT} =critical yield stress at elevated temperature, T

σ_y = yield stress at ordinary room temperature

F_e = factor of safety, elastic design

F_p = factor of safety, plastic design

Z_e = elastic section modulus

Z_p = plastic section modulus

For $0<T\leq 600°C$,

$$\sigma_{yT} = \left[1 + \frac{T}{900 \ln\left(\frac{T}{1750}\right)}\right] \sigma_{y0} \tag{87}$$

$$E_T = \left[1 + \frac{T}{2000 \ln\left(\frac{T}{1100}\right)}\right] E_0 \tag{88}$$

For T>600°C,

$$\sigma_{yT} = \left[\frac{340 - 0.34T}{T - 240}\right] \sigma_{y0} \tag{89}$$

$$E_T = \left[\frac{690 - 0.69T}{T - 53.5}\right] E_0 \tag{90}$$

$$\alpha_T = \left(0.004T + 12\right) \times 10^{-6} \tag{91}$$

where

σ_{yT} =yield strength at temperature T (MPa) (psi)
σ_{y0} = yield strength at 20°C (68°F) (MPa) (psi)
E_T = modulus of elasticity at temperature, T (MPa) 9psi)
E_0 = modulus of elasticity at 20°C (68°F) (MPa) (psi)
α_T = coefficient of thermal expansion at temperature T (m/m°C)
T = steel temperature (°C)

Critical Load

The critical load is defined as the minimum applied load that will result in failure if the structural member is heated to a temperature, T. The critical load can be expressed as a point load or distributed load. Critical load calculations can be performed using algebraic equations or a computer program.

Residual Strength

Residual strength is the load or force that a damaged object or material can still carry without failing. Material toughness, fracture size and geometry as well as its orientation all contribute to residual strength. Residual strength is a measure of the platforms ability to sustain damage without failure.

Find the system members that are most critical to maintaining structural integrity. Detect potential weak links in the system and understand how progressive collapse may occur. Residual strength analysis may be carried out using the AFGROW software.

Steel jacket-type offshore platform can realize considerably greater natural fire endurance solely through the application of fundamental design considerations. Specifically, the fire resistance of unprotected steel members, and the endurance of the overall system capacity may be improved by limiting D/t and kL/r ratios of critical above-water braces to a maximum of 30 and 60 respectively, and optimizing structural configurations of the framing system using X-frame configurations to maximize residual strength (Bea & Williamson, 1991).

Failure Criteria

1. **Ultimate Strength:** The simplest failure criterion assumes that failure occurs at the ultimate (or yield) strength of the material. Thus, failure occurs when

$$\sigma_f = F_{tu} \tag{92}$$

where

σ_f = the fracture stress
F_{tu} = the ultimate strength

This criterion is applicable primarily to uncracked structures.

2. **Fracture Toughness - Abrupt Fracture:** Fracture toughness is the ability of a material (e.g., steel pipe) to resist stress and prevent cracks in the material from spreading.

Technically speaking, fracture toughness is the ability of a material to deform under increasing tensile stress in the presence of a defect or crack without exhibiting rapid and extensive fracture propagation. Materials that have high fracture toughness can absorb larger amounts of energy (i.e., can withstand higher pressures or levels of stress) before an existing crack spreads.

The fracture toughness is a measure of the material's resistance to unstable cracking. In a fractured structure, the stress intensity factor (*K*) correlates the local stresses in the region of the crack tip with crack geometry, structural geometry, and the level of load on the structure. When the applied load level increases, the *K* value also increases and reaches a critical value at which time the crack growth becomes unstable.

The Irwin's failure criterion states that abrupt fracture occurs when the crack-tip stress-intensity factor reaches or exceeds the fracture toughness of the material. The corresponding applied stress at failure is called the fracture strength. The failure occurs when

$$K \geq K_{cr} \tag{93}$$

where

$$K = \sigma\sqrt{\pi a}$$

For plane stress

$$K_{cr} = \sqrt{EG_c} \tag{94}$$

For plane strain

$$K_{cr} = \sqrt{\frac{EG_c}{1-\upsilon^2}} \tag{95}$$

K_{cr} = the material's fracture toughness
υ =Poisson's ratio
G = strain energy release rate
σ = applied stress
a = half the crack length
E = young modulus

Robustness

Robustness is defined according to EN 1991-1-1, Part 1-7, as the ability of a structure to withstand events like fire, explosions, impact or the consequences of human error, without being damaged to an extent disproportionate to the original cause. A robust structure has the ability to redistribute load when a load carrying member experience a loss of strength or stiffness and exhibits ductile rather than brittle global failure modes (Szarka, 2015).

In Eurocode EN 1990:2002 (CEN 2002), the basic requirement to robustness is given in clause 2.1 4(P): "A structure should be designed and executed in such a way that it will not be damaged by events such as explosion, impact, and the consequences of human errors, to an extent disproportionate to the original cause."

Robustness can be measured using the following three approaches:

1. Risk-based robustness index
2. Reliability – based robustness index: also known as Probabilistic robustness index

3. Deterministic robustness indexes

Offshore platforms should be provided with adequate robustness level in order to avoid progressive collapse. There must be adequate thermal robustness in the design to provide the time (endurance) required to accomplish critical tasks such as platform shutdown, fire-fighting response, disembarkation, and so on (Bea & Williamson, 1991).

Risk-Based Robustness Index

The risk-based robustness index is defined as

$$I_{rob} = \frac{R_{dir}}{R_{dir} + R_{ind}} \tag{96}$$

Where, I_{rob} is the robustness index, R_{dir} and R_{ind} are the direct and indirect risk respectively.

The values of I_{rob} can be between zero (0) and one (1), with larger values signifying larger robustness. I_{rob} can have values very close to 1.0 with relatively large direct and small indirect risks (Szarka, 2015).

Reliability-Based Robustness Index

Reliability – based robustness index is also known as probabilistic robustness index

1. **Redundancy Index:** Redundancy index (RI) is defined as

$$RI = \frac{P_{f(damaged)} - P_{f(intact)}}{P_{f(intact)}} \tag{97}$$

where, $P_{f(damaged})$ and $P_{f(index)}$ are probability of failure for a damaged and intact system.

The values of RI can be between zero (0) and infinity (∞), the smaller values meaning more robustness (i.e. $P_{f\,(damaged)}$ is not much higher than $P_{f\,(intact))}$.

2. **Redundancy Factor:** Redundancy factor is defined as

$$\beta_R = \frac{\beta_{intact}}{\beta_{intact} - \beta_{damaged}} \tag{98}$$

where, β_R is the redundancy factor, $\beta_{intact}\ and\ \beta_{damaged}$ are the reliability indexes of the intact and damaged system (Szarka, 2015).

The value of β_R changes between 1 and ∞, and the higher the redundancy factor (β_R), the more robust the structure is.

3. **Reliability Index:** Reliability index is defined as

$$\beta = \varnothing\left(1 - P\left(F\right)\right)^{-1} \tag{99}$$

where, β is the reliability index, Φ is the cumulative normal distribution function, and P (F) is the failure probability.

Deterministic Robustness Indexes

1. **Reserve Strength Ratio:** According to Faber (2007), the reserve strength ratio (RSR) is calculated as

$$RSR = \frac{R_c}{S_c} \tag{100}$$

Where, R_c and S_c are the base shear capacity and design value in ULS, the R_c value coming from a pushover analysis. The value of RSR can be between 1 and ∞, where the bigger number denotes more unaccounted capacity (Szarka, 2015).

2. **Residual Influence Factor:** A simple and practical measure of structural redundancy (and robustness) used in the offshore industry is based on the so-called RIF-value (Residual Influence Factor), (ISO 19902: 2008). In order to measure the effect of damage (or loss of functionality) of structural member, i, on the structural capacity, the so-called RIF-value can be defined sometimes referred to as the Damaged Strength Ratio. The RIF's values can be between zero (0) and one (1), with larger values indicating larger robustness or larger redundancy (Szarka, 2015).

To specify better, the effect of losing one particular member (I) the RIF value *(damaged strength ratio)* is defined as

$$RIF_i = \frac{RSR_{fail,i}}{RSR_{intact}} \tag{101}$$

where, $RSR_{fail,i}$ is the RSR value of the platform given that member '*i*' has failed.

3. **Stiffness Based Robustness Measure:** Robustness can also be measured by using the determinant of the static stiffness matrix of the structural system

$$R_s = \min \frac{\det K_j}{\det K_o} \tag{102}$$

where, R_s is the stiffness based robustness measure, K_0 is the stiffness matrix of the intact structure, and K_j is the stiffness matrix of the structure with the given member removed.

This expression needs further normalization in order for it to give a value between 0 and 1, as written in Haberland (2008).

Energy-Based Structural Robustness Criterion

This is a new structural robustness index and is used to analyse the structure in terms of energy balance. The general expression of the structural robustness index, I_R, is given by (André et al, 2015) as

$$I_R\left(A_L \mid H\right) = \frac{Damages\ up\ to\ unavailable\ collapse\ state\ for\ hazard\ h}{Damages\ up\ to\ collapse\ state\ for\ hazard\ h} \tag{103}$$

$$I_R\left(A_L \mid H\right) = \frac{D_{uc} - D_{1st\ failure}}{D_c - D_{1st\ failure}}\ with \begin{cases} 0 \leq I_R \leq 1 \\ D_c - D_{1st\ failure} = 0 \rightarrow I_R = 1 \end{cases} \tag{104}$$

where

A_L represents the leading action

$H = \{h_1, h_2 \ldots.h\ldots h_n\}$ is a set of hazard scenarios. For example, a set of different actions with determined values applied in a given sequence

$D_{1st\ failure}$ = damage energy of the structure when the 'first failure' state takes place for hazard scenario considered

D_{uc} =the damage energy corresponding to the state where collapse is unavoidable, the "unavoidable collapse" state, for the hazard scenario considered.

D_c =the damage energy corresponding to the collapse state for the hazard scenario considered

If the value of the structural robustness index (I_R) is equal to 1.0, then it means the structure is very robust. For the hazard scenario considered, if the structural robustness index is zero it means the structure completely lacks optimization in terms of structural robustness or is less robust.

Ductility

Ductility is the ability of the material to endure and resist after yielding energy and thus allowing energy to be dissipated in a stable manner and stresses to be reallocated without substantial deterioration of the structure's performance. Mild steel is an example of a ductile material that can be bent and twisted without rupture. Use sections with low width-thickness ratios and adequate lateral bracing. High strength steels are generally less ductile (lower elongations) and generally have a higher yield ratio. High strength steels are generally undesirable for ductile elements. Provide connections that are stronger than members. Recognize that compression member buckling is non-ductile.

The material ductility is attained by material strain-hardening and /or by material deformation capacity.

There is a need to design the offshore structure to have redundant elements appropriately placed, be able to withstand increased loadings, and have the ability to redistribute loadings (i.e. ductility).

In ductility design, consider strain rate sensitivity, strength and stiffness degradation, cyclic behaviour, and low cyclic fatigue of materials and joints.

If ductile members are used to form a structure, the structure can undergo large deformations before failure. This is beneficial to the users of the structures, as in case of overloading, if the structure is to collapse, it will undergo large deformations before failure and thus provides warning to the occupants. This gives a notice to the occupants and provides sufficient time for taking preventive measures. This will reduce loss of life.

Ductility permits redistribution of internal stresses and forces, increases strength of members, connections and structures, permits design based on simple equilibrium models, results in more robust structures, provides warning of failure, permits structure to survive severe earthquake loading.

The jacket must be sufficiently robust in terms of capacity, redundancy, and ductility, to transfer the environmental and deck loads to the pile foundation without loss of serviceability over a wide range of conditions. In the case of fire, failure is realized as a result of progressive ductile collapse (Bea & Williamson, 1991).

Ductility Ratios

All structural steel has a minimum strain capacity of 17% at low strain rates. The offshore platform should have sufficient toughness against brittle fracture not to limit strain capacity significantly at the high strain rates associated with blast response.

The strain limits for evaluating the effect of temperature may depend on the class of steel used for the design. The limiting strains for different classes of steel sections are specified in design codes.

It is important to know the shape and plastic hinges that form in order to allow the maximum strain to be calculated. Reduce the concept of strain limit to a limiting deformation (i.e is ductility ratio). Ductility ratio is defined as

$$Ductility\,ratio = \frac{total\,deformation}{deflection\,at\,elastic\,limit} \tag{105}$$

The deflection at elastic limit (γ_{el}) is the deflection at which bending behaviour can be assumed to change from elastic to plastic. Transition from elastic to plastic does not occur at a specific deflection. The following assumptions are made to define deflection at elastic limit (γ_{el})

- Type of loading
- Beam fixity
- Shape of stress-strain curve
- Rate of loading and hence hinge formation

Ductility ratio is an important parameter for designing structures against explosion- induced forces. The ductility ratio (μ) of the steel member can also be defined as the ratio of the ultimate displacement at failure (Δu) to the displacement at the yield point (Δy).

$$\mu = \frac{\Delta_u}{\Delta_y} \tag{106}$$

The ductility reduction factor is calculated as

$$R_\mu = F_c / F_y \tag{107}$$

where

Fc = the ultimate base shear in the linear elastic behaviour
Fy = the ultimate base shear in the non-linear Elasto-plastic response.

For the duration of hydrocarbon explosions on offshore installations, the ductility of joints is the key safety element which relates to structural performance as well as the level of damage. Primary members of the installation must not collapse and provide safe escape after the event; all main connections should not have yield strength much higher than expected which can overload and prevent yielding of adjacent members. Therefore, in blast engineering where safety is the main concern, requisite of ductility for materials and the responses against the accidental loads are critical issues, which can only be achieved through a better understanding of fracture characteristics, both brittle and ductile.

The ductility ratios currently being used in design of structural components are obtainable from design codes. The structural member is considered to fail if the ductility ratio has exceeded 20 as given by TM 5 1300.

Duration of the Load and Natural Period of Vibration

It might be expected that the damage produced in a confined explosion would depend simply on the relative magnitude of the peak pressure generated and the pressure required to fail the confining structure. In reality, it is more complex. An explosion produces a pressure loading which varies with time and the response of the structure or structural component is time-dependent.

Structure response depends on:

$$\frac{t_d}{T} = \frac{duration\ of\ imposed\ load}{natural\ period\ of\ vibration} \tag{108}$$

Ie the ratio between the duration of the imposed load and the natural period of vibration of the structure There are three (3) basic types of response

1. $t_d > T$: Here the loading experienced will effectively be equivalent to a static load equal to the peak explosion overpressure.

2. $t_d \approx T$: Here the loading experienced will effectively be equivalent to a static loading of a magnitude greater than the peak overpressure. The equivalent static overpressure can be up to $\frac{\pi}{2}$ times the incident overpressure
3. $t_d < T$: Here the pressure is effectively partially absorbed and the loading experienced will be equivalent to a static loading lower than the explosion peak overpressure ie a structure can withstand a higher dynamic pressure than static load necessary to cause failure.

From the above, gas explosions will produce structural response type (i) ie (t_d > T)

Thus, gas explosions could be considered approximately as static loadings equal to the peak overpressure.

Fracture Criteria

A fracture criterion is a standard against which the expected fracture behaviour of a structure can be judged. In general terms, fracture criteria are related to the three levels of fracture performance, namely plane strain, elastic plastic, or fully plastic.

An offshore platform rig is an example of a structure where complex loads are presents in the junctions. It is possible to carry fracture tests on these structures with a very large and highly sophisticated equipment. However, this is very expensive and time consuming. Radon (1974), proposed a simplest approach of this problem. Gilles work in 1977 reported not less than thirty-five (35) mode of fracture criterion; some of these fracture criterions are

1. Sih's criteria
2. Mandel's criterion
3. Strifors' criterion

The Strifors's criterion seems to be the best fracture criteria (Radon, 1974).

Table 4. Typical periods for structural elements and explosion durations

Item	Period/Duration (msec)
Concrete floors	10 – 30
Concrete walls	10 - 15
Brick walls	20 - 40
Confined gas explosions	100 – 300
Detonation of explosive charge	1 - 10

(Harris, 1983)

Fire Partitioning

Topside and other parts of an offshore platform consist of load-bearing, non-load-bearing, and partitioning components. The load-bearing components are to withstand service loads during fire; partitioning components are to prevent the spread of fire to adjoining spaces. Fire resistance is quantified by strength, integrity and insulation. Strength is the ability of a structural member such as column, beam, load bearing wall, slab etc, to withstand the service loads during fire. It applies to any load bearing member. Integrity and insulation are prescribed for partitioning components of the platform such as walls, doors (e.g. firewalls, fire-rated doors, fire resisting ducts or dampers etc). The ability of a partitioning element to limit the rise of temperature on its unexposed side is termed as insulation; whiles the ability to prevent hot gases to reach unexposed side through cracks/fissures is called integrity. Fire resistance is measured by the time to which a structural member satisfies all three criteria (ie. strength, integrity and insulation) as applicable in fire; this is termed the fire rating of the member.

A partition is defined in British Standards as an "internal, dividing, non-loadbearing, vertical construction". In European (CEN) standards, it is defined as a non-loadbearing wall, and EOTA European Technical Approval Guideline for partitions (ETAG 003), is entitled "Internal partition kits for use as non-loadbearing walls". A partition may be used for space division within the offshore platform, to separate areas of different floors, or used as an independent lining to an external wall.

Safety guidelines have been set out to ensure that the spread of fire remains limited during an incident. Further, that the safety of those present is guaranteed as much as possible, that they can escape if necessary and that the fire can be fought. The necessary provisions can be included when the platform is designed and built. The guidelines are stated in ISO, EN standards, MODU code for drilling platforms, and the IMO FSC (Fire Safety Code) code for ships (Nutec, 2008).

A platform can be divided into a number of fire compartments. A production platform and an accommodation platform can be 'separated' from each other in this way. The intention of the compartmentalisation is to limit the spread of fire as much as possible. Compartmentalisation is achieved by placing partitions that are classified as class A, class B, class C, and class H. Partitions of different strengths can be placed depending on the degree of compartmentalisation. In class A, B, and H the number after the letter shows how long (in minutes) the partitions are fire resistant or fire-retardant. After the stated number of minutes, the rise in temperature on the side not exposed to fire is so high that the fire will spread by means of fire transport. The fire will no longer be resisted/ delayed by the partition. H-120 is used especially as a partition wall between production platform and accommodation. This wall is resistant to the extreme heat of a liquid fire for 2hours. The start or the spread of a fire on an offshore installation should be prevented in the earliest stage possible (Nutec, 2008).

Use a fire-rated partition or fire resisting partition for which the fire resistance performance has been determined according to the appropriate British or European standards. Similarly, the reaction to fire performance of the exposed surfaces should also be determined by the appropriate fire test standards. It is important to determine the fire resistance and the reaction to fire performance of a partition. The fire resistance of non-loadbearing partitions is evaluated by:

1. BS 476: Part 20:1987: "Method for determination of the fire resistance of elements of construction", which details the general principles of fire resistance testing,
2. BS 476: Part 22:1987: "Methods for determination of the fire resistance of elements of non-loadbearing elements of construction", which details the procedures for testing partitions.

3. BS EN 1363-1, Fire Resistance Tests - Part 1 - General Requirements and
4. BS EN 1364-1, Fire Resistance Tests for Non-Loadbearing Elements: Part 1: Non-Loadbearing Walls

These test method measures two criteria of the partition's behaviour in the fire test: insulation and integrity.

A fire-rated partition will not allow hot gases to pass from the fire compartment to the surroundings by creating a structure (i.e., a compartment) which does not collapse and contains the fire for a given period. The fire resistance of such partition should range from 30 to 240 minutes (or more). In the event of fire, we have to ensure that surface materials are difficult to burn; the fire performance of boards, and wall coverings that comprise the outer faces of partitions should be subjected to the guidance of appropriate regulations.

Care must be taken to ensure that there are no adverse effects in performance between the fire door and the partition. If the ductwork is not fire rated, it must be fitted with a fire resistant damper(s). If the duct is fire-rated, a fire penetration seal must be incorporated to both faces of the partition between the duct and the partition. Electrical cables need to be sealed into the partition by a fire penetration seal compatible with both cables and partition. The fire penetration seal for pipes will need to be flexible to allow for structural movement and/or thermal expansion of the pipe(s). Some pipes may be hot; others may be cold, dependent on their use. Lagging is often applied to conserve heat, and any penetration sealant must be applied onto the pipe and not onto the lagging. The temperature of the pipe must be taken into account when choosing the penetration seal material. Strength and robustness classification should comply with BS 5234:1987.

Practical Ways of Achieving Fire Resistance of Steel Structures

There are three main methods of doing this:

1. **Oversizing:** For bare steel members, the fire resistance time can be increased by oversizing the members (i.e increasing the wall thickness), by maintaining the member size but using a higher strength steel, by utilizing the restraining effects of connections, or by a combination of these methods (see Figure 10).
2. **Insulation:** For protected steel members, the thickness of the insulation must be such that the temperature of the steel at the required fire resistance time (taking into account its section factor) does not exceed the critical (or limiting) temperature.

Fire resistive coating or Intumescent Coating may be applied at an appropriate thickness. The required thickness of insulation for a structural steel member may be determined by using a nomogram, which relates critical temperature, applied load, section factor and fire resistance (ESDEP, 2017).

Intumescent fireproofing is a layer of paint, which is applied along with the coating system on the structural steel members. The thickness of this intumescent coating is dependent on the steel section used. Intumescent coatings are paint like substances which are inert at low temperatures but which provide insulation by swelling to provide a charred layer of low conductivity materials at temperatures of approximately 200-250°C. At these temperatures, the properties of steel will not be affected. Most intumescent coatings can traditionally provide up to 60-120 minutes' fire resistance economically (e.g. thick film epoxy intumescent).

It should be noted that in the eventuality of a fire, the steel structure will collapse once the steel attains the critical core temperature (around 550 degrees Celsius or 850 degrees Fahrenheit). The Passive Fire Protection (PFP) system will only delay this by creating a layer of char between the steel and fire. Depending upon the requirement, PFP systems can provide fire ratings in excess of 120 minutes. PFP systems are highly recommended in infrastructure projects as they can save lives and property. Passive fire protection materials (PFP) such as fire resistant insulation products can be used either to envelope individual structural members, or to form fire walls that contain or exclude fire from compartments, escape routes, and safe areas.

Active fire protection (AFP) should be provided by water deluge and, in some instances by fire suppressing gas that is delivered to the site of the fire by dedicated equipment preinstalled for that purpose (AkzoNobel, 2017). Refer to API 2218, UL1709 and ISO 834.

For longer periods of fire resistance for floors with high loading and long spans, additional reinforcement may be necessary. Reinforce the walls with internal and/or external stiffeners. Fire and blast resistant walls should be used.

The fire resistance of tubular members can be improved by utilizing the hollow interior to cool the load-bearing steelwork. Filling such members with water gives extremely high fire resistance when circulation is maintained.

Flexible/blanket system could be used. Concrete encasement also provides fire resistance. Spray protection materials can also be used and some could be suitable for situations where the threat is from hydrocarbon fires. Fire doors should be encouraged for offshore applications.

Other areas of the topside that should be considered for fire resistance are nuts and bolts used in flanges (one of the weakest areas of any platform). Typical fire protection, which covers the complete flange, will not allow easy inspection of the units. By protecting only, the nuts, regular inspection can be performed, reducing installation time and overall weight. Using molded rubber-based material on just the flange nuts protects the stud bolts from elongating and the flange from breaking the seal during a fire.

3. **Dissipation:** Ensuring that heat applied to an assembly is rapidly dissipated to other materials or to the air, so that the temperature of the assembly is not raised to a critical level.

FIRE DESIGN OF STRUCTURAL CONNECTIONS

The design of connections for fire safety involves calculation of the heat transfer to them and a determination of the response of the structural elements of the connections. PrEN 1993-1-2 (prEN 1993-1-2: 2003) gives two methods for the design of steel connections. The first method involves applying a fire protection to the member and its connections. The level of protection is based on that applied to the connected members taking into account the different level of utilisation that may exist in the connection compare to the connected members (Moore & Wald, 2003). For the second method, we use an application of the component approach in prEN 1993-1-8 together with a method for calculation the behaviour of welds and bolts at elevated temperature. By using this methodology, the connection moment, shear and axial capacity can be calculated at elevated temperature (Moore & Wald, 2003).

For the fire design of structural connections, the following need to be considered

Figure 10. Fire resistance of bare steel beams as a function of section factor, different load levels and different resultant emissivity values
(ESDEP, 2017)

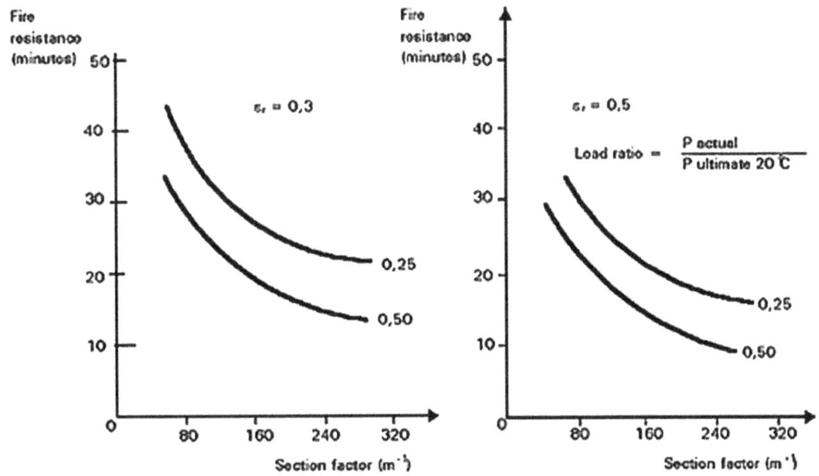

1. Bolt resistance at high temperature
2. Weld resistance at high temperature
3. Temperature distribution with time within a joint
4. Fire resistance of joints

The methodology for the fire design of structural connections is obtainable from prEN 1993-1-2, prEN 1993-1-8.

FIRE, EXPLOSION, AND BLAST EFFECT ANALYSIS

The overall structural response of the fixed offshore platform to explosion loads may be determined by any of the following two methods

1. Non-linear dynamic finite element analysis
2. Simple calculation models based on Single Degree Of Freedom (SDOF) analogies and elastic-plastic methods of analysis

Refer to Section 6.0 of DNV-RP-C204 for the design of offshore platform against explosive loads. Also in chapter 18.0 of API RP 2A, an assessment procedure for fire and blast is shown in Figure 18.2-1.

Single Degree of Freedom Analysis

Single-degree-of-freedom (SDOF) system is a system whose motion is defined just by a single independent co-ordinate (or function). SDOF systems are often used as a very crude approximation for a generally much more complex system.

When the blast response of individual components or assemblages is characterised by a dominant deflection mode, simplified analysis based on an equivalent single degree of freedom (SDOF) system can often be undertaken to evaluate blast resistance.

The single degree of freedom approximation can provide an accurate assessment of the explosion resistance of individual components within a structure, the key requirements being the realistic representation of the boundary conditions, the strain-rate effect and the beam-column action. For the assessment of the overall structural response, the use of nonlinear finite element analysis has become essential, particularly for modelling complex interactions between the structural components (Zzuddin, Lloyd, 1997).

Equation of Motion

The parameters for the simplified model are the mass (m), stiffness or spring constant (k), external force F (t), structural resistance R, and displacement, y.

The blast load can also be idealized as a triangular pulse having a peak force F_m and positive phase duration t_d (see Figure 11-12), (Corr &Tam, 1998). The forcing function is given as

$$F(t) = F_m\left(1 - \frac{t}{t_d}\right) \quad (109)$$

The blast impulse is approximated as the area under the force-time curve, and is given by

$$I = \frac{1}{2}F_m t_d \quad (110)$$

The equation of motion of the undamped elastic single degree of freedom system for a time ranging from zero (0) to the positive phase duration, t_d, is given by Biggs (1964) as

$$my + ky = F(t) \quad (111)$$

$$my + ky = F_m\left(1 - \frac{t}{t_d}\right) \quad (112)$$

The general solution is expressed as

Displacement:

$$y(t) = \frac{F_m}{k}\left(1 - \cos\omega t\right) + \frac{F_m}{kt_d}\left(\frac{\sin\omega t}{\omega} - t\right) \quad (113)$$

Velocity:

$$y(t) = \frac{dy}{dt} = \frac{F_m}{k}\left[\omega \sin \omega t + \frac{1}{t_d}\left(\cos \omega t - 1\right)\right] \tag{114}$$

In which ω is the natural circular frequency of vibration of the structure and *T* is the natural period of vibration of the structure, which is calculated as

$$\omega = \frac{2\pi}{T} = \sqrt{\frac{k}{m}} \tag{115}$$

$$k = \frac{3EI}{L^3} \tag{116}$$

From the natural frequency, the natural period can be calculated as

$$T = 2\pi / \omega \tag{117}$$

$$T = \frac{2\pi}{\omega} = 2\pi\left(m / k\right)^{0.5} \tag{118}$$

The maximum response is defined by the maximum dynamic deflection y_m, *which* occurs at time t_m. The maximum dynamic deflection y_m can be evaluated by setting dy/dt in Equation above equal to zero, i.e. when the structural velocity is zero.

where

ω = the natural frequency of SDOF system
T = natural period
t_d = duration or time taken for the overpressure to be dissipated
m = actual mass (total mass)
k = effective spring constant (or stiffness)

Duration 't_d' is related directly to the time taken for the overpressure to be dissipated (Corr &Tam, 1998).

The energy dissipated by damping is very small in a system exposed to a very short pulse such as an explosion load (Chopra, 2007). Thus, the damping effect is usually ignored when studying the structural response under a gas explosion load (Ki-Yeob, Kwang, JaeWoong, YongHee, & Jae-Myung, 2016).

Figure 11. Single degree of freedom (SDOF) system

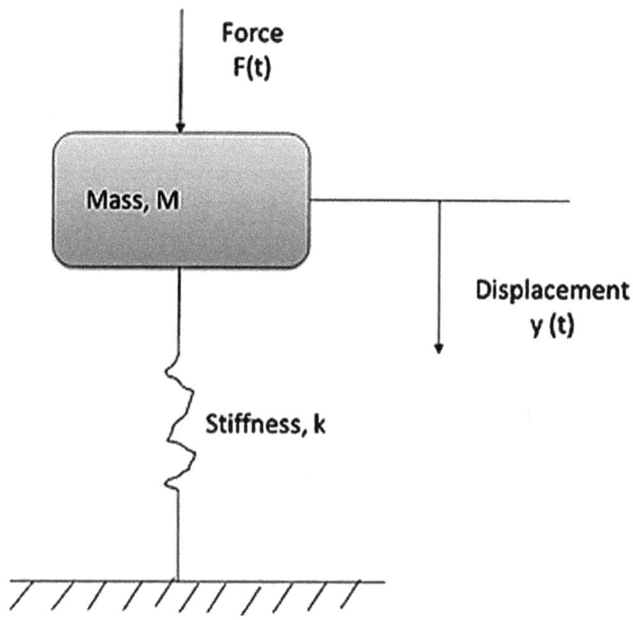

Figure 12. Blast loading

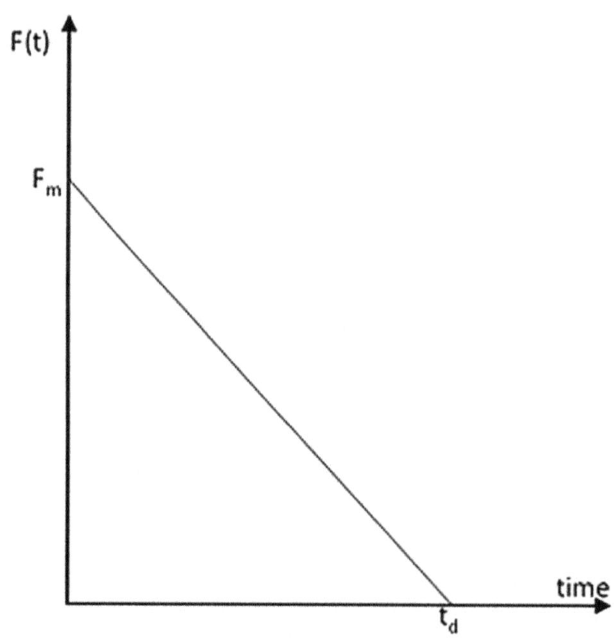

Dynamic Load Factor

The dynamic load factor (DLF) is defined as the ratio of the maximum dynamic deflection (y_m) to the static deflection (y_{st}) which would have resulted from the static application of the peak load F_m, which is shown as follows

$$DLF = \frac{y_{\max}}{y_{st}} = \frac{y_{\max}}{F_m / k} = \varphi\left(\omega t_d\right) = \varphi\left(\frac{t_d}{T}\right) \tag{119}$$

The structural response to blast loading is considerably influenced by the ratio $\left(\frac{t_d}{T}\right)$ or $\left(\omega t_d\right)$

$$\frac{t_d}{T} = \frac{\omega t_d}{2\pi} \tag{120}$$

Three loading regimes are categorized as follows:

1. Impulsive loading regime

$$\omega t_d < 0.4$$

2. Quasi-static loading regime

$$\omega t_d < 0.4$$

3. Dynamic loading regime

$$0.4 < \omega t_d < 40$$

The increase in the effect of a dynamic load is given by the dynamic amplification factor (DAF) or dynamic load factor (DLF). The dynamic amplification factor (DAF) is equal to the dynamic load factor (DLF).

Hinge Rotation

Hinge rotation is another measure of member response, which relates maximum deflection to span and indicates the degree of instability present in critical areas of the member. Hinge rotation has a symbol θ (see figure 13 below). Hinge rotation is also known as support rotation (Stolz, Klomfass & Millon, 2016).

In Figure 13, θ_a is hinge rotation at support, and θ_b is the hinge rotation at centre $\approx 2\,\theta_a$

Figure 13. Hinge rotation

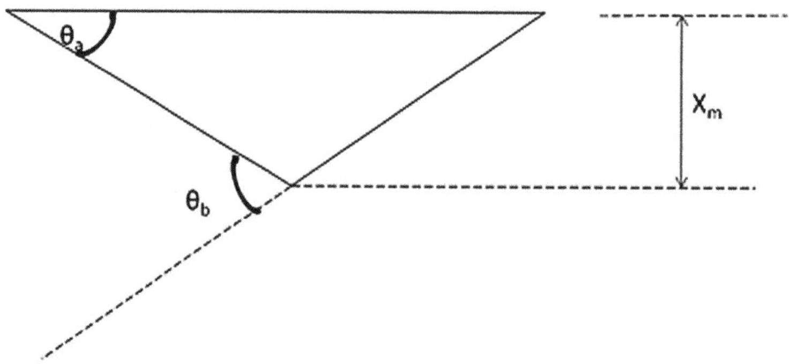

Elastic-Plastic Methods of Analysis

The nonlinear finite element approach is an accurate, even though computationally expensive, method for predicting the large displacement inelastic response of structures subject to explosion loading. Computational fluid dynamics (CFD) analysis is a very useful method that is able to model mechanical phenomena in complicated flow geometries (Ki-Yeob, Kwang, JaeWoong, YongHee, & Jae-Myung, 2016). The Flame Acceleration Simulator (FLACS) is a general tool used in the oil and gas industries; it is used to develop the potential overpressure caused by explosions in offshore facilities (Qiao, 2010). SACS or ABAQUS software can also be used for the blast load analysis.

Effects of Gas Explosions on Structures

Most analysis have been based on explosives. Although peak pressure magnitudes may be similar the very different pressure/time profiles of these and of gas explosions result in different levels and types of damage.

FIRE PROTECTION MATERIALS

Types of Fire Protection Materials

There are two basic types of fire protection: active and passive.

Active Fire Protection Material

Fire protection system, which, in event of fire, can function only after its operation has been manually or automatically, activated (ASFP, 2015). Active fire protection consist of

1. The means to deliver the large volume of water required to tank cooling and firefighting purposes.
2. Alarms, fire and gas detection systems within all process areas of the platform including living quarters

3. Firefighting equipment
4. Dry powder and sometimes foam fire suppression equipment or extinguishers, automatic fire sprinklers and water deluge systems. Refer to API 2218, UL1709 and ISO 834.
5. Effective communications systems so that the assistance of the emergency services can be quickly brought in.

Passive Fire Protection Material

Fire protection system that performs its functions without the need for manual or automatic start of its operation in the event of fire (ASFP, 2015).

Passive fire protection materials protect steel structures from the effects of the high temperatures that may be produced in fire. They can be divided into two types, non-reactive, of which the most common types are boards and sprays, and reactive, of which thick film intumescent coatings are the most common example. The intumescent coatings can be applied either on-site or off-site (AkzoNobel, 2017). The techniques used to protect steel are:

1. Fire resisting boards
2. Vermiculite cement sprays
3. Fibre sprays
4. Dry linings
5. Mineral wool
6. Intumescent coatings

Choosing the right places, members, joints, vessels, and pipes along with the right material in the appropriate thickness in offshore platforms can result in cost-effective use of Passive Fire Protection.

OPTIMIZATION

Integrity and cost are the most concerned problems in the design and manufacture of offshore platforms. The design optimization of a platform is mainly concentrated on structural size optimization. The structural properties of an offshore platform is enhanced by adjusting size parameters of the structural members. The structural shape and topology not only determine the weight and cost of the structures but also directly affect the structural characteristic and responses when loads act on them. Thus, shape and topology optimization of platforms can obtain more remarkable benefits comparing with size optimization.

For the layout of the platform, effective procedures should be developed for designing synthetically the structural size parameters (such as, cross-sections of components), shape and topology of the platform structures. The load conditions considered should include weight, wind, wave and/or ice loads, accidental loads and their combinations. The design computation can be completed by use of the software such as ABAQUS, SACS, and JIFEX and so on.

A fully probabilistic approach to fire safety assessment and optimal design of fire protection on offshore topsides is achieved by integrating:

1. Quantitative Risk Analysis techniques
2. Fire and explosion models
3. Heat transfer models
4. Non-linear structural analysis methods
5. Structural System Reliability Analysis techniques, and
6. Reliability-Based Design Optimization methods.

This integration have been achieved by enhanced event-trees where the final failure events in an enhanced event-tree being: loss of escape ways, loss of Temporary Refuge, loss of evacuation systems and structural collapse of the topside (Alessanciro et al, 1995).

The (conditional) probabilities of the above events are calculated using structural reliability methods by taking into account the uncertainties in the fire loading parameters (exit size, flow-rates, fuel properties, fire models), thermal properties (insulation thickness} thermal properties of structural steel and the insulation), structural properties (yield strength, expansion coefficient, etc.). In this way, dominant accident scenarios leading to loss of TR, EER or structural collapse can be identified and their probabilities quantified. The optimization of passive fire protection on structural systems is performed by adopting the above-mentioned method.

Optimized Fire Safety of Offshore Structures (OFSOS) software system is used for fire *safety assessment and optimal* design of offshore structures. The OFSOS software system can be applied to the topsides of various types of offshore structures such as jackets, semi-submersibles and Tension Leg Platforms. The methodology and the software can be used for the fire safety assessment and optimization during the conceptual phase and the detailed design phase of a new platform or for an existing platform (Alessanciro et al, 1995).

REFERENCES

AISC. (2011). *Steel Construction* (14th ed.). AISC.

AkzoNobel. (2017, June 20). *Protective coatings*. Retrieved from AkzoNobel: http://www.international-pc.com/products/fire-protection/fire-protection-technical-information.aspx

Ali, R. M. (2007). *Performance Based Design Of Offshore Structures Subjected To Blast Loading*. London: Imperial College.

Anatol Longinow, F. A. (2003). Blast resistant design with structural steel - common questions answered. *Modern Steel Construction*, 61 - 66.

André, J., Beale, R., & Baptista, A. M. (2015). New indices of structural robustness and structural fragility. *Structural Engineering and Mechanics*, *56*(6), 1063–1093. doi:10.12989em.2015.56.6.1063

API. (2000). *API recommended practice 2a-WSD (RP 2A-WSD)*. American Petroleum Institute.

ASFP. (2015). *Types of fire protection materials*. Retrieved October 14, 2017, from http://www.asfp.org.uk/Technical%20Services/types_of_fire_protection.php

Baker, M. S. (2007). *Assessment of robustness*. Academic Press.

Bea, R. W. (1991). *Design and characterize the offshore fire problem. Improved means of offshore platform fire resistance.* Berkeley, CA: University of California.

Biggs, J. M. (1964). *Introduction to structural dynamics.* New York: Mac Graw-Hill.

Bresler, B., & Iding, R. (n.d.). *Effect of Fire Exposure on Structural Response and Fireproofing Requirements of Structural Steel Frame Assemblies.* Wiss, Janney, Elstner Associates, Inc.

Brode, H. L. (1955). Numerical solution of spherical blast waves. *Journal of Applied Physics, American Institute of Physics.*

Chopra, A. (2007). Dynamic of structures: theory and applications to earthquake engineering (3rd ed.). Pearson Education, Inc.

Corr, R. B., & Tam, V. H. Y. (1998). Gas explosion generated drag loads in offshore installations. *Journal of Loss Prevention in the Process Industries, 11*(1), 43–48. doi:10.1016/S0950-4230(97)00054-5

Eknes, M. L. T. M. (1994). Escalation of gas explosion event offshore. *Offshore Structural Design, Hazard, Safety and Engineering.* London: ERA Report No 94-0730, 3.1.1-3.1.16.

ESDEP. (n.d.). *ESDEP 15A.* Retrieved July 5, 2017, from Dynamic analysis: http://fgg-web.fgg.uni-lj.si/%7E/pmoze/esdep/master/wg15a/l0100.htm

Farid Alfawakhiri, L. A. (2003). *Blast resistant design with structural steel - common questions answered.* Modern Steel Construction.

Harris, R. J. (1983). *The investigation and control of gas explosions in buildings and heating plant.* The University of Michigan.

ISO19902. (2007). *Petroleum and natural gas industries — Fixed steel offshore structures.* International Standard Organization.

Izadifard, M. M. (2010). Ductility effects on the behaviour of steel structures under blast loading. *Iranian Journal of Science & Technology*, 49 - 62.

Izzuddin, D. L. (1997). Response of Offshore Structures to Explosion Loading. *International Journal of Offshore and Polar Engineering*, 212.

James, M. A. (1988). Analytical Methods for Determining Fire Resistance of Steel Members. In SFPE Handbook of fire engineering. Quincy, MA: Society of fire protection Engineers.

Jodin, P. G. P. (1980). Experimental and theoritical study of cracks in mixed mode conditions. In J. C. Radon (Eds.), Fracture and Fatigue: Elasto-plasticity, thin sheet and micromechanisims problems (pp. 350 - 352). London: Imperial College, London.

Ki-Yeob Kang, K.-H. C.-M. (2016). Dynamic response of structural models according to characteristics. *Ocean Engineering*, 174 - 190.

Lloyds. (2014). *Guidance notes for risk based analysis: fire loads and protection.* Lloyd's Register.

M, B. J. (1964). *Introduction of Structural Dynamics.* New York: McGraw-Hill.

Mills, C. A. (1987). The design of concrete structure to resist explosions and weapon effects. In *Proceedings of the first Int. Conference on concrete hazard protections*, (pp. 61 - 73). Edinburgh, UK: Academic Press.

Moore, D. B. F. W. (2003). Design of structural connections to Eurocode 3. Watford: Leonardo da Vinci.

Ngo, P. M. (2007). Blast Loading and Blast Effects on Structures – An Overview. *EJSE*, 79-80.

Nutec, F. (2008). *Basic Safety Offshore*. Falck Nutec.

Paik, J. K. (2011). Quantitative assessment of hydrocarbon explosion and fire risks in offshore installations. Elsevier.

Pitaluga. (1992 - 1995). *Optimized Fire Safety of Offshore Structures (OFSOS).* Brite-Euram Project.

Profire. (2014). *Section factors*. Retrieved June 20, 2017, from W/D, A/P, M/D Calculation Method: http://www.profire.com.tr/eng/222-page-wd-ap-md-calculation-method.aspx

Qiao, S. Z. (2010). Advanced CFD modeling on vapour dispenser and vapour cloud explosion. *Journal of Loss Prevention*, 843-848.

Robertson. (1959). Proposed criteria for defining load failure of beam floors and roof constructions during fire test. *Journal of Research of the National Bureau of Standards*, *63C*, 121.

Starossek, U. M. H. (2008). Approaches to measures of structural robustness. Seoul, South Korea: Academic Press.

Stolz, A. K. (2016). A large blast simulator for the experimental investigation of explosively loaded building components. *Chemical Engineering Transactions*, 151–156. doi:10.3303/CET1648026

Szarka, I. (2015). *Structural Integrity Management ensuring robustness and barriers*. Stavanger: University of Stavanger.

TM5. (1990). *Design of structures to resist the effects of accidental explosions.* Department of Army - TM5-1300.

Tam, V. H. Y. (1990). Modelling of missile energy from gas explosions offshore. *International conference on the management and engineering of fire safety and loss prevention onshore and offshore.*

TMR. (2009). *Resistance to accidental and catastrophic fires: General principles*. MTS.

Vasilis, K. (2013). Calculation of blast loads for application to structural components. European Commission - Joint Research Centre.

ADDITIONAL READING

Chakrabarti, S. (2005). Handbook of offshore engineering. In K. Demir (Ed.), *Fixed offshore platform design* (pp. 279–401). Elsevier.

Dag Bjerketvedt, J. R. (1992). *Gas Explosion Handbook. Christian Michelsen Research's*. CMR.

El-Reedy, M. A. (2012). Offshore Structures: Design, Construction and Maintenance. In *Historical review of major North Sea incidents* (p. 460). Gulf Professional Publishing.

Holen, J., Hekkelstrand, B., & .... (June28, 1991). Modelling Of Hydrocarbon Fires Offshore. In *Final Report, STF25 A91029*. Trondheim, Norway: SINTEF.

Itkin, A. (2017). *Major Offshore Accidents of the 20th and 21st Century*. Retrieved October 13, 2017, from http://www.oilrigexplosionattorneys.com/Oil-Rig-Explosions/History-of-Offshore-Accidents.aspx

Okyere, M. S. (2018). *Fixed Offshore Platforms: Structural Design for Fire Resistance*. United Kingdom: CRC Press.

Raabe, D. (n.d.). *Dramatic failure of materials in drilling platforms*. Retrieved August 12, 2017, from http://www.dierk-raabe.com/dramatic-material-failure/oil-drilling-platforms/

Sakha, G. S. (2015). *Design, modeling, analysis and calculation of offshore module structure*. Stavanger: University of Stavanger.

KEY TERMS AND DEFINITIONS

Deflection: Is the degree to which a structural element is displaced under a load. It may refer to an angle or a distance.

Design Optimization: Is the process of finding the best design parameters that satisfy project requirements. Engineers typically use design of experiments (DOE), statistics, and optimization techniques to evaluate trade-offs and determine the best design.

Explosion: Is a sudden and violent release of energy the violence of which depends on the rate at which the energy is released.

Fire: Is the rapid oxidation of a material in the exothermic chemical process of combustion, releasing heat, light, and various reaction products.

Fire Resistance: Is the ability of construction or its element to satisfy for a stated period of time load bearing capacity, integrity, and insulation when exposed to fire.

Fire Resistant Material: Is one that is designed to resist burning and withstand heat.

Fixed Offshore Platform: Is a type of offshore platform used for the production of oil or gas in shallow waters. It extends above the water surface and supported at the seabed by means of piling or shallow foundation with the intended purpose of remaining stationary over an extended period.

Slenderness Ratio: Is the ratio of the effective length of a column (L) and the least radius of gyration (r) about the axis under consideration. It is given by the symbol "λ" (lambda).

Chapter 5
Wind Loads on Structures, and Energy Dissipation Systems Optimization

Aboubaker Gherbi
Constantine 1 University, Algeria

Mourad Belgasmia
Setif 1 University, Algeria

ABSTRACT

Wind has a great impact on civil structures. It is considered a dynamic and random phenomena and it plays an important role in the design of tall structures. Existing buildings with certain height must resist wind effect. Many researchers have developed theories and schemes that consider more thoroughly wind components and the influence of its turbulence on buildings. It is known that any structure inherently dissipates and absorbs energy due to external loads thanks to its inherent damping. In order to improve this capacity and limit structural damage, fluid viscous dampers are commonly used for structural protection; they have confirmed their efficiency and reliability. Many researchers have investigated their effect by inserting them in the structure; some of the optimization methods for the design of these dampers previously used will be discussed. Finally, an effective method for optimal design of additional dampers will be illustrated by an example and discussion.

SCOPE AND OBJECTIVES

The aims of this chapter is to illustrate and show essential concepts in wind engineering. Relying on some brief theoretical background about wind loads on structures, its nature and components, which will be presented in such an illustrative way that engineers and young researchers may understand with ease. Along with time series generation which can be seen as an important task in wind engineering.

In the same manner, discussion and description of an energy dissipation system is presented in this chapter, historical use, advantages and mainly the design methods which can be seen as a main objective of this chapter; various attempts of an optimized design process will be shown and it will be concen-

DOI: 10.4018/978-1-5225-7059-2.ch005

trated on an effective and simple procedure. Finally, all the methodology discussed will be illustrated with some examples on an RC structure, along with results and comments that gives directions towards further research.

INTRODUCTION

Even if the subjects treated in this chapter are fully investigated, the majority of available literature can be considered as hard to assimilate for students, and their application in engineering field can be challenging. The need of simplifying certain concepts can induce a lack of important information related to the governing theory. This work, proposes plenty of materials i.e. references, along the chapter in order to allow the reader to access further details. Wind have a great deal of importance on civil structures, and a thorough characterization of tis components is of utmost importance. In the other hand, engineers and researchers always try to enhance the resistance of these structures using many approaches, and energy dissipation systems are increasingly applied.

It is a common practice for academics the ability of having a two side view of a given problem i.e. time and frequency domains, in order to valid their work. In the next section, the authors describe the means and methods that facilitate this task, presenting the proper tools for a time domain analysis i.e. generation of wind time series (for a Monte Carlo analysis), and illustrate the spectral analysis method for the frequency domain analysis.

The second main objective of this work is to investigate the effect of additional dampers, and highlight an effective method that allows an optimal design. Optimization is a large concept, that modern engineering is applying in numerous fields, aerospace, electronics, computer and structural engineering. This work, also simplifies this concept, where it is applied in the design of supplemental damping devices, it is attempted to figure out their optimal location, number, and an optimum time for the design process.

WIND LOADS AND SIMULATION

Wind have a great impact on civil structures and their design, a satisfactory representation of wind load must deal with several random parameters i.e. stochastic process, as wind climate, terrain roughness and structures geometry. These parameters require the use of statistical methods in order to adapt them to each studied structure.

Wind load vary in time and space and over the height and surface of structures, and it is subjected to two types of aerodynamic forces: drag forces i.e. along wind, that acts on the direction of the mean flow, and lift force i.e. across-wind, which acts perpendicularly to that direction (Simiu & Scanlan, 1986). The use of codes and standards is of utmost importance in the design phase, since it simplifies widely the complex nature of wind. However, it is crucial to understand the methodology behind the rules and clauses in order to make good use of the codes.

Mean Wind Speed

It is common practice in wind engineering to define wind velocity as a stationary stochastic process. It is described by a mean speed and turbulent components (Dyrbye & Hansen, 1997). The mean wind speed varies over height, and it is mainly conditioned by the surface roughness:

$$U(z) = U(z_{ref})\,(z/z_{ref})^{\alpha} \quad (1)$$

In which Z_{ref} and the index α are functions of the ground roughness, Table 1 suggests values of these quantities for different types of terrain.

As explained by Davenport, the wind is retarded by surface friction, and some of its kinetic energy is dissipated in turbulence. Thus, the mean speed increases from zero at ground level to a maximum value above ground where it is no longer affected by surface friction (Figure 1), this height is called gradient height or reference height, it depends on surface roughness and varies from approximately 200 to 500 meters.

Table 1. Influence of surface roughness on wind related parameters

Type of Surface	Power Law Exponent α	Gradient Height Z_{ref} (meters)
Open terrain with very few obstacles, e.g. open grass or farmland with few trees	0.16	270
Terrain uniformly covered with obstacles 10-20 m in height, e.g. small towns.	0.28	400
Terrain with large and irregular objects, e.g. centers of large cities	0.40	520

(Davenport, 1961)

Figure 1. Variation of mean wind velocity profile with surface roughness

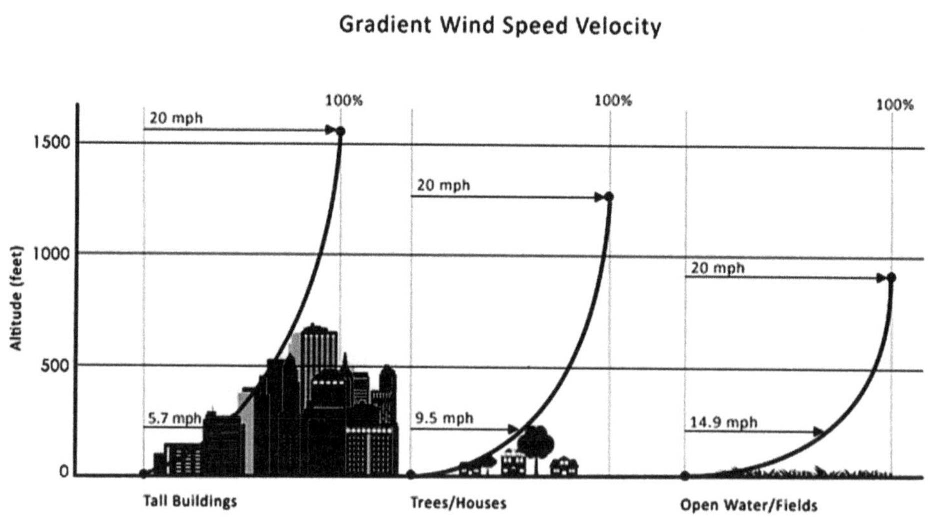

Fluctuating Wind Speed

Wind is always turbulent, and it fluctuates randomly, its properties need to be visualized in a statistical manner. It also means that the flow is chaotic with random periods ranging from fractions of seconds to few minutes. The turbulent components are described by: their standard deviation, time scales, integral length scale and power spectral density functions that define the frequency distribution and normalized co-spectra that specify the spatial correlation (see Dyrbye & Hansen, 1997).

In the need to analyze structures under wind loads, engineers are most of the time forced to simulate wind time series. Digitally generated samples become more essential for time domain analysis of structures i.e. Monte Carlo method, since it is easier to get artificial samples of loading, these samples are introduced into the governing equation to be solved for each specific problem. These simulations are mostly based on models describing the process to be simulated, and it can be an exact theoretical model, phenomenological or empirical model or observed data (Kareem, 2008). The distribution of turbulent energy in wind as a function of frequency can be described by the power spectrum of the fluctuating component of wind speed (Das, 1988). A typical power density spectrum, referred to hereafter by PSD, has a peaked form as indicated in Figure 2.

Wind Simulation

Many researchers have proposed a wind PSD model, the Davenport wind spectrum reads,

$$S_f(n) = \frac{n\frac{2}{3}\left(\frac{L}{U}\right)^2 \sigma^2}{\left(1+\left(\frac{nL}{U}\right)^2\right)^{4/3}} \tag{2}$$

Figure 2. Wind power spectral density

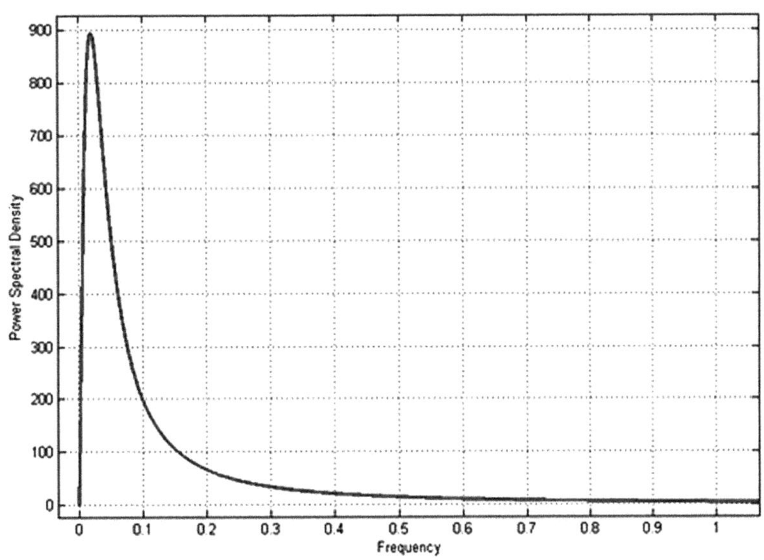

As for Von Karman, the power spectral density it is formulated:

$$S_f(n) = \frac{4L\sigma^2}{\left(1 + 70.8\left(\frac{nL}{U}\right)^2\right)^{5/6}} \tag{3}$$

where, σ^2 is the mean square of the wind turbulence, L is a turbulent length scale and n is the frequency. For the next sections, Davenport's spectrum will be taken into consideration for the simulation of wind time series. Wind generation methods did not cease developing, from approaches that uses a superposition of trigonometric functions (Shinozuka, 1971), to the use of Fast Fourier Transform to improve the computational efficiency, and the cholesky decomposition of the cross-spectral density matrix and many other schemes.

In the following, the Fourier series for sample generation will be shown; these time histories are obtained for any point in space and should satisfy Davenport's power spectral density function.

The fundamental characteristic of a power spectral density for a random process is that the integration over the frequency range corresponds to the variance of the process (Denoël, 2005):

$$\int_0^{+\infty} S_f(n) = \sigma^2 \tag{4}$$

Based on the ergodicity hypothesis, the power spectral density of a stationary process can be obtained from a single sample (Denoël, 2005). The generation could be done simply by choosing:

$$X(w,T) = \sqrt{\frac{T}{2\pi}}\sqrt{S_f(w_i)}\,e^{i\varphi_j} \tag{5}$$

where φ_j is a random phase angle taken between zero and 2π.

Based on this theoretical background briefly presented (see Denoël,2005, Newland, 2012, and Roberts & Spanos, 2003 for more useful detail), a Matlab code has been implemented to generate artificial time histories for later use (illustrations). These samples are simply obtained by applying the inverse Fourier Transform on equation (5). From the generated fluctuating wind speed, the determination of the fluctuating wind force can be done as follows:

$$F(t) = \rho C_d A U u(t) \tag{6}$$

where ρ is the air density, A is the projected area, U, u(t) are the mean wind speed and fluctuating wind speed at a specific height and C_d is the drag coefficient, which can be determined for any given structure, many relations are proposed in codes and standards depending on the structure's geometry and the construction material. This method is simple and precise, even if it is seen as time consuming, however, it is efficient for a time domain analysis of any problem i.e. solving the equation of motion. Using any suitable code, the generation is done following steps shown previously, based on a power spectral density denoted by *SF,* and for a number of values for the generated sample *N*:

```
for j=1:N/2+1
   x(j) = sqrt (T/(2*pi))* sqrt(SF(j))*exp(1i*2*pi*rand(1,1));
end
x(1) = 0;
x(N/2+1) = real(x(N/2+1));
x(N/2+2:N) = conj(x(N/2:-1:2));
x = ifft(x);
```

where the function *rand()* generates random numbers, their multiplication by π generates the random phase angles in equation (5), and *ifft()* is a function that applies the inverse Fourier Transform after the normalization of the time series. An example of the simulated sample will be presented below.

It is known that the Fourier Transform is considered as a link between time and frequency domains, this relation can be understood by applying the Fourier Transform on the governing equation of motion:

$$\int_{-\infty}^{+\infty}\left(\left[M\right]\left\{\ddot{x}\left(t\right)\right\}+\left[C\right]\left\{\dot{x}\left(t\right)\right\}+\left[K\right]\left\{x\left(t\right)\right\}\right)e^{-2j\pi nt}dt=\int_{-\infty}^{+\infty}\left\{p\left(t\right)\right\}e^{-2j\pi t}dt \tag{7}$$

where M, C and K are the system's characteristics, and p(t) is the excitation (e.g. the generated sample), we obtain:

$$4\pi^2n^2\left(\left[M\right]+2j\pi n\left[C\right]+\left[K\right]\right)\left\{X\left(n\right)\right\}=\left\{P\left(n\right)\right\} \tag{8}$$

where the new unknown is the function {X(n)}, with the variable changing from time to frequency. Equation (8) is the formulation of the equation of motion in the frequency domain. The relationship between the two domains can be illustrated by schematic representation:

For a resolution in the frequency domain, a spectral analysis is the best fit, thanks to its simplicity in case of an excitation of a stochastic nature. It is based on the PSD of the loading, from which the PSD of the response can be easily obtained. In the following an example is given, upon which the spectral analysis is illustrated.

Figure 3. Relationship between the frequency and time domain

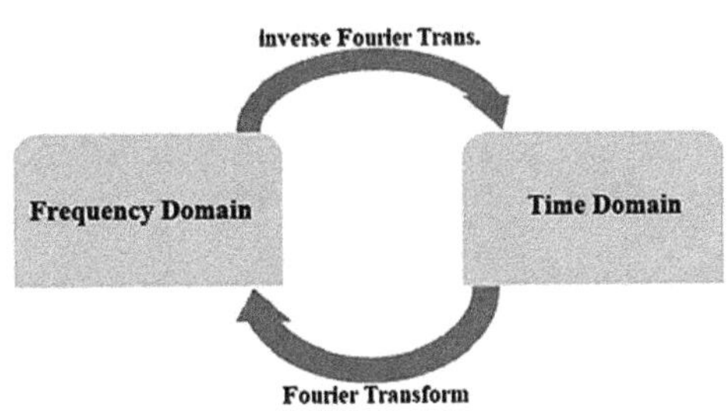

Spectral Analysis

The spectral analysis is well suited to determine the response of a deterministic structure under a stochastic excitation, considering that both the response and the loading are stochastic processes. A single degree of freedom system is considered for this example, subjected to a wind PSD. The power spectral density function of the response (displacement of the SDOF) is obtained as follows:

$$S_x(n) = \int_{-\infty}^{+\infty} |H(n)|^2 S_f(n)\,dn = \sigma_x^2 \quad (9)$$

σ_x^2 present the variance of the displacement of the SDOF system, and *H(n)* is the transfer function:

$$|H(n)|^2 = \frac{1}{k^2}\frac{1}{\left(1-\left(\frac{n}{\bar{n}}\right)^2\right)^2+\left(2\xi\frac{n}{\bar{n}}\right)^2} \quad (10)$$

where k is the stiffness, ξ is the damping ratio and $\bar{n}$ is the natural circular frequency. In the same manner the variance of the velocity:

$$\sigma_{\dot{x}}^2 = \int_{-\infty}^{+\infty} n^2 |H(n)|^2 S_f(n)\,dn \quad (11)$$

As well as for the variance of the acceleration:

$$\sigma_{\ddot{x}}^2 = \int_{-\infty}^{+\infty} n^4 |H(n)|^2 S_f(n)\,dn \quad (12)$$

As mentioned above, Matlab or any other similar tool can be used to perform such calculus. After defining the appropriate input, the transfer function is implemented in the code as follows:

```
for l = 1:(Nmax/DN)
    n = Z(l);
    h = (1/(K).^2)*1/((1-(n/f).^2).^2+(2*ksi*n/f).^2);
    H(l) = h;
end
plot (Z,H)
```

where *Nmax* and *DN* present the maximum frequency considered and frequency step respectively, *H* is used for storing purposes in every step of the *for* loop, and *Z* stores the ensemble of frequencies *n*.

Figure 4. Transfer function of an SDOF system

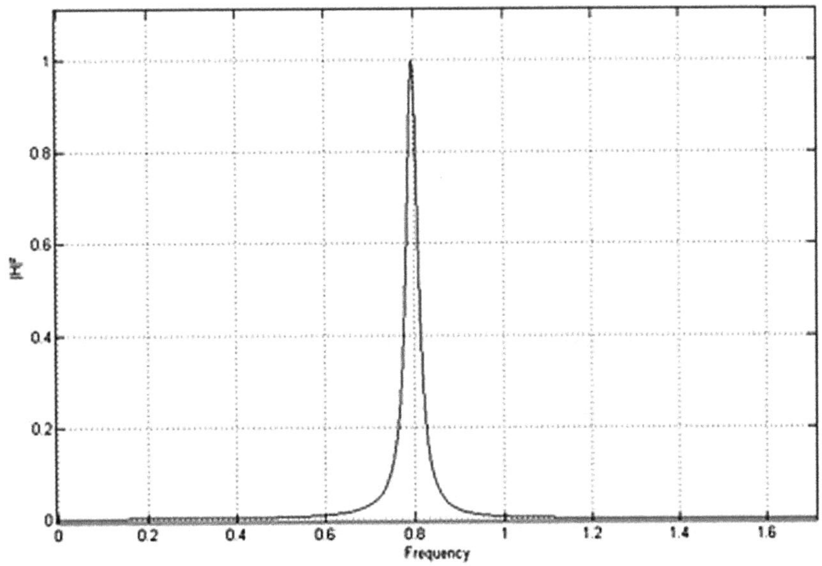

Multiplication of the power spectral density of the loading by $\left|H\left(n\right)\right|^2$, and using the same concept of a *for* loop in the code, the response is shown in Figure 5.

The simplicity of these presented concepts, allows an academic comparison between the time domain analysis and the frequency one, applying the concept of generation of the equivalent time series then the equation of motion is solved using step by step methods i.e. Newmark (Clough & Penzien, 2003). The variance of the displacement presents the area under the response spectrum curve, which can be calculated for comparison purposes.

As discussed previously, using Fourier Transform, one can switch from a domain to another, based on this definition, both the loading and the response of the time domain analysis will be presented in

Figure 5. The response of an SDOF system

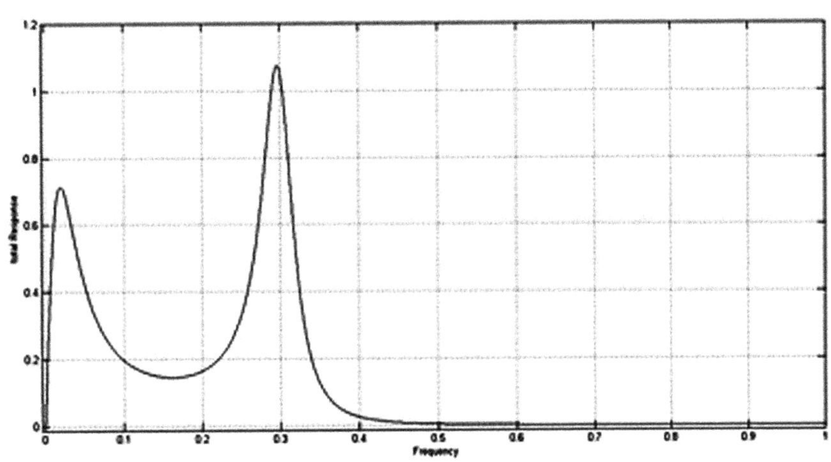

the frequency domain along with the results already obtained. First, the generated time history will be transformed for a frequency representation (using a Matlab code).

It is obvious that the generated time history is well matched with the equivalent Davenport PSD, and thus, a good match is expected when comparing the results, a linear Newmark method is used for this example.

From Figure 7 and as expected, a good match is achieved, proving the efficiency of the used method; this academic view of the problem (even for an SDOF) can be a powerful tool for further research in this field.

ENERGY DISSIPATION SYSTEMS

When a structure is subjected to a dynamic excitation, a quantity of energy is diffused into the structure. This latter absorbs and dissipates this energy (through heat) by transforming it to kinetic and potential energy, this inherent damping which consists in a combination of strength, deformability and flexibility

Figure 6. a) Generated time history b) Spectral representation of the loading

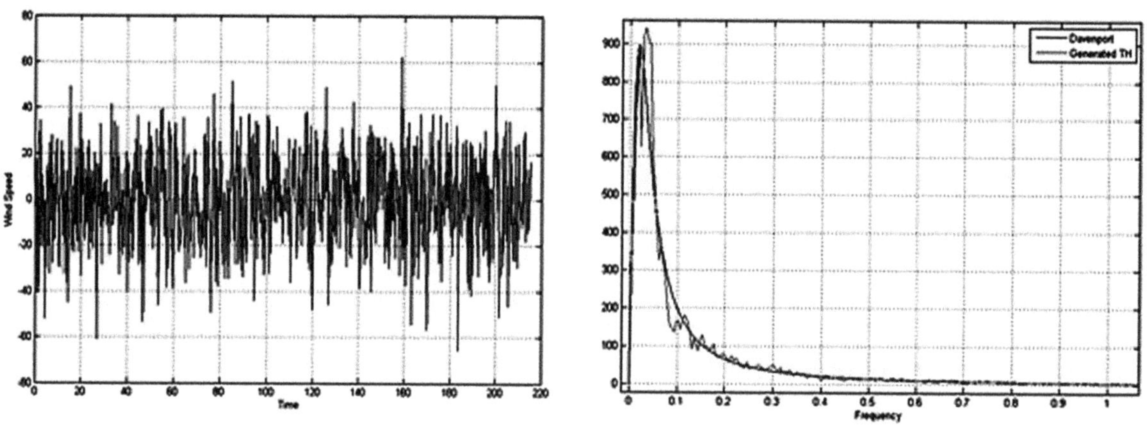

Figure 7. a) Displacement of the SDOF system b) Spectral representation of the response

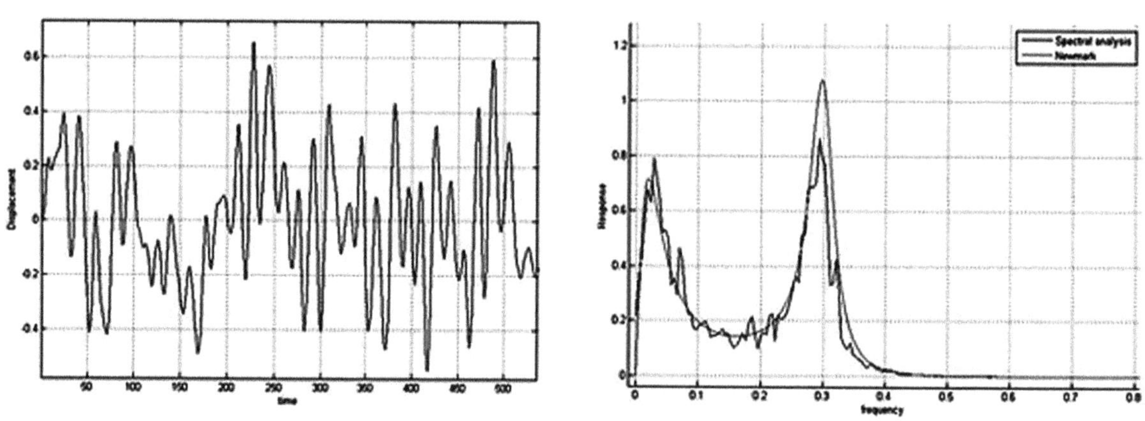

allows the input energy to be extinguished. Additional energy dissipation devices may achieve a better structural behavior and improvement. In the last decade, structural control devices caught a lot of attention from the civil engineering community, thanks to their proven ability to reduce the damage induced by wind and seismic excitations. So much efforts was done in order to make the concept of additional damping as a workable technology, nowadays, numerous civil structures are equipped with such devices. Many researchers have studied these systems, and since a large variation of these systems is available, a classification was of utmost importance, three major classes are distinguished, base isolation, passive control systems and active control systems.

The principle of an isolation system hides in the flexibility in the device in the horizontal direction, and a very stiff capacity in the vertical direction in such way to increase the rocking stability and thus, the isolation device absorbs a part of the input energy, and their implementation could be done in different location of the structure. This type of systems is more suitable to buildings with medium to short height; however, it has a low efficiency in the case of wind loading due to the flexibility in the horizontal direction (Marko & Thambiratnam, 2004, Saaed & Nikolakopoulos, 2015).

Active control systems are considered as a more advanced technology, their concept is based on a real time processing that allows them to act accordingly to the excitation, thanks to the sensors integrated within the device. These sensors collect information about the excitation and the structural response, then adapts the device's behavior based on the collected data (Saaed & Nikolakopoulos, 2015, Soong & Spencer, 2002).

Passive control systems consists in materials and devices that enhance damping, strength and stiffness, the most common devices of this class operate generally on principles such as deformation of fluids and fluid orificing (the case of this chapter), deformation of viscoelastic solids, frictional sliding and yielding of metals (Soong & Spencer, 2002). A passive dissipation device utilizes the motion of the structure to generate the control force (no external power source needed). In the following, it will be mainly focused on fluid viscous dampers i.e. FVD, and their design optimization, which is seen as intensive task, all illustrated with an example. Figure.8 shows some of the passive dissipation devices followed by a brief definition and a summary of their advantages and disadvantages.

Friction damper consists of four links located at the intersection of cross bracings (tension brace and compression brace)(Canstantinou & Spencer, 1992). A slippage is permitted for one of the braces then the other slips, which allow the device to dissipate energy in both braces. Filiatrault & Cherry (1990) tested and confirmed that this type of devices can increase substantially the capacity of dissipation in one cycle and reduce the amount of drifts. However Symans & al (2008), cited some reliability issues that concerns this sliding option, which can change the interface condition with time, and since their behavior is highly nonlinear, it may provoke undesirable structural behavior by exciting higher modes.

Figure 8. Passive energy dissipation systems.

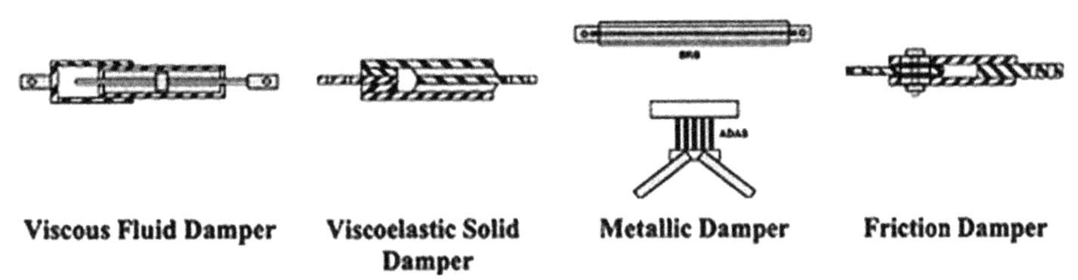

Metallic damper or also called yielding steel elements, many concepts are proposed, it can consist on round steel bar integrated in the bracing frame (Canstantinou & Symans, 1992, Tyler, 1985 and Whittaker & al, 1991), with their stable hysteretic behavior and reliability for long periods, they also present an insensitivity to temperature. One must mention that after a severe excitation, this device may be severely damaged and must be replaced.

Viscoelastic damper initially developed for wind protection, consisting of bonded viscoelastic layer, suitable also for earthquake resistance (Canstantinou & Symans, 1992, Chang & al, 1991 and Aiken & al, 1990), which mean that they are activated at low displacements and easily modeled due to their linear behavior. Some of the inconvenient when using this type of dampers, is their temperature and frequency dependency, along with their low capacity for deformation and some reliability concerns (Symans & al, 2008).

Fluid viscous damper, which will make a subject for this chapter, have a proven record of performance in the military applications and present a large independency to frequency and temperature, also activated at low displacements with the only disadvantage of a possible fluid seal leakage (Symans & al, 2008).

Fluid Viscous Damper

FVDs operate on the principle of fluid flow through orifices. A stainless steel piston travels through chambers that are filled with silicone oil (inert, non flammable, non toxic and stable), when the damper is excited the fluid is forced to flow either around or through the piston. The input energy is transformed into heat when the silicone oil flows through orifices between the two chambers (Duflot & Taylor, 2008). Canstantinou & al (1993) demonstrated that energy dissipation systems are capable of producing significant reduction of inter-story drift to structures and column bending moments to which they are installed. Thus, they are all suitable in the design of new or existing buildings i.e. retrofitting. Constantinou & Symans (1992) conducted an experimental and analytical study of this device, which evaluated their behavior in a range of frequencies, amplitudes and temperatures. The dampers where installed in a one and three-story model structures, it was concluded that the reduction in story drifts reached a percentage of 70%, also, they were capable of achieving a better benefits offered by active control devices with the advantage of low cost, no power is needed and reliability for long period of time. Figure 9 shows a typical fluid viscous damper and his different components (Canstantinou & al, 1993).

Figure 9. A typical fluid viscous damper (FVD)
(Canstantinou & al, 1993)

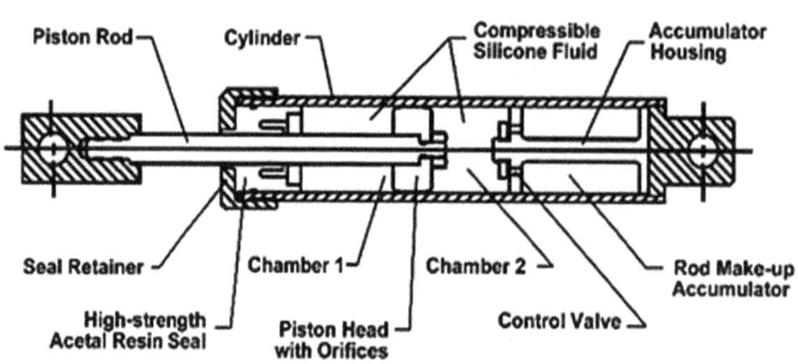

As mentioned above, these devices can be included in the structures in the preliminary stage of construction or for structural rehabilitation. Different inclination possibilities are available, engineer may choose which is suitable for any given structure, respecting the architectural and technical requirements. Some of these inclination possibilities are presented in the next figure.

Many manufacturers supply a wide variety of devices, operating at high or low fluid pressure and generally having nonlinear force-velocity relationship as follows (Constantinou & Symans, 1992)):

$$F = C_D \left|\dot{x}\right|^{\alpha} \operatorname{sgn}\left(\dot{x}\right) \tag{13}$$

where C_D is the damping coefficient, α is a damping exponent in the range of 0.3 – 2, and *sgn(.)* is a signum function . The value of α depends on the shape of the piston head, when $\alpha = 2$ it means that the orifices are cylindrical, and presents an unacceptable performance. For $\alpha = 0.3$ an effective value for the case of high velocities, as for the case of near fault earthquakes (Soong & Constantinou, 2014). A linear behavior is achieved by a value of $\alpha = 1$ which is frequently used in case of seismic and wind energy dissipation and will be the subject of further developments. While the damping coefficient can be determined by the damper diameter and the orifice area (Haskell & Lee, 1996). Figure 11 shows the damper's force-velocity relationship.

In order to represent the effect that the additional dampers have on a structure, the SDOF treated previously will be considered for a spectral analysis, using Davenport's PSD and taking arbitrary values for the mass and stiffness of the SDOF. With a natural frequency of approximately 0.5 Hz, this system was equipped with a fluid viscous damper which damping coefficient was increased in each step. The

Figure 10. FVD inclination: a- diagonal ; b- chevron ; c- toggle ; d-scissor
(Guo & al, 2014)

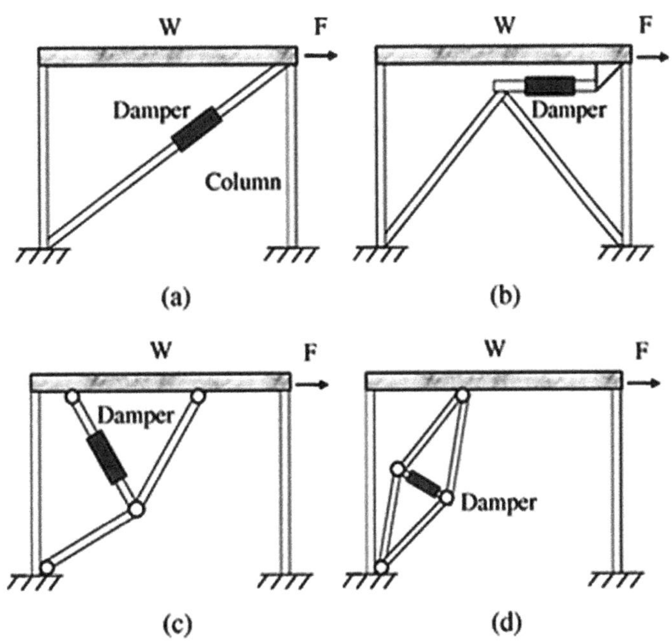

Figure 11. Force-velocity relationship of FVD
(Ras & Boumechra, 2016)

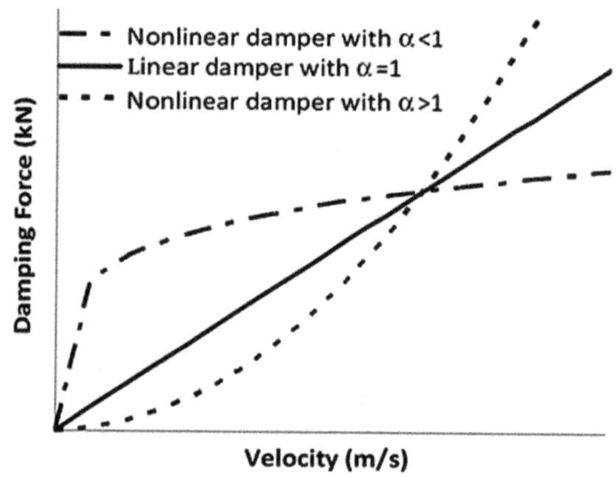

result shown below (figure 12), confirm the efficiency of such devices, since the variance (area under the graph) of the displacement presents a reduction after each iteration i.e. increased damping coefficient.

Since this chapter aims towards simplification and initiations, only dampers with linear behavior will be considered. As for the design of the typical structural members, the additional dampers should be optimally designed. The choice of an optimum damping coefficient value and an optimal configuration could be seen as intensive and time consuming task, since the majority of the proposed methods are based on iterations and analysis concept (a first value is considered, then the analysis is undertaken, and this action is repeated until a better structural behavior is observed).

Figure 12. The reduction in the response with an increasing value of damping coefficient C_D

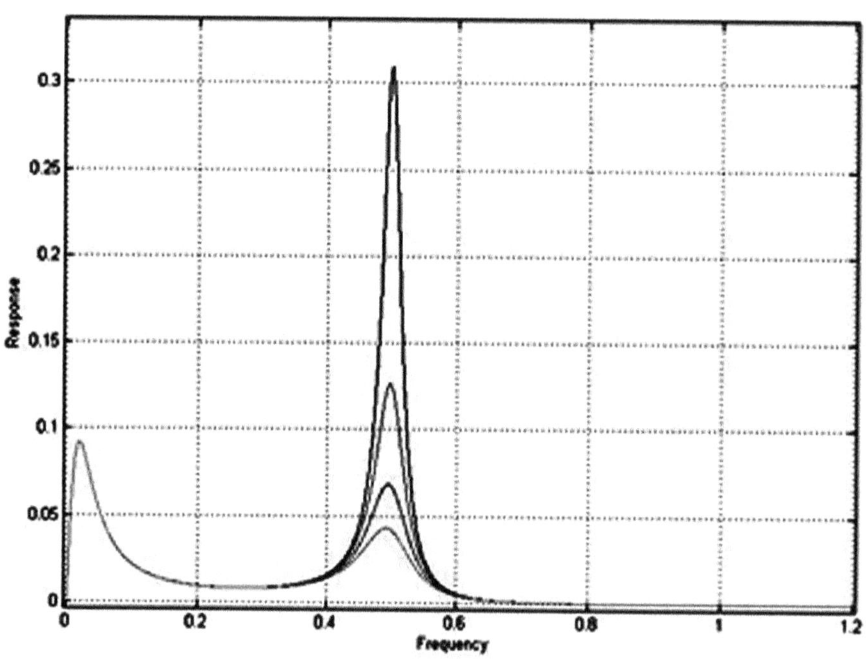

Many researchers attempted to optimize the design process of additional dampers, Moreschi (2000) used two techniques to achieve an optimum design, a Gradient projection method for the selection of a damping value, and a genetic algorithm which deals with the choice of their configuration (the location of dampers over the height of a structure). Martinez & Romero (2003) proposed a strategy to an optimum retrofitting option with added dampers (linear and nonlinear), based on a numerical optimization process, which takes into account two indexes: the performance index that achieves the damage control (DC) performance point; and the force index used to calculate the reduction in the damper force. Kandemir & Mazanoglu (2017) investigated an optimum damper capacity and number installed between two adjacent buildings, depending on a parametric study. This brief bibliography shows the variation of both the design and the application of fluid viscous dampers. Despite the effectiveness of the numerous design method, it is still seen as a time consuming task. In the following, an optimization in the design and time is satisfied by a method proposed by Duflot & al (2017), which proved to be both efficient and simple, and gives a wide range of flexibility for the designer.

METHODOLOGY FOR THE PRELIMINARY DESIGN

For a an MDOF structure, the governing equation of motion can be written in the matrix form as follows:

$$M\ddot{x} + C\dot{x} + Kx = F\left(t\right) \tag{14}$$

M, C and *K* are the mass, damping and stiffness matrices of the MDOF structure, *F(t)* is the dynamic excitation. Using Finite Element methods, system's characteristics can be determined by assembling element matrices; in the case of added damping devices, the damping matrix can be expressed as:

$$C = C_s + C_d + C_a \tag{15}$$

C_s is the structural damping, C_a is the aerodynamic damping (in case of wind loading) and C_d is the damping in the additional FVDs, the assembling of each added device's matrix constitutes the C_d matrix, in a local frame:

$$C_d = \sum_{j=1}^{nd} l_j \begin{bmatrix} c_j & -c_j \\ -c_j & c_j \end{bmatrix} l_j \tag{16}$$

c_j is the viscosity of damper *j*, n_d is the number of additional dampers and l_j is an $N \times 2$ vector symbolizing the rotation localization. It should be mentioned that the only non-zero elements in C_d matrix corresponds to the added dampers ends. A projection in the modal basis is necessary in the preliminary design phase, since the appearing of the mode shape matrix

ϕ simplifies the upcoming steps (diagonalized matrices M^*, K^* $and\, C^*$). The diagonal element of the added dampers matrix is:

$$C^*_{d_{n,n}} = \sum_{j=1}^{nd} c_j \Delta\phi^2_{n,j} \tag{17}$$

where $\Delta\phi^2_{n,j}$ is the relative longitudinal displacement of both ends of damper *j* in mode *n*, and it is more naturel to express this quantity as a function of the inter-story drift $\Delta\phi = \varphi\Delta u_{n,j}$, with φ is called an efficiency factor (depending on the inclination of the FVD). The non-diagonal elements of this matrix could be non-zero especially when mode shapes are similar, sometimes, a load correction is used to compensate for the modal coupling (Denoël & Degée, 2009). The damping ratio of the added dampers can be expressed as follows:

$$\zeta_d = \frac{C^*_{d_{n,n}}}{2M^*_{n,n}\Omega_{n,n}} \tag{18}$$

where $\Omega_{n,n}$ is the matrix of the natural circular frequencies. This method is based on the idea of a target damping, which could be defined in the preliminary stage of the design process, or could be required by the client of the building for example.

The modal damping ratio in mode *n* can be expressed $\zeta = \zeta_s + \zeta_d + \zeta_a$ and since the target damping ζ_t is initially known (replacing the modal damping ratio by the target damping ratio) it can be deduced that $\zeta_d = \zeta_t - \zeta_s - \zeta_a$, the minimal viscosity for all dampers can be obtained:

$$c_j \geq \frac{2M^*_{n,n}\Omega_{n,n}\left(\zeta_t - \zeta_s - \zeta_a\right)}{\sum_{j=1}^{nd} c_j \Delta\phi^2_{n,j}} \tag{19}$$

This formula gives a flexibility in the design from different points of view, number of the dampers, their location and their inclination (Duflot & al, 2017). All these parameters are left for the designer to consider, and some iterations maybe needed to obtain the optimal values after comparing few scenarios.

ILLUSTRATION

In this section, the preliminary design method will be applied on a 5-story RC building, the structural scheme consists in a column-beam system, figure 13 presents a plan view. With a total height of 18.28 meters and concrete grade of 25 MPA, this structure can be designed using commercially available finite element-based software package. In order to apply a time history analysis, and since wind loads changes along the height of the building, the structure was divided into three representative section along its height, where the wind load is applied as a dynamic force. Using the method explained above for a multivariate time history generation. Based on the interstory drift, the additional dampers are installed in the building; few scenarios will be proposed to finally settle with the optimal design.

Since the approach adopted in this chapter is to simplify the comprehension, and in order to illustrate the flexibility obtained by the described method, two scenarios will be considered:

1. Additional dampers will be installed in every floor, in the middle bay in lines (A) and (E).
2. In the second scenario, the dampers will be installed where the structure presents higher values of interstory drift, since their use is to reduce the drift.

To illustrate the design methodology, Table 2 summarizes the properties used for this purpose, choosing a target damping ratio for this structure to be 8%, also, the structural damping ratio is supposed to be 2%, then, $\zeta_d = \zeta_t - \zeta_s = 6\%$ (these values where taken arbitrarily for illustration purpose). As for the aerodynamic damping, it is usually taken equal to 1%, however, due to a certain reduction of the damper's contribution that can be caused during their installation, the aerodynamic damping is neglected, and thus insuring that the target damping is reached.

Before proceeding with the proposed methodology, some assumption must be made in order to restrain the various design possibilities.

- The dampers are installed in the building diagonally (see figure 10). This means that the efficiency factor φ. in equation (17) is equal to $\varphi = \cos\theta$., with the inclination of $\theta = 45°$, φ is assumed to be equal to $\sqrt{2}\,/\,2$.
- For the first scenario, the damper's viscosity is assumed to be the same for all dampers (10 dampers).

Figure 13. Plan view of the analyzed structure

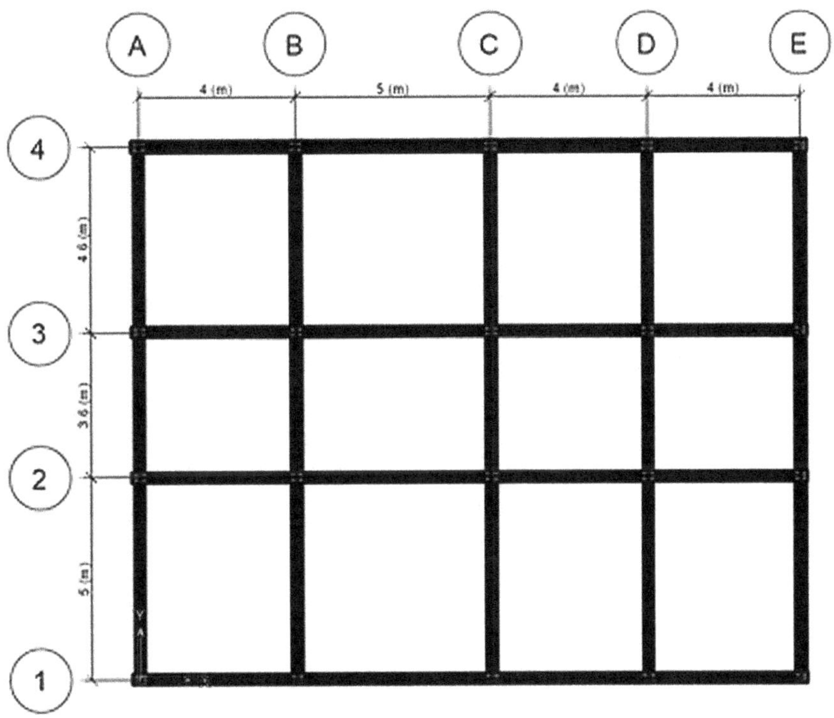

Figure 14. Inter-story drift for the structure (no additional dampers)

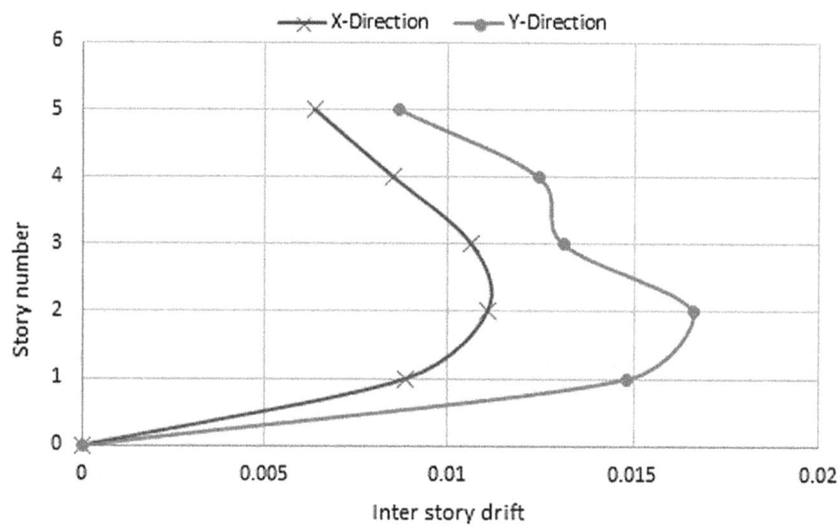

Table 2. Structural properties for target damping

Mode	Frequency (Hz)	Modal Mass (ton.)	Additional Damping ζ_d	Target Viscosity (KN.s / m)
1 - Y direction	0.694	1371.7	6%	720
2 – X direction	0.833	1340	6%	840

- As stated before in equation (17), the relative longitudinal displacement is expressed as a function of the inter-story drift, $\Delta\phi = \varphi\Delta u_{n,j}$.

r the modal analysis, the required modal viscosity is easily defined using equation (18). It wnd that for the first mode (Y direction), the target modal viscosity needed to achieve the target damping ratio is equal to a total of 720 KN.s / m. Dividing the target viscosity by the number of added dampers, we find that the average contribution of a single damper is 71 KN.s / m. From figure 14, the inter-story drift is $\Delta u_n^2 = 0.016$., and $\varphi = \cos 45$. (damper inclination). From equation (17), the obtained value is $0.51.6.10^{-3} c_j$, it is found that to reach a damping ratio of 6%, each damper should have a viscosity approximately equal to 9000 KN.s / m.

In the same manner for the X direction, and for a value of interstory drift of $\Delta u_n^2 = 0.011$, with ten dampers diagonally installed in the middle bay (between (B) and (C)) of lines (1) and (4). The average contribution of each damper is 83 KN.s/m; following the same procedure, it is found that the viscosity of each damper is 13000 KN.s/m.

Figure 15 shows that the maximum inter-story drift was reduced by more than 30% when adding FVDs, which is considered as a major improvement and proves the efficiency of these devices.

For the second scenario, the focus will be on the Y direction, since it represents higher values of drift and the same procedure can be followed for the other direction. The designer can obviously see that

Figure 15. Inter-story drift reduction after the first scenario

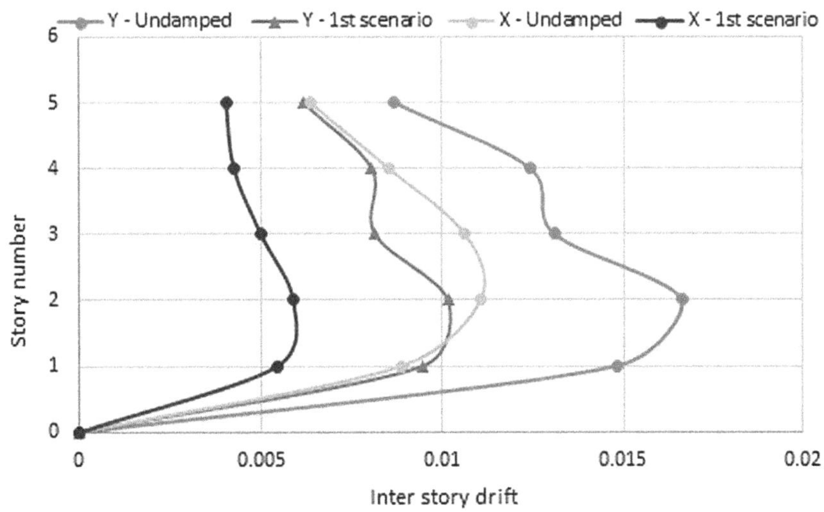

Figure 16. Inter-story drift reduction after the second scenario

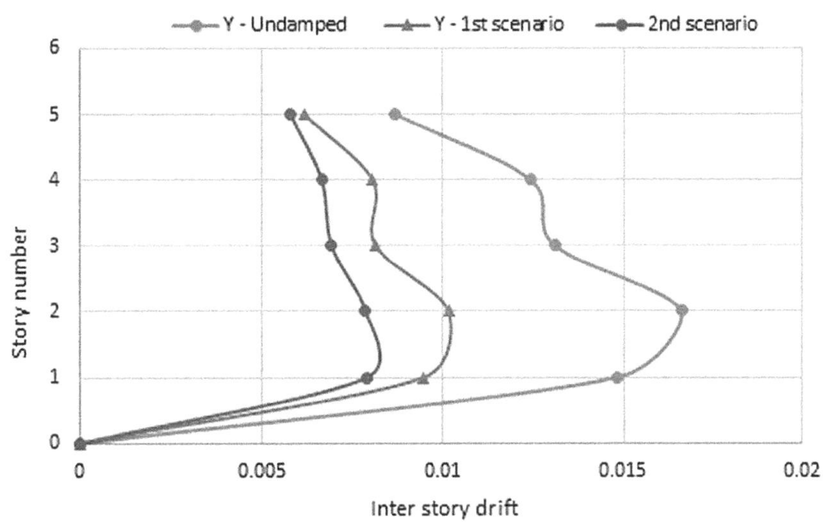

floors from first to fourth present high inter-story drift, and even after the addition of dampers, it may be demanded or for better behavior to reduce the drift in first and second floor. Thus, the flexibility given by this method permits to adjust the viscosity of the dampers installed in those floors, or increase the number of dampers. The latter proposition may be seen as a costly solution; so, the damping coefficient is increased for this case.

Figure.16 illustrates the optimized solution, after a few iteration where the viscosity of the dampers installed in the first and second floors is adjusted, respecting the total target viscosity. In addition dampers installed in the fifth story where removed, and an optimization of time, damping value and cost is achieved. Many other optimization proposition can be adopted for this case, but a reduction of over 50% in inter-story drift is a major improvement.

CONCLUSION

The concept of wind loads on civil structure was simplified in this work, introducing its components, and presenting effective method for the generation of wind time histories matched with a target power spectral density, that can be implemented on any available tool i.e. Matlab. The relation between time and frequency domain have been shown, this relation that is held by the Fourier Transform have also been implemented along with a spectral analysis of a SDOF.

Structural protection against any type of excitation is discussed, using energy dissipation systems, with the variety of such devices, it is only concentrated on viscous fluid dampers. Fluid viscous dampers are increasingly used in modern engineering, with their proved efficiency in achieving better structural behavior. Many design optimization were briefly mentioned with their respective references, however, a practical method for the preliminary design of the added dampers was used in this study, it has obviously simplified the task of predicting optimal values of the damping coefficient for any given structure, and it allows a flexible choice in the configuration of dampers respecting architectural requirements.

A time history analysis was performed, using the discussed method for the generation, on a five-story RC building, the rapidity of the application of this approach has been illustrated for a target damping of 8%. Based on the results obtained, a reduction of approximately 30% for the inter-story drift was successfully achieved when only ten linear FVDs where introduced in the first scenario. For the second scenario, the inter-story drift gives a good initial estimate for the location of the added dampers (even for more complicated and tall structures), the viscosity of the dampers in those location have been increased, to finally achieve more than 50% reduction for the inter-story drift and with less dampers.

Considering the references is recommended for readers who wish to get more details, since this chapter have an illustrative purpose.

REFERENCES

Aiken, I. D., Kelly, J. M., & Mahmoodi, P. (1990, May). The application of viscoelastic dampers to seismically resistant structures. In *Proceedings of Fourth US National Conference on Earthquake Engineering* (*Vol. 3*, pp. 459-468). Academic Press.

Chang, K., Soong, T. T., Oh, S. T., & Lai, M. L. (1991). *Seismic response of a 2/5 scale steel structure with added viscoelastic dampers*. Academic Press.

Constantinou, M. C., & Symans, M. D. (1992). *Experimental and analytical investigation of seismic response of structures with supplemental fluid viscous dampers*. Buffalo, NY: National Center for Earthquake Engineering Research.

Constantinou, M. C., Symans, M. D., Tsopelas, P., & Taylor, D. P. (1993). Fluid viscous dampers in applications of seismic energy dissipation and seismic isolation. *Proceedings ATC*, *17*(1), 581–592.

Das, N. K. (1988). *Safety analysis of steel building frames under dynamic wind loading* (Doctoral dissertation). Texas Tech University.

Davenport, A. G. (1961). The application of statistical concepts to the wind loading of structures. *Proceedings - Institution of Civil Engineers*, *19*(4), 449–472. doi:10.1680/iicep.1961.11304

Denoël, V. (2005). *Application des méthodes d'analyse stochastique à l'étude des effets du vent sur les structures du génie civil* (Doctoral dissertation). University of Liège.

Denoël, V., & Degée, H. (2009). Asymptotic expansion of slightly coupled modal dynamic transfer functions. *Journal of Sound and Vibration, 328*(1-2), 1–8. doi:10.1016/j.jsv.2009.08.014

Duflot, P., & Taylor, D. (2008). Experience and Practical Considerations in the Design of Viscous Dampers. *Third international conference.*

Duflot, P., Vigano, M. G., & Denoël, V. (2017). Method for Preliminary Design of a System of Viscous Dampers Applied to a Tall Building, *European African Conference on Wind Engineering.*

Filiatrault, A., & Cherry, S. (1990). Seismic design spectra for friction-damped structures. *Journal of Structural Engineering, 116*(5), 1334–1355. doi:10.1061/(ASCE)0733-9445(1990)116:5(1334)

Guo, T., Xu, J., Xu, W., & Di, Z. (2014). Seismic Upgrade of Existing Buildings with Fluid Viscous Dampers: Design Methodologies and Case Study. *Journal of Performance of Constructed Facilities, 29*(6), 04014175. doi:10.1061/(ASCE)CF.1943-5509.0000671

Haskell, G., & Lee, D. (1996). *Fluid viscous damping as an alternative to base isolation (No. CONF-960706).* New York: American Society of Mechanical Engineers.

Kandemir-Mazanoglu, E. C., & Mazanoglu, K. (2017). An optimization study for viscous dampers between adjacent buildings. *Mechanical Systems and Signal Processing, 89*, 88–96. doi:10.1016/j.ymssp.2016.06.001

Kareem, A. (2008). Numerical simulation of wind effects: A probabilistic perspective. *Journal of Wind Engineering and Industrial Aerodynamics, 96*(10-11), 1472–1497. doi:10.1016/j.jweia.2008.02.048

Marko, J., Thambiratnam, D., & Perera, N. (2004). Influence of damping systems on building structures subject to seismic effects. *Engineering Structures, 26*(13), 1939–1956. doi:10.1016/j.engstruct.2004.07.008

Martinez-Rodrigo, M., & Romero, M. L. (2003). An optimum retrofit strategy for moment resisting frames with nonlinear viscous dampers for seismic applications. *Engineering Structures, 25*(7), 913–925. doi:10.1016/S0141-0296(03)00025-7

Moreschi, L. M. (2000). *Seismic design of energy dissipation systems for optimal structural perfromance* (Doctoral dissertation). Virginia Tech.

Newland, D. E. (2012). *An introduction to random vibrations, spectral & wavelet analysis.* Courier Corporation.

Ras, A., & Boumechra, N. (2016). Seismic energy dissipation study of linear fluid viscous dampers in steel structure design. *Alexandria Engineering Journal, 55*(3), 2821–2832. doi:10.1016/j.aej.2016.07.012

Roberts, J. B., & Spanos, P. D. (2003). *Random vibration and statistical linearization.* Courier Corporation.

Saaed, T. E., Nikolakopoulos, G., Jonasson, J. E., & Hedlund, H. (2015). A state-of-the-art review of structural control systems. *Journal of Vibration and Control, 21*(5), 919–937. doi:10.1177/1077546313478294

Shinozuka, M. (1971). Simulation of multivariate and multidimensional random processes. *The Journal of the Acoustical Society of America*, *49*(1B), 357–368. doi:10.1121/1.1912338

Simiu, E., & Scanlan, R. H. (1986). *Wind effects on structures: an introduction to wind engineering*. John Wiley.

Soong, T. T., & Costantinou, M. C. (Eds.). (2014). *Passive and active structural vibration control in civil engineering* (Vol. 345). Springer.

Soong, T. T., & Spencer, B. F. Jr. (2002). Supplemental energy dissipation: State-of-the-art and state-of-the-practice. *Engineering Structures*, *24*(3), 243–259. doi:10.1016/S0141-0296(01)00092-X

Symans, M. D., Charney, F. A., Whittaker, A. S., Constantinou, M. C., Kircher, C. A., Johnson, M. W., & McNamara, R. J. (2008). Energy dissipation systems for seismic applications: Current practice and recent developments. *Journal of Structural Engineering*, *134*(1), 3–21. doi:10.1061/(ASCE)0733-9445(2008)134:1(3)

Tyler, R. G. (1985). Test on a Brake Lining Damper for Structures. *Bulletin of the New Zealand National Society for Earthquake Engineering*, *18*(3), 280–284.

Whittaker, A. S., Bertero, V. V., Thompson, C. L., & Alonso, L. J. (1991). Seismic testing of steel plate energy dissipation devices. *Earthquake Spectra*, *7*(4), 563–604. doi:10.1193/1.1585644

ADDITIONAL READING

Genta, G. (2009). *Vibration dynamics and control* (pp. 280–306). New York: Springer. doi:10.1007/978-0-387-79580-5

Lee, D., & Taylor, D. P. (2001). Viscous damper development and future trends. *Structural Design of Tall and Special Buildings*, *10*(5), 311–320. doi:10.1002/tal.188

Preumont, A. (2013). *Random Vibration and Spectral Analysis/Vibrations aléatoires et analyse spectral* (Vol. 33). Springer Science & Business Media.

Robert, C. P. (2004). *Monte Carlo methods*. John Wiley & Sons, Ltd.

Sockel, H. (Ed.). (2014). *Wind-excited vibrations of structures* (Vol. 335). Springer.

KEY TERMS AND DEFINITIONS

Correlation: Dependence or association in any statistical relationship, whether causal or not, between two random variables.

Ergodicity: An ergodic process is a process which its statistical properties can be deduced from a single, sufficiently long, random sample of the process.

Fast Fourier Transform: (FFT): Is an algorithm that samples a signal over a period of time and divides it into its frequency components used to improve the computational efficiency.

Stochastic Process: A stochastic or random process is a phenomenon usually defined as a collection of random variables.

Variance: In probability theory, it is the expectation of the squared deviation of a random variable from its mean, and can be estimated as the area of the PSD graph (response).

Chapter 6
Optimization of Condensed Stiffness Matrices for Structural Health Monitoring

Kong Fah Tee
University of Greenwich, UK

ABSTRACT

This chapter aims to develop a system identification methodology for determining structural parameters of linear dynamic systems, taking into consideration practical constraints such as insufficient sensors. Based on numerical analysis of measured responses (output) due to known excitations (input), structural parameters such as stiffness values are identified. If the values at the damaged state are compared with the identified values at the undamaged state, damage detection and quantification can be carried out. To retrieve second-order parameters from the identified state space model, various methodologies developed thus far impose different restrictions on the number of sensors and actuators employed. The restrictions are relaxed in this study by a proposed method called the condensed model identification and recovery (CMIR) method. To estimate individual stiffness coefficient from the condensed stiffness matrices, the genetic algorithms approach is presented to accomplish the required optimization problem.

INTRODUCTION

Civil engineering structures such as buildings, bridges and offshore platform continuously accumulate damage during their service life due to natural and man-made actions. Damage in a structure is often translated into change in physical properties of its structural components. When an element of the structure contains damage, the stiffness of the damage element as well as the load-carrying capacity of that element will change (Tee et al, 2013). If not monitored and rectified, damage would increase maintenance cost and render structures unserviceable. In extreme event, damage may even cause structures to collapse catastrophically, resulting in loss of lives and assets. The only way to safeguard the safety of human life and to reduce loss of wealth is to carry out regular monitoring for early detection of structural damage.

DOI: 10.4018/978-1-5225-7059-2.ch006

It is therefore essential to detect the existence, location and extent of damage in the structure early and to carry out remedial work if necessary.

The science of monitoring (continuous or periodic) of the condition of a structure using built-in or autonomous sensory systems is now called Structural Health Monitoring (SHM). Some of the noteworthy efforts in SHM are reported in special issues in Journal of Engineering Mechanics, ASCE in July 2000 (Ghanem and Sture, 2000) and January 2004 (Bernal and Beck, 2004) and in Computer-Aided Civil and Infrastructure Engineering in January 2001 (Adeli, 2001). For civil engineering structures, the current methods used by practicing engineers are mainly visual inspection (Moore, 2001) and localized on-site methods such as acoustic or ultrasonic methods, magnetic field methods, radiography, eddy-current methods and thermal field methods (Doherty, 1987). All these latter on-site methods require that the vicinity of the damage is known *a priori* and that the portion of the structure being inspected is readily accessible. These experimental methods can usually be used to detect damage on or near the surface of the structure and are thus limited in application.

The need for quantitative global damage detection methods that can be applied to complex structures has led to research into SHM methods that examine changes in the vibration characteristics of the structure. Vibration-based inspection is currently an active area of research in SHM, on the basis of examining changes in the characteristics of a structure before and after damage occurrence based on analysis of input and output signals due to dynamic excitation (Tee et al, 2009; Koh et al, 2002). The general idea is that changes in the physical properties (i.e., stiffness, mass, and or damping) of the structure will, in turn, alter the dynamic characteristics (i.e., natural frequencies, modal damping and mode shapes) of the structure. A monitoring system can provide invaluable insight into the accuracy of these structural models and not only can assist engineers in refining them but also can verify design assumptions and parameters for future construction.

For the purpose of SHM, the use of vibration-based inspection or system identification provides a non-destructive means to quantify structural parameters based on measured structural response due to dynamic excitation. Using a monitoring system to measure structural responses, a damage detection strategy is then employed to monitor the structural health and to provide information for facilitating the planning of inspection and maintenance activities. Any health monitoring and damage detection methodology often involves some kind of system identification algorithm. Therefore, structural system identification will be briefly reviewed, and its correlation to structural damage identification will be highlighted in the following sections.

System identification, in a broad sense, can be described as the identification of the conditions and properties of mathematical models that aspire to represent real phenomena in an adequate fashion. System identification originally began in the area of electrical engineering and later extended to the field of mechanical and control engineering, and civil engineering. The underlying philosophy of most system identification discussions attempts at addressing two important questions: (a) Choosing a mathematical model that is characterized by a finite set of parameters and (b) Identifying these parameters based on collected data. System Identification techniques to study the actual states of civil engineering structures have received considerable attention in recent years. The application of system identification techniques has increasingly become an important research topic in connection with damage assessment and safety evaluation of structures (Qin, 2006; Tee et al, 2003). To properly identify a structure means to create a mathematical model that represents the real structure in an appropriate way. The primary measure of the effectiveness of the system identification is how well the identified mathematical model produces an output which matches the measured output for a given input signal. Hence, such a model must, with a certain degree of accuracy, (a) represent the dynamic characteristics of the structure (i.e. natural frequencies, damping factors, etc.), (b) provide some information on the mechanical properties of the structure (mass, stiffness, etc), and (c) be able to estimate the structure response in the case of future excitations (Wang et al, 2011). System identification methods have been shown to be effective in producing models which exactly or closely match the true system.

From the viewpoint of system identification, civil engineering applications present unique and challenging features such as the large size of the structure, difficulty and high cost in field experiments, limited number of sensing devices and high level of measurement noise. Full-scale experiments of civil engineering structures are expensive and difficult to conduct due to the fact that many structural elements may not be accessible (Chen et al, 2012). In this respect, system identification techniques can be used for structural identification based on dynamic response of structures subjected to low intensity excitations. With the development of data acquisition technology and enhanced computational resources, structural assessment by means of system identification techniques has become a viable option.

Considering repeated experiments corresponding to the damaged and undamaged configurations can detect the location and extent of damage in structural systems using identification algorithm. It is possible to determine, somewhat rigorously, where and how much structural damage has occurred between these two states by comparing the changes in various structural parameters. To do so, it is necessary to use an identification algorithm that provides a reliable and accurate physical model of the structural system. Thus, structural system identification technique has become increasingly popular to study numerically the undamaged and damaged states of existing structures.

Based on the amount of information provided regarding the damage state, these methods can be classified as providing four levels of damage detection. The four levels are (Rytter, 1993):

1. Identify that damage has occurred
2. Identify that damage has occurred and determine the location of damage
3. Identify that damage has occurred, locate the damage, and estimate its severity
4. Identify that damage has occurred, locate the damage, estimate its severity and determine the remaining useful life of the structure

Generally, system identification techniques can be classified under various categories, such as frequency and time domains, parametric and nonparametric models, deterministic and stochastic approaches, classical and non-classical methods and online and offline identifications. Further information can be found in the literature on the application of system identification in structural engineering reviewed by several investigators including Lin et al. (1990), Agbabian et al. (1991), Ghanem and Shinozuka (1995), Hjelmstad and Banan (1995), Lus (2001) and Chen and Tee (2014).

The identification of Markov parameters has been studied in the literature. The Markov parameters can be defined as the coefficients in the convolution sum for the state difference equation, and analogously, as the pulse response sequence of a discrete time system. Under ideal test conditions these parameters are obtained using FFT. However, this procedure requires a very rich input to ensure computational accuracy. The use of time domain methods for the determination of Markov parameters may also be problematic in the sense that their results are unsatisfactory and numerically ill-conditioned. Therefore, in the work by Phan et al. (1992), Observer/Kalman filter Identification (OKID), which incorporated observer-based identification concept, was developed to improve the stability of the system and to make the problem better conditioned numerically. This approach was extended to include also observers with complex eigenvalues, and both the system and its associated observer could be identified simultaneously (Phan et al. 1993). This observer would converge to an optimal Kalman filter when both the length of the record and the order of the identified input-output model approached infinity. To this end, Phan et al. (1995) proposed an improvement of OKID using residual whitening, which uses an auto-regressive model with the moving average terms to model the noise dynamics. The OKID approach is used to obtain Markov parameters, which are pre-requisites for the Eigensystem Realization Algorithm (ERA) based algorithms.

One of the most important theoretical concepts in the control theory is that of 'minimal realizations'. Ho and Kalman (1965) showed that the problem of minimal realization was equivalent to identifying the first order system matrices in the state space formulation. Then, a more practical algorithm, namely Eigensystem Realization Algorithm (ERA) was developed by Juang and Pappa (1985). They have further developed the Ho-Kalman algorithm to include the singular-value decomposition and applied it to modal parameter identification problems. ERA is one of the most widely used and studied algorithms. Juang and Pappa (1985) conducted numerical studies on the Galileo spacecraft test data for modal parameter identification and discussed some accuracy indicators such as the modal amplitude coherence and modal phase collinearity. Effects of noise in the data were studied by Juang and Pappa (1986) using Monte Carlo simulations, and Longman and Juang (1987) attempted to develop a confidence criterion for the ERA identified modal parameters. Their numerical studies show that ERA performs better for most cases considered. Later, Juang et al. (1988) proposed ERA with Data Correlations (ERA/DC), which was refined to better handle the effects of noise and structural nonlinearities. ERA/DC uses data correlations rather than response values and their numerical studies demonstrated that ERA/DC performed better when noise characteristics were significantly large.

Identification of stiffness, mass and damping in a second-order matrix differential equation has also received considerable attention. However, for identification of physical parameters of the second-order model from the results of first-order model, issues such as non-uniqueness of the solution have to be considered. The existing literature imposes restrictions on the number of sensors and actuators employed in order to retrieve the second-order model parameters. Yang and Yeh (1990) required full sensors and full actuators. This requirement was relaxed by Alvin and Park (1994) that only require full sensors, with one co-located sensor-actuator pair. Tseng et al. (1994a, b) presented a further generalization where the number of actuators is equal to the number of second order modes, with one co-located sensor-actuator pair. For structural damage assessment, however, it is impractical to use full measurement to identify the unknown structural parameters. DeAngelis et al. (2002) utilized mixed type information, thereby enabling one to treat the information from a sensor or an actuator in an analogous fashion. This conceptual "input–output equivalence" helps relax the necessity of either full sensors or full actuators. However, it is still not practical for real life engineering application. The focus in this study is on recovering the stiffness value from the identified condensed stiffness matrices. This approach allows fewer numbers of sensor and actuator than those required in previously discussed approaches. This technique was identified as being of practical importance because this could provide an alternative to the problem of insufficient sensors in structural system identification.

A review of published literature shows that the first major step toward a method of reducing or condensing the dimension of the eigenproblem of a structural dynamic system (model condensation methods) appeared in the paper published by Guyan (1965). The well-known method of Guyan is based on static condensation of unwanted or dependent coordinates in order to reduce the stiffness matrix of the system. Since the dynamic effects were ignored in this method, the error can be large for dynamic problems. Hence many methods have subsequently been proposed to improve the accuracy. The inertia terms were considered partially by Kidder (1975) and Miller (1980). The inertia terms are also considered statically in the Improved Reduced System (IRS). A method of reduction that may be considered an extension of the static condensation method has been proposed, namely dynamic condensation (Paz, 1989). Many other algorithms for dynamic condensation have been developed. Among them, the iterative methods are the most accurate ones because dynamic condensation is updated repeatedly until a convergent value is obtained. Qu and Fu (1998) proposed a new iterative method for dynamic condensation of finite element models. Two constraint equations for the dynamic condensation matrix are derived directly from the modified eigenvalue equation. Most recently, a dynamic condensation approach applicable to non-classically damped structures was proposed by Rivera et al. (1999). O'Callahan et al. (1989) proposed a new model reduction technique, which requires the full system eigenvectors corresponding to the set of modes of interest, and this is called System Equivalent Reduction Expansion Process (SEREP). Later, an approach using the eigenvectors from the reduced model is proposed by Papadopoulos et al. (1996) to avoid using the full system eigenvectors. Three different model condensation methods are considered in this study, i.e. static condensation, dynamic condensation and SEREP to eliminate the requirement of complete measurement.

STRUCTURAL IDENTIFICATION FOR HEALTH MONITORING

The equations of motion for an N-DOF time invariant structural system can be written in a system of second-order differential equations as

$$\mathbf{M}\ddot{\mathbf{q}}(t) + \mathbf{L}\dot{\mathbf{q}}(t) + \mathbf{K}\mathbf{q}(t) = \mathbf{B}_{\mathrm{f}}\,\mathbf{u}(t) \tag{1a}$$

$$\mathbf{y}(t) = \begin{bmatrix} \mathbf{C}_{\mathrm{p}}\mathbf{q}(t) \\ \mathbf{C}_{\mathrm{v}}\dot{\mathbf{q}}(t) \\ \mathbf{C}_{\mathrm{a}}\ddot{\mathbf{q}}(t) \end{bmatrix} \tag{1b}$$

where $\mathbf{M}$, $\mathbf{L}$ and $\mathbf{K}$ are symmetric positive definite mass, damping, and stiffness matrices (size $N\times N$) of the structure, respectively, $\mathbf{B}_{\mathrm{f}}$ is the input matrix ($N\times r$) containing the location of r external excitations acting on the structure, $\mathbf{q}(t)$ is the displacement vector ($N\times 1$), overdot denotes differentiation with respect to time t, and $\mathbf{u}(t)$ is the input excitation vector ($r\times 1$). The matrix $[\mathbf{C}_p^T\ \mathbf{C}_v^T\ \mathbf{C}_a^T]^T$ is the output matrix ($m\times N$) that incorporates displacement, velocity and acceleration measurements, with m denoting the total number of outputs. The $m\times 1$ output vector $\mathbf{y}(t)$ can be displacement, velocity or acceleration measurements or a combination of them. $\mathbf{q}(t)$ has a chosen physical meaning, and therefore one knows these input and output matrices based on the type and placement of the sensors and actuators, i.e. in the case of the input matrix $\mathbf{B}_{\mathrm{f}}$, the coefficient in the ith row (i =1, 2, …, N) and j^{th} column (j = 1, 2, …, r) of $\mathbf{B}_{\mathrm{f}}$ is 1 if the jth actuator is placed on the ith DOF, and this coefficient is 0 if the jth actuator is not placed on the ith DOF. Similarly, the coefficient in the ith row (i =1, 2, …, m) and j^{th} column (j = 1, 2, …, N) of the output matrix $\mathbf{C}_{\mathrm{a}}$ is 1 if the ith sensor (acceleration) is placed on the jth DOF, and this coefficient is 0 if the ith sensor is not placed on the jth DOF.

By defining a state vector $\mathbf{x}(t) = [\mathbf{q}(t)^T \ \dot{\mathbf{q}}(t)^T]^T$, the equations of motion can be written in first-order difference equations form (state space form) with N DOFs, m number of observations and r number of input forces as

$$\dot{\mathbf{x}}(t) = \mathbf{A}_c\mathbf{x}(t) + \mathbf{B}_c\mathbf{u}(t) \tag{2a}$$

$$\mathbf{y}(t) = \mathbf{C}\mathbf{x}(t) + \mathbf{D}\mathbf{u}(t) \tag{2b}$$

where $\mathbf{A}_c$ $(2N\times 2N)$, $\mathbf{B}_c$ $(2N\times r)$, $\mathbf{C}$ $(m\times 2N)$ and $\mathbf{D}$ $(m\times r)$ are the time invariant continuous time system matrices, while $\mathbf{x}(t)$ is the $n\times 1$ state vector and $\dot{\mathbf{x}}(t)$ is its first derivative with respect to time. The equivalent discrete time representation of the system in Eqs. (2) is given by

$$\mathbf{x}(k+1) = \mathbf{A}\mathbf{x}(k) + \mathbf{B}\mathbf{u}(k) \tag{3a}$$

$$\mathbf{y}(k) = \mathbf{C}\mathbf{x}(k) + \mathbf{D}\mathbf{u}(k) \tag{3b}$$

where **A** and **B** are the discrete time system matrices. The sampling time is represented by ΔT, and k is an integer (> 0) that indicates the time step, i.e. $\mathbf{x}(k)$ actually implies $\mathbf{x}(k\,\Delta T)$.

Consider a pulse input denoted by $u_i(0) = 1$ $(i = 1,\ 2,\ \ldots,\ r)$ and $u_i(k) = 0$ $(k = 1,\ 2,\ \ldots)$ as well as $x(0) = 0$. Substituting into Eq. (3) yield the following $m\times r$ pulse-response matrices **Y** (Markov parameters):

$$\mathbf{Y}_0 = \mathbf{D},\quad \mathbf{Y}_1 = \mathbf{CB},\quad \mathbf{Y}_2 = \mathbf{CAB},\quad \cdots,\quad \mathbf{Y}_k = \mathbf{CA}^{k-1}\mathbf{B} \tag{4}$$

Observer Kalman Filter Identification (OKID)

The usual practice uses the Fast Fourier Transform (FFT) of the inputs and measured outputs to compute the frequency response function, and then use the Inverse Fast Fourier Transform (IFFT) to compute the sampled pulse response histories. Rather than identifying the system Markov parameters in frequency domain, one can use an OKID to identify in time domain. In practice, the primary purpose of introducing an observer is to compress the data and improve system identification results.

Introducing the term $\mathbf{Gy}(k)$ to the right-hand side of the state equation in Eq. (3a) yields

$$\mathbf{x}(k+1) = \mathbf{A}\mathbf{x}(k) + \mathbf{B}\mathbf{u}(k) + \mathbf{G}\mathbf{y}(k) - \mathbf{G}\mathbf{y}(k)$$

$$= (\mathbf{A} + \mathbf{GC})\mathbf{x}(k) + (\mathbf{B} + \mathbf{GD})\mathbf{u}(k) - \mathbf{G}\mathbf{y}(k)$$

or

$$\mathbf{x}(k+1) = \bar{\mathbf{A}}\mathbf{x}(k) + \bar{\mathbf{B}}\mathbf{v}(k) \tag{5}$$

where

$$\bar{\mathbf{A}} = \mathbf{A} + \mathbf{GC}$$
$$\bar{\mathbf{B}} = [\mathbf{B} + \mathbf{GD}, -\mathbf{G}] \qquad (6)$$
$$\mathbf{v}(k) = \begin{bmatrix} \mathbf{u}(k) \\ \mathbf{y}(k) \end{bmatrix}$$

and $\mathbf{G}$ is an $n \times m$ observer gain matrix. Although Eq. (5) is mathematically identical to Eq. (3a), it is expressed using different system matrices and has a different input. In fact, Eq. (5) is an observer equation if the state $\mathbf{x}(k)$ is considered as an observer state vector. Therefore, the Markov parameters of the system will be referred to as the observer Markov parameter. The input-output description in matrix form becomes

$$\mathbf{y} = \bar{\mathbf{Y}}\ \mathbf{V} \qquad (7)$$

$$(m \times l)\ (m \times [(r+m)p + r])\ ([(r+m)p + r] \times l)$$

where

$$\mathbf{y} = [\mathbf{y}(0)\ \mathbf{y}(1)\ \mathbf{y}(2) \ldots\ \mathbf{y}(p) \ldots\ \mathbf{y}(l-1)]$$

$$\bar{\mathbf{Y}} = [\,\mathbf{D}\ \mathbf{C}\bar{\mathbf{B}}\ \mathbf{C}\bar{\mathbf{A}}\bar{\mathbf{B}} \ldots\ \mathbf{C}\bar{\mathbf{A}}^{p-1}\bar{\mathbf{B}}\,]$$

$$\mathbf{V} = \begin{bmatrix} \mathbf{u}(0) & \mathbf{u}(1) & \mathbf{u}(2) & \cdots & \mathbf{u}(p) & \cdots & \mathbf{u}(l-1) \\ \mathbf{0} & \mathbf{v}(0) & \mathbf{v}(1) & \cdots & \mathbf{v}(p-1) & \cdots & \mathbf{v}(l-2) \\ \mathbf{0} & \mathbf{0} & \mathbf{v}(0) & \cdots & \mathbf{v}(p-2) & \cdots & \mathbf{v}(l-3) \\ \vdots & \vdots & \mathbf{0} & \ddots & \vdots & \cdots & \vdots \\ \mathbf{0} & \mathbf{0} & \cdots & \mathbf{0} & \mathbf{v}(0) & \cdots & \mathbf{v}(l-p-1) \end{bmatrix}$$

The matrix $\mathbf{y}$ is an output data matrix $m \times l$ where l is the number of time steps in the data. The matrix $\bar{\mathbf{Y}}$ of size $m \times [(r+m)p + r]$ where p is an integer such that $\mathbf{CA}^k\mathbf{B} \approx 0$ for $k \geq p$, contains all the observer Markov parameters to be determined. The matrix $\mathbf{V}$ is an input matrix of size $[(r+m)p + r] \times l$. If the data have a realization in the form of Eq. (3), then the first p observer Markov parameters approximately satisfy $\mathbf{Y} = \mathbf{yV}^{+}$ where $\mathbf{V}^{+}$ is the pseudo-inverse of the matrix $\mathbf{V}$, and the approximation error decreases as p increases.

The Markov parameters include the system Markov parameters and the observer gain Markov parameters. The system Markov parameters are used to compute the system matrices $\mathbf{A}$, $\mathbf{B}$, $\mathbf{C}$ and $\mathbf{D}$. To recover the system Markov Parameters in $\mathbf{Y}$ from the observer Markov parameters in $\bar{\mathbf{Y}}$, $\bar{\mathbf{Y}}$ is partitioned such that

$$\bar{\mathbf{Y}} = [\bar{\mathbf{Y}}_0 \quad \bar{\mathbf{Y}}_1 \quad \bar{\mathbf{Y}}_2 \quad \cdots \quad \bar{\mathbf{Y}}_p] \tag{8}$$

where

$$\bar{\mathbf{Y}}_0 = \mathbf{D}$$

$$\bar{\mathbf{Y}}_k = \mathbf{C}\bar{\mathbf{A}}^{k-1}\bar{\mathbf{B}} \tag{9}$$

$$= [\mathbf{C}(\mathbf{A}+\mathbf{GC})^{k-1}(\mathbf{B}+\mathbf{GD}) \quad -\mathbf{C}(\mathbf{A}+\mathbf{GC})^{k-1}\mathbf{G}]$$

$$= [\,\bar{\mathbf{Y}}_k^{(1)} \; -\bar{\mathbf{Y}}_k^{(2)}\,];\; k = 1,\, 2\,, 3, \ldots.$$

The minus sign used for $\bar{\mathbf{Y}}_k^{(2)}$ in the last equality of Eq. (9) is chosen so that $\bar{\mathbf{Y}}_k^{(2)} = \mathbf{C}(\mathbf{A}+\mathbf{GC})^{k-1}\mathbf{G}$. Note that the identified observer Markov parameter $\bar{\mathbf{Y}}_0$ has a smaller size than the remaining Markov parameters.

Having identified the observer Markov parameters, the true system's Markov parameters can be retrieved using the recursive formula. By induction, the general relationship between the actual system Markov parameters and the observer Markov parameters is

$$\mathbf{D} = \mathbf{Y}_0 = \bar{\mathbf{Y}}_0$$

$$\mathbf{Y}_k = \bar{\mathbf{Y}}_k^{(1)} - \sum_{i=1}^{k} \bar{\mathbf{Y}}_i^{(2)}\mathbf{Y}_{k-i} \text{ for } k = 1, \ldots, p \tag{10}$$

$$\mathbf{Y}_k = -\sum_{i=1}^{p} \bar{\mathbf{Y}}_i^{(2)}\mathbf{Y}_{k-i} \text{ for } k = p+1, \ldots, \infty$$

Equation (10) shows that $\mathbf{Y}_k$ for $k \geq p+1$ is a linear combination of its past p system Markov parameters, i.e., $\mathbf{Y}_{k-1}$, $\mathbf{Y}_{k-2}$, …, $\mathbf{Y}_{k-p}$. In other words, there are only p independent system Markov parameters.

Most of the methods are estimated under the ideal assumption that the system is perfectly linear, the process and measurement noises are Gaussian, white and zero-mean, and the data length is sufficiently long. In practice, this ideal assumption can rarely, if ever, be satisfied, because of unexpected disturbances, system non-linearities, non-whiteness of the process and measurement noises, insufficient data records, and an incorrect system model. However, OKID produces an observer with minimum residual in the least-squares sense for the given input-output data record (Chang and Pakzad, 2014). There are several advantages for this approach. First, the number of independent Markov parameters has been compressed by using the observer. This allows one to use a smaller Hankel matrix and thus reduce the computational effort in the identification algorithm. Second, one can identify the number of independent system Markov parameters with multiple inputs and multiple outputs. This is a result of increased stability produced by adding an observer gain.

Eigensystem Realization Algorithm (ERA)

Once the system's Markov parameters have been identified, they can be used in the ERA formulation for the identification of the dynamic structural characteristics. Realization refers to the computation of a triplet $[\mathbf{A}, \mathbf{B}, \mathbf{C}]$ from the Markov parameters, for which the discrete-time model, Eq. (3) is satisfied. Any system has an infinite number of realizations, which will predict the identical response for any particular input. Minimum realization means a model with the smallest state-space dimension among all realizable systems that have the same input-output relations. All minimum realizations have the same set of eigenvalues, which are modal parameters of the system.

System realization begins by forming the generalized Hankel matrix ($\alpha m \times \beta r$), composed of the Markov parameters:

$$\mathbf{H}(k-1) = \begin{bmatrix} \mathbf{Y}_{k} & \mathbf{Y}_{k+1} & \cdots & \mathbf{Y}_{k+\beta-1} \\ \mathbf{Y}_{k+1} & \mathbf{Y}_{k+2} & \cdots & \mathbf{Y}_{k+\beta} \\ \vdots & \vdots & \ddots & \vdots \\ \mathbf{Y}_{k+\alpha-1} & \mathbf{Y}_{k+\pm} & \cdots & \mathbf{Y}_{k+\alpha+\beta-2} \end{bmatrix} \tag{11}$$

For the case when $k = 1$,

$$\mathbf{H}(0) = \begin{bmatrix} \mathbf{Y}_{1} & \mathbf{Y}_{2} & \cdots & \mathbf{Y}_{\beta} \\ \mathbf{Y}_{2} & \mathbf{Y}_{3} & \cdots & \mathbf{Y}_{1+\beta} \\ \vdots & \vdots & \ddots & \vdots \\ \mathbf{Y}_{\alpha} & \mathbf{Y}_{1+\alpha} & \cdots & \mathbf{Y}_{\alpha+\beta-1} \end{bmatrix}$$

Note that $\mathbf{Y}_0 = \mathbf{D}$ is not included in $\mathbf{H}(0)$. If $\alpha \geq n$ and $\beta \geq n$ (the order of the system), the matrix $\mathbf{H}(k-1)$ is of rank n. If the order of the system is n, then the minimum dimension of the state matrix is $n \times n$. Therefore, the Hankel matrix is of rank n by Eq. (11).

The basic development of the state-space realization is attributed to Ho and Kalman (1965) who introduced the important principles of minimum realization theory. The Ho-Kalman procedure uses the Hankel matrix, Eq. (11), to construct a state-space representation of a linear system from noise-free data. The methodology has been modified and substantially extended to develop the Eigensystem Realization Algorithm to identify modal parameters from noisy measurement data. The ERA algorithm begins by forming a Hankel matrix.

The ERA process continues with the factorization of the Hankel matrix, for $k = 1$, using singular value decomposition,

$$\mathbf{H}(0) = \mathbf{R}\Sigma\mathbf{S}^{T} \tag{12}$$

where the columns of matrices $\mathbf{R}$ and $\mathbf{S}$ are orthonormal and **£** is a rectangular matrix

$$\Sigma = \begin{bmatrix} \Sigma_{\mathrm{n}} & \mathbf{0} \\ \mathbf{0} & \mathbf{0} \end{bmatrix}$$

with

$$\Sigma_{\mathrm{n}} = diag[\sigma_1, \sigma_2, \cdots, \sigma_i, \sigma_{i+1}, \cdots, \sigma_n]$$

and monotonically non-increasing σ_i $(i = 1, 2, \cdots, n)$

$$\sigma_1 \geq \sigma_2 \geq \cdots \geq \sigma_i \geq \sigma_{i+1} \geq \cdots \geq \sigma_n > 0$$

Next, let $\mathbf{R}_{\mathrm{n}}$ and $\mathbf{S}_{\mathrm{n}}$ be the matrices formed by the first n columns of $\mathbf{R}$ and $\mathbf{S}$, respectively. Hence, the matrix $\mathbf{H}(0)$ becomes

$$\mathbf{H}(0) = \mathbf{R}_{\mathrm{n}}\Sigma_{\mathrm{n}}\mathbf{S}_{\mathrm{n}}^{\mathrm{T}} \text{ where } \mathbf{R}_{\mathrm{n}}^{\mathrm{T}}\mathbf{R}_{\mathrm{n}} = \mathbf{I}_{\mathrm{n}} = \mathbf{S}_{\mathrm{n}}^{\mathrm{T}}\mathbf{S}_{\mathrm{n}} \tag{13}$$

This is the basic formulation of realization for the ERA. The triplet

$$\hat{\mathbf{A}} = \Sigma_{\mathrm{n}}^{-1/2}\mathbf{R}_{\mathrm{n}}^{\mathrm{T}}\mathbf{H}(1)\mathbf{S}_{\mathrm{n}}\Sigma_{\mathrm{n}}^{-1/2},\ \hat{\mathbf{B}} = \Sigma_{\mathrm{n}}^{1/2}\mathbf{S}_{\mathrm{n}}^{\mathrm{T}}\mathbf{E}_{\mathrm{r}},\ \hat{\mathbf{C}} = \mathbf{E}_{\mathrm{m}}^{\mathrm{T}}\mathbf{R}_{\mathrm{n}}\Sigma_{\mathrm{n}}^{1/2} \tag{14}$$

is a minimum realization. Here the quantities with ^ mean estimated quantities to distinguish from the true quantities. $\mathbf{E}_r = [\mathbf{I}_{r\times r} \quad \mathbf{0}_{r\times r} \quad \mathbf{0}_{r\times r} \quad \cdots \quad \mathbf{0}_{r\times r}]_{r\times\beta r}^{T}$ with $\mathbf{I}$ denoting an identity matrix and $\mathbf{0}$ denoting a matrix whose elements are all zeros, and $\mathbf{E}_m$ is defined similarly. A major benefit of this approach is that there is no requirement for *a priori* knowledge of the order of the system. The order of the matrix $\hat{\mathbf{A}}$ is n which is the order of the system for sufficiently low-noise data. The singular value decomposition of $\mathbf{H}(0)$ reveals exactly n non-zero singular values when there is no noise. Due to measurement noise, nonlinearity, and numerical round off, the Hankel matrix $\mathbf{H}(k)$ may be full rank which does not, in general, equal the true order of the system under test. It should not be the aim to obtain a system realization which exactly reproduces the noisy sequence of data. A realization which produces a smoothed version of the sequence, and which closely represents the underlying linear dynamics of the system, is more desirable.

Some singular values, say $\sigma_{i+1}, \cdots, \sigma_n$, may be relatively small and negligible in the sense that they contain more noise information than system information. In other words, the directions determined by the singular values $\sigma_{i+1}, \cdots, \sigma_n$, have less significant degrees of controllability and observability relative to the noise. It would be unwise to require a realization to include these directions. The reduced model of order i after deleting singular values $\sigma_{i+1}, \cdots, \sigma_n$ is then considered as the robustly controllable and observable part of the realized system. The modal frequencies and modal damping ratio may then be computed from the eigenvalues of the estimated continuous-time state matrix, which is converted from the realized discrete time system matrix. In order to reduce the bias due to noise in the data, an alternative formulation of the ERA, called the ERA with data correlations (ERA/DC), can be used (Juang et al, 1988).

Identification of Mass, Stiffness and Damping Matrices

For the purposes of this section, methods of determining a mass-stiffness-damping model from identified state space realization are presented. A state space model of the dynamic system is identified from input-output data using the OKID/ERA approach (Tee et al, 2004), which is presented in previous sections. Two methods are presented here, i.e. Method 1 (Identification with full set of sensors and actuators) and Method 2 (Identification with mixed sensors and actuators).

Method 1 requires either a full set of sensors or a full set of actuators with at least one DOF containing a collocated sensor-actuator pair. The Method 2 has one main advantage over the Method 1, in the sense that the Method 2 has more general theoretical implications about the number of sensors or actuators that can be used in dynamic testing. The requirement for Method 2 is that all DOFs should have either a sensor or an actuator, with at least one DOF containing a collocated sensor-actuator pair (hence $r + m = N + 1$). For structural damage assessment, it is impractical to use huge sensors for measurement to identify the unknown structural parameters. The methods presented in the subsequent sections are proposed to remove the limitations of the two methods in this section.

The two methods (Lus et al., 2003) are chosen because (1) they all start from a state space realization of the second-order structural system; (2) they can handle all types of measurements (i.e., displacements, velocities, and/or accelerations); and (3) they are inherently capable of handling the nonproportional viscous damping case.

Method 1: Identification With Full Set of Sensors and Actuators

The first method to be discussed was proposed initially by Tseng et al. (1994a, b). Equation (1a) is an equation of motion for an *N*-DOF time invariant structural system. The so-called normal modes for the system represented by Eq. (1a) are obtained by solving the following generalized eigenvalue problem:

$$\mathbf{M}\Phi_i\Omega_i^2 = \mathbf{K}\Phi_i \tag{15}$$

where Ω_i is the ith undamped natural frequency and Φ_i the corresponding i^{th} undamped mode shape. Although scaling of the eigenvectors is arbitrary, the most common choice in modal analysis is adopted, i.e.

$$\Phi^T\mathbf{M}\Phi = \mathbf{I} \qquad \Phi^T\mathbf{K}\Phi = \Omega^2 \tag{16}$$

Accordingly, the equation of motion in Eq. (1) can be expressed in terms of modal coordinates via a transformation of the form $\mathbf{q}(t) = \Phi\eta(t)$ to yield

$$\ddot{\eta}(t) + \mathbf{E}\dot{\eta}(t) + \Omega^2\eta(t) = \bar{\mathbf{B}}_c\mathbf{u}(t) \tag{17a}$$

$$\mathbf{y}(t) = \begin{bmatrix} \bar{\mathbf{C}}_{\mathrm{p}}\eta(t) \\ \bar{\mathbf{C}}_{\mathrm{v}}\dot{\eta}(t) \\ \bar{\mathbf{C}}_{\mathrm{a}}\ddot{\eta}(t) \end{bmatrix} \tag{17b}$$

where **E** denotes the damping matrix in the modal coordinate system. If **E** is diagonal, the system is referred to as classically damped (or it is said to have modal damping). Otherwise, the system is referred to as nonclassically damped. By defining a state vector as $\mathbf{x}_1 = [\eta^T \ \dot{\eta}^T]^T$, Eq. (17) can be transformed to

$$\dot{\mathbf{x}}_1(t) = \mathbf{A}_1\mathbf{x}_1(t) + \mathbf{B}_1\mathbf{u}(t) \tag{18a}$$

$$\mathbf{y}(t) = \mathbf{C}_1\mathbf{x}_1(t) + \mathbf{D}_1\mathbf{u}(t) \tag{18b}$$

where

$$\mathbf{A}_1 = \begin{bmatrix} \mathbf{0} & \mathbf{I} \\ -\Omega^2 & -\mathbf{E} \end{bmatrix};\ \mathbf{B}_1 = \begin{bmatrix} \mathbf{0} \\ \bar{\mathbf{B}}_{\mathrm{c}} \end{bmatrix};\ \mathbf{C}_1 = [\bar{\mathbf{C}}_{\mathrm{p}} \quad \mathbf{0}] \tag{19}$$

Equations (2) are equations of motion with *N*-DOF structural system in the state space form which is obtained from the OKID/ERA. This state space realization is compared with Eqs. (18) to determine similarity transformation as explained below. If the matrix $\mathbf{A}_{\mathrm{c}}$ is diagonalized by a square matrix Ψ such that $\Psi^{-1}\mathbf{A}_{\mathrm{c}}\Psi = \Gamma$, then by considering the transformation $\mathbf{x} = \Psi\theta$, the continuous time system of Eqs. (2) can also be written in modal coordinates as

$$\dot{\theta}(t) = \Gamma\theta(t) + \Psi^{-1}\mathbf{B}_{\mathrm{f}}\mathbf{u}(t) \tag{20a}$$

$$\mathbf{y}(t) = \mathbf{C}\Psi\,\theta(t) \tag{20b}$$

where the matrix Γ contains the continuous time eigenvalues of the identified state space model, and Ψ of order $2N \times 2N$ is the matrix of the corresponding eigenvectors.

The model in Eq. (20) can be expressed in a new set of coordinates with the transformation to the McMillan normal form. Then reordering the sequence of state variables appearing in the McMillan normal form produces the following realization ($\mathbf{A}_2$, $\mathbf{B}_2$, $\mathbf{C}_2$), which is in terms of real valued quantities only.

$$\dot{\mathbf{z}}_{\mathrm{r}}(t) = \mathbf{A}_2\mathbf{z}_{\mathrm{r}}(t) + \mathbf{B}_2\mathbf{u}(t) \tag{21a}$$

$$\mathbf{y}(t) = \mathbf{C}_2\mathbf{z}_{\mathrm{r}}(t) \tag{21b}$$

with the following system matrices

$$\mathbf{A}_2 = \begin{bmatrix} \mathbf{0} & \mathbf{I} \\ -\Omega_{\mathrm{B}}^2 & -\Sigma \end{bmatrix}; \ \mathbf{B}_2 = \begin{bmatrix} \mathbf{B}_{1\mathrm{M}} \\ \mathbf{B}_{2\mathrm{M}} \end{bmatrix}; \ \mathbf{C}_2 = \begin{bmatrix} \mathbf{C}_{1\mathrm{M}} & \mathbf{C}_{2\mathrm{M}} \end{bmatrix} \tag{22}$$

where Σ is a diagonal matrix of the damping factors $-2\sigma_j$, and Ω_{B}^2 is a diagonal matrix of the $(\sigma_j^2 + \omega_j^2)$ which reduces to the undamped natural frequencies only when there is no damping present. The realization (Eq. 22), which we have obtained from data, and the desired realization (Eq. 18) are representations of the same system. Therefore, there must exist a similarity transformation $\bar{\mathbf{N}}$.

$$\bar{\mathbf{N}}\mathbf{A}_2\bar{\mathbf{N}}^{-1} = \mathbf{A}_1; \ \bar{\mathbf{N}}\mathbf{B}_2 = \mathbf{B}_1; \ \mathbf{C}_2\bar{\mathbf{N}}^{-1} = \mathbf{C}_1 \tag{23}$$

The determination of this similarity transformation $\bar{\mathbf{N}}$ is quite cumbersome and the detailed procedure is presented in Tee (2004). Once the mass normalized eigenvector is determined, the mass matrix $\mathbf{M}$, damping matrix $\mathbf{L}$, and stiffness matrix $\mathbf{K}$ are all uniquely determined.

Method 2: Identification With Mixed Sensors and Actuators

A method, which does not require a full set of actuators or a full set of sensors, has been proposed by DeAngelis et al. (2002). Unlike Method 1, by defining a state vector $\mathbf{x}(t) = [\mathbf{q}(t)^T \ \dot{\mathbf{q}}(t)^T]^T$, Eqs. (1) also can be conveniently written as

$$\begin{bmatrix} \mathbf{L} & \mathbf{M} \\ \mathbf{M} & \mathbf{0} \end{bmatrix} \dot{\mathbf{x}}(t) + \begin{bmatrix} \mathbf{K} & \mathbf{0} \\ \mathbf{0} & -\mathbf{M} \end{bmatrix} \mathbf{x}(t) = \begin{bmatrix} \mathbf{B}_{\mathrm{f}} \\ \mathbf{0} \end{bmatrix} \mathbf{u}(t) \tag{24a}$$

$$\mathbf{y}(t) = [\mathbf{C}_{\mathrm{p}} \quad \mathbf{0}]\, \mathbf{x}(t) \tag{24b}$$

The advantage of rewriting Eq. (1) into Eq. (24) is that now the associated eigenvalue problem is kept symmetric and can be written in a matrix form as:

$$\begin{bmatrix} \mathbf{L} & \mathbf{M} \\ \mathbf{M} & \mathbf{0} \end{bmatrix} \begin{bmatrix} \mathbf{P} \\ \mathbf{P}\,\Gamma \end{bmatrix} \Gamma = \begin{bmatrix} -\mathbf{K} & \mathbf{0} \\ \mathbf{0} & \mathbf{M} \end{bmatrix} \begin{bmatrix} \mathbf{P} \\ \mathbf{P}\,\Gamma \end{bmatrix} \tag{25}$$

where $\mathbf{P}$ is the matrix containing the eigenvectors of the complex eigenvalue problem and $\Gamma_{2N\times 2N}$ is the diagonal matrix of complex eigenvalues.

Once the symmetric eigenvalue problem (Eq. 25) has been solved, we can now conveniently rewrite Eq. (24) by using the transformation $\mathbf{z}(t) = [\mathbf{P}^{\mathrm{T}} \ (\mathbf{P}\,\Gamma)^{\mathrm{T}}]^{\mathrm{T}} \zeta(t)$ so that

$$\dot{\zeta}(t) = \Gamma\zeta(t) + \mathbf{P}^{\mathrm{T}}\mathbf{B}_{\mathrm{f}}\mathbf{u}(t) \tag{26a}$$

$$\mathbf{y}(t) = \mathbf{C}_{\mathrm{p}}\mathbf{P}\zeta(t) \tag{26b}$$

Like Method 1, if the first order system of Eq. (20) was identified using data that actually came from the second order model of Eq. (26), one can look for a transformation matrix, $\hat{\mathbf{N}}$, that relates these two representations, i.e.:

$$\hat{\mathbf{N}}^{-1}\,\Gamma\,\hat{\mathbf{N}} = \Gamma,\ \hat{\mathbf{N}}^{-1}\Psi^{-1}\mathbf{B}_{\mathrm{c}} = \mathbf{P}^{T}\mathbf{B}_{\mathrm{f}},\ \mathbf{C}\Psi\hat{\mathbf{N}} = \mathbf{C}_{\mathrm{p}}\mathbf{P} \tag{27}$$

Once the properly scaled eigenvector matrix $\mathbf{P}$ is evaluated, the mass, damping, and the stiffness matrices of the finite element model can be obtained:

$$\mathbf{M} = (\mathbf{P}\Gamma\mathbf{P}^{\mathrm{T}})^{-1}, \quad \mathbf{L} = -\mathbf{M}\mathbf{P}\,\Gamma^{2}\mathbf{P}^{\mathrm{T}}\mathbf{M}, \quad \mathbf{K} = -(\mathbf{P}\,\Gamma^{-1}\mathbf{P}^{\mathrm{T}})^{-1} \tag{28}$$

Damage Detection

Once the second order models corresponding to the undamaged and damaged configurations have been determined, it is possible to compare the physical and mechanical characteristics of the two models in order to locate and quantify the structural damage. We seek to locate and quantify structural damage by inspecting the relative changes in the stiffness of each storey identified for the undamaged and damaged models. To this end, a simple but effective damage indicator (stiffness integrity index) is defined as

$$D_{i} = \frac{K_{d}(i)}{K_{ref}(i)} \tag{29}$$

where $K_{d}(i)$ is the stiffness of the ith member for the damaged state, and $K_{ref}(i)$ refers to the stiffness of the ith member for the reference (presumably undamaged) state. The stiffness integrity index is introduced such that $D_{i} = 1$ for no loss in stiffness (presumably no damage) and $D_{i} = 0$ for complete loss of storey stiffness (presumably 100% damage).

OPTIMIZATION OF CONDENSED STIFFNESS MATRICES

In practice, the number of sensors for measurement is often limited making the identification of many unknown parameters difficult, particularly when one attempts to identify the full system in one go. Alternatively, a reduced or condensed system is identified corresponding to the number of sensors used. This, however, does not necessarily give information on all unknown parameters. As such, the objective of this chapter is to determine individual stiffness parameters of the full system from the identified condensed system. In this respect, model condensation to reduce the structural stiffness matrix is involved.

In the discretization process, it is sometimes necessary to divide a structure into a large number of elements. When the elements are assembled for the entire structure, the number of unknowns may be prohibitively large. Identification of large stiffness, mass and damping matrices becomes computationally difficult, if not impossible. To overcome this, it may be necessary to condense the size of these matrices, for example by means of model condensation method. In addition, full measurement is often impractical due to limited sensors and other practical constraints in instrumentation. As such, a condensed model based on incomplete measurement can only directly identify the condensed stiffness matrices. A robust method is needed to estimate the stiffness of each DOF indirectly from the identified condensed stiffness matrices. The method proposed herein is thus called Condensed Model Identification and Recovery (CMIR) Method (Koh et al, 2006; Tee et al, 2005).

To illustrate the method, consider a twelve-storey shear building idealized as a 12-DOF system. The storey stiffness values are contained in the stiffness matrix. If only 5 sensors are available, it is not possible to identify the entire stiffness matrix, but it is possible to identify a 5×5 stiffness matrix, which in fact, is a condensed matrix corresponding to the five measured DOFs. In order to estimate individual stiffness coefficients from the condensed stiffness matrices, GA is adopted to recover stiffness of individual storey in a numerically efficient way.

Implementation of the GA approach to recover the storey stiffness values from the reduced stiffness matrices is relatively straightforward, depending heavily on the forward analysis and without having to reformulate into an inverse problem. The forward analysis referred to is the numerical simulation to predict the condensed eigenvalues and eigenvectors based on a trial set of stiffness parameters generated by the GA approach. The forward analysis is carried out by using the model condensation methods described as below which is commonly used to solve the condensed eigenvalues and eigenvectors. Two sets of condensed eigenvalues and eigenvectors are obtained and termed the identified values and analytical values. The identified values are obtained from the measurements through the proposed approach discussed in previous sections whereas the analytical values are estimated from the GA forward analysis. The CMIR method is then used to check if the identified and analytical values match. The GA approach is used in this sense to provide an optimal solution for the set of unknown parameters of the structural system to be identified (Koh et al, 2003).

The procedure for recovering the storey stiffness values from the condensed stiffness matrices is highly nonlinear and GA is found to be suitable in this study. Genetic Algorithm Optimization Toolbox (GAOT), designed by Houck et al. (1995) for the implementation of GA within Matlab is adopted. The software provides basic subroutines such as various mutation methods, crossover methods and selection methods. Most importantly Matlab is used for the following reasons: it provides many built in auxiliary functions useful for function optimization; it is completely portable; and it is efficient for numerical computations especially solving a series of non-linear equations.

CMIR-Static Condensation Method (CMIR-SC)

As a first attempt, a simple and well-known method of static condensation by Guyan (1965) to condense the structural stiffness matrix is considered. This method ignores dynamic effects completely. It is therefore important to study the accuracy of parameters deduced from the condensed system assuming that static condensation is applicable despite ignoring the dynamic effects in the static condensation process. To condense the structural stiffness matrix, it involves first defining the DOFs to be condensed as dependent, or secondary, DOFs, and expressing them in terms of the remaining independent, or primary, DOFs. The relationship between the secondary and primary DOFs is found by establishing the static equilibrium relationship between them. This approach of identification is named CMIR-SC. For illustration, the DOFs are re-arranged in the system such that secondary DOFs are the first s elements in the displacement vector, and the primary DOFs are the last p elements. Assuming that no load (for simplicity of illustration) is applied at any of the secondary DOFs, the equilibrium equations may be written in the following partition form

$$\begin{bmatrix} \mathbf{K}_{ss} & \mathbf{K}_{sp} \\ \mathbf{K}_{ps} & \mathbf{K}_{pp} \end{bmatrix} \begin{Bmatrix} \mathbf{y}_{s} \\ \mathbf{y}_{p} \end{Bmatrix} = \begin{Bmatrix} \mathbf{0} \\ \mathbf{F}_{p} \end{Bmatrix} \tag{30}$$

where $\mathbf{y}_s$ is the displacement vector corresponding to the s DOFs to be condensed and $\mathbf{y}_p$ is the vector corresponding to the remaining p independent DOFs. Equation (30) can be re-written in two equations as follows:

$$\mathbf{K}_{ss}\mathbf{y}_{s} + \mathbf{K}_{sp}\mathbf{y}_{p} = \mathbf{0} \tag{31}$$

$$\mathbf{K}_{ps}\mathbf{y}_{s} + \mathbf{K}_{pp}\mathbf{y}_{p} = \mathbf{F}_{p} \tag{32}$$

Equation (31) yields

$$\mathbf{y}_{s} = \bar{\mathbf{T}}\,\mathbf{y}_{p} \tag{33}$$

where $\bar{\mathbf{T}}$ is the transformation matrix given by

$$\bar{\mathbf{T}} = -\mathbf{K}_{ss}^{-1}\,\mathbf{K}_{sp} \tag{34}$$

Substituting Eq. (33) into Eq. (32) and using Eq. (34) results in the condensed stiffness equation relating forces and displacements at the primary DOFs, that is,

$$\bar{\mathbf{K}}\,\mathbf{y}_{p} = \mathbf{F}_{p} \tag{35}$$

where $\bar{\mathbf{K}}$ is the condensed stiffness matrix given by

$$\bar{\mathbf{K}} = \mathbf{K}_{pp} - \mathbf{K}_{ps}\,\mathbf{K}_{ss}^{-1}\,\mathbf{K}_{sp} \tag{36}$$

To recover stiffness of each storey from the identified condensed stiffness matrices, the objective function, in terms of the differences between the element in the identified condensed stiffness matrix and the analytical condensed stiffness matrix, is defined as follows

$$\varepsilon = \sum_{i=1}^{n} \left(1 - \frac{\bar{\mathbf{K}}(i,i)}{\mathbf{K}(i,i)} \right)^2 \tag{37}$$

where $\bar{\mathbf{K}}(i,i)$ and $\mathbf{K}(i,i)$ are the ith row and i^{th} column of the analytical (Guyan) and identified condensed stiffness matrix, respectively. The off-diagonal zero terms are not included in the objective function. The problem of recovering the stiffness values is transformed into minimization of Eq. (37). Figure 1 shows the flowchart for identification of the stiffness of each storey using the CMIR-CS method.

CMIR-Dynamic Condensation Method (CMIR-DC)

Static condensation may introduce modeling error when applied to structural dynamic problems, since it does not include inertia and damping forces associated with the secondary DOFs that are "condensed out". To resolve this problem, the concept of dynamic condensation (Paz 1985) is adopted.

The algorithm for this method starts by assigning an approximate value (e.g. zero) to the first eigenvalue Ω_1^2, applying dynamic condensation to the dynamic matrix of the system $\mathbf{D}_1 = \mathbf{K} - \Omega_1^2\,\mathbf{M}$, and then solving the condensed eigenproblem to determine the first eigenvalues Ω_1^2 and estimate the other eigenvalues. Next, dynamic condensation is applied to the dynamic matrix $\mathbf{D}_2 = \mathbf{K} - \Omega_2^2\,\mathbf{M}$ to condense the problem and calculate the second eigenvalues Ω_2^2 and others. The process continues in this manner, where basically one virtually exact eigenvalue and an approximation of the next eigenvalue are calculated at each step.

It is proposed in this study to modify the dynamic condensation method to be *non- iterative* in combination with CMIR and is named the CMIR-DC method. Unlike the CMIR-SC method, the CMIR-DC method uses the eigenvalues from the identified condensed stiffness and mass matrices to find the analytical condensed stiffness and mass matrices. The method requires neither matrix inversion nor series expansion. To demonstrate this, consider the eigenvalue problem of a discrete structural system for which it is desired to eliminate the secondary DOFs $\mathbf{y}_s$ and retain the primary DOFs $\mathbf{y}_p$. In this case, the equations of motion may be written as

$$\begin{bmatrix} \mathbf{M}_{ss} & \mathbf{M}_{sp} \\ \mathbf{M}_{ps} & \mathbf{M}_{pp} \end{bmatrix} \begin{Bmatrix} \ddot{\mathbf{y}}_s \\ \ddot{\mathbf{y}}_p \end{Bmatrix} + \begin{bmatrix} \mathbf{K}_{ss} & \mathbf{K}_{sp} \\ \mathbf{K}_{ps} & \mathbf{K}_{pp} \end{bmatrix} \begin{Bmatrix} \mathbf{y}_s \\ \mathbf{y}_p \end{Bmatrix} = \begin{Bmatrix} 0 \\ 0 \end{Bmatrix} \tag{38}$$

Substitution of $\mathbf{y} = \Phi \sin \Omega_i t$ in Eq. (38) results in the generalized eigenproblem

$$\begin{bmatrix} \mathbf{K}_{ss} - \Omega_i^2\mathbf{M}_{ss} & \mathbf{K}_{sp} - \Omega_i^2\mathbf{M}_{sp} \\ \mathbf{K}_{ps} - \Omega_i^2\mathbf{M}_{ps} & \mathbf{K}_{pp} - \Omega_i^2\mathbf{M}_{pp} \end{bmatrix} \begin{Bmatrix} \mathbf{y}_s \\ \mathbf{y}_p \end{Bmatrix} = \begin{Bmatrix} 0 \\ 0 \end{Bmatrix} \tag{39}$$

where Ω_i^2 is the ith eigenvalue and Φ_i is the corresponding mode shape or eigenvector. Consequently, the ith eigenvector Φ_i can be expressed as

Figure 1. Flowchart for identification of storey stiffness values using CMIR-SC

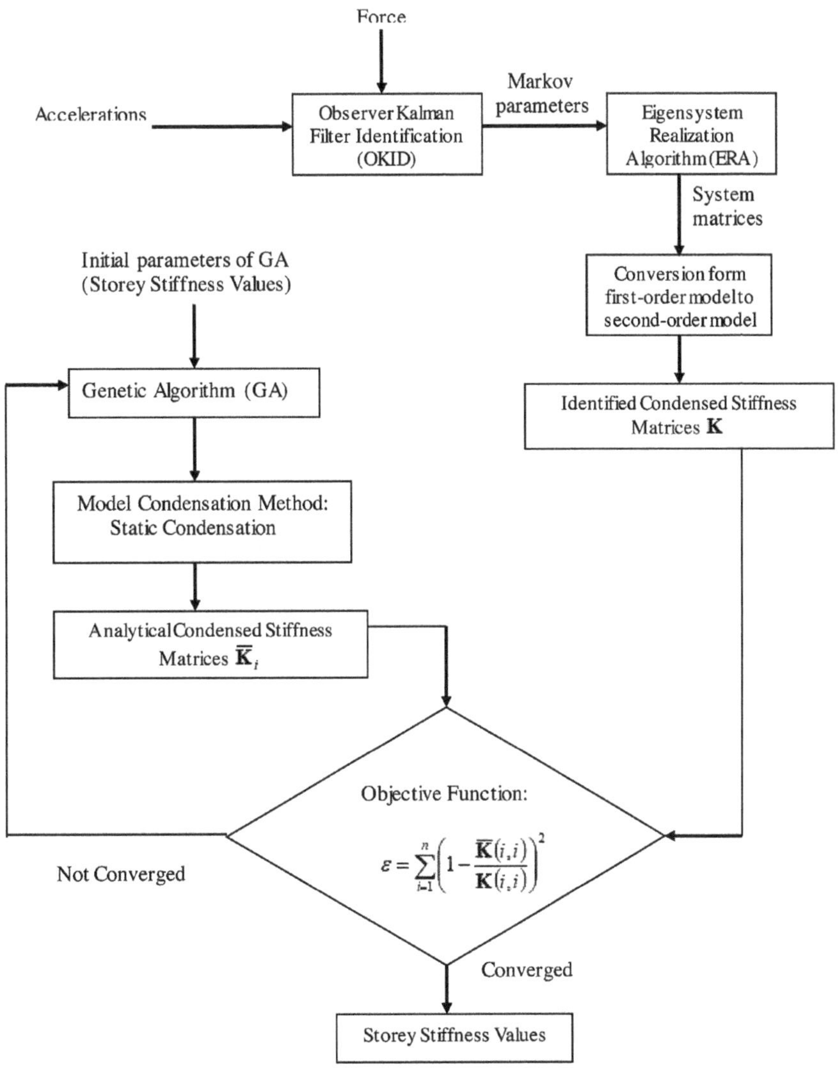

$$\Phi_i = \begin{Bmatrix} \Phi_{si} \\ \Phi_{pi} \end{Bmatrix} \tag{40}$$

where Φ_{pi} is the ith eigenvector for the condensed system.

The following two steps are executed to calculate the condensed stiffness and mass matrices.

Step 1: Gauss-Jordan elimination of the secondary DOFs Φ_s is used to reduce Eq. (39) to the following form:

$$\begin{bmatrix} \mathbf{I} & -\bar{\mathbf{T}}_i \\ \mathbf{0} & \bar{\mathbf{D}}_i \end{bmatrix} \begin{Bmatrix} \Phi_s \\ \Phi_p \end{Bmatrix} = \begin{Bmatrix} \mathbf{0} \\ \mathbf{0} \end{Bmatrix} \tag{41}$$

The first equation in Eq. (41) can be written as

$$\Phi_{s} = \bar{\mathbf{T}}_{i}\Phi_{p} \tag{42}$$

The ith eigenvector of the system is determined as

$$\Phi_{i} = \mathbf{T}_{i}\Phi_{p} \tag{43}$$

where

$$\mathbf{T}_{i} = \begin{bmatrix} \bar{\mathbf{T}}_{i} \\ \mathbf{I} \end{bmatrix} \tag{44}$$

Step 2: The condensed mass matrix $\bar{\mathbf{M}}_{i}$ and condensed stiffness matrix $\bar{\mathbf{K}}_{i}$ are calculated as

$$\bar{\mathbf{M}}_{i} = \mathbf{T}_{i}^{T}\mathbf{M}\mathbf{T}_{i} \tag{45}$$

and

$$\bar{\mathbf{K}}_{i} = \bar{\mathbf{D}}_{i} + \Omega_{i}^{2}\bar{\mathbf{M}}_{i} \tag{46}$$

Consider the corresponding characteristic equation

$$(-\mathbf{M}\Omega_{i}^{2} + \mathbf{K})\Phi_{i} = 0 \tag{47}$$

Rearranging Eq. (47) and defining the ith eigenvalue, Λ_{i}, as the ith square of the frequency yields i^{th} Rayleigh Quotient, $\Re_{i}$ as,

$$\Re_{i} = \Lambda_{i} = \frac{K_{i}^{*}}{M_{i}^{*}} = \Omega_{i}^{2} \tag{48}$$

where M_{i}^{*} is the ith identified modal mass, and K_{i}^{*} is the ith identified modal stiffness, respectively, given by

$$M_{i}^{*} = \Phi_{i}^{T}\mathbf{M}\Phi_{i} \qquad K_{i}^{*} = \Phi_{i}^{T}\mathbf{K}\Phi_{i} \tag{49}$$

On normalizing the eigenvectors with respect to the modal mass, Eq. (48) reduces to

$$\Re_i = \Lambda_i = K_i^* \tag{50}$$

In dynamic condensation, Eq. (48) will be

$$\bar{\Re}_i = \bar{\Lambda}_i = \frac{\bar{K}_i^*}{\bar{M}_i^*} \tag{51}$$

where $\bar{M}_i^*$ is the ith analytical condensed modal mass, and $\bar{K}_i^*$ is the ith analytical condensed modal stiffness, respectively, given by

$$\bar{M}_i^* = \{\Phi_p\}_i^T \bar{\mathbf{M}}_i \{\Phi_p\}_i \qquad \bar{K}_i^* = \{\Phi_p\}_i^T \bar{\mathbf{K}}_i \{\Phi_p\}_i \tag{52}$$

Rayleigh quotients based on the identified and analytical reduced matrices to be used in the objective function are as follows

$$\Re_i = \frac{\{\Phi_p\}_i^T \mathbf{K} \{\Phi_p\}_i}{\{\Phi_p\}_i^T \mathbf{M} \{\Phi_p\}_i} \qquad \bar{\Re}_i = \frac{\{\Phi_p\}_i^T \bar{\mathbf{K}}_i \{\Phi_p\}_i}{\{\Phi_p\}_i^T \bar{\mathbf{M}}_i \{\Phi_p\}_i} \tag{53}$$

where Φ_{pi} is the ith eigenvector for the condensed model and $\bar{\mathbf{K}}_i$, $\bar{\mathbf{M}}_i$ and **K**, **M** are the ith analytical condensed stiffness and mass matrices (dynamic condensation) and identified condensed stiffness and mass matrix, respectively. The objective function, in terms of the differences between the Rayleigh quotients of the identified condensed model and the analytical condensed model, is as follows

$$\varepsilon = \sum_{i=1}^{n} (1 - \frac{\bar{\Re}_i}{\Re_i})^2 \tag{54}$$

where $\Re_i$ and $\bar{\Re}_i$ is the ith identified and analytical Rayleigh Quotient, respectively. The problem of recovering the stiffness values is thus transformed into minimization of ε as defined in Eq. (54). Figure 2 shows the flowchart for identification of the stiffness of each storey using the CMIR-DC method.

CMIR-System Equivalent Reduction Expansion Process (CMIR-SEREP)

The third condensation method considered here makes use of the System Equivalent Reduction Expansion Process (SEREP) which was first proposed by O'Callahan et al. (1989). It requires the full system eigenvectors corresponding to the set of modes of interest. However, an approach using the eigenvectors from the condensed model is later proposed by Papadopoulas et al. (1996) to avoid using the full system eigenvectors. The latter modification of SEREP is adopted in the CMIR, leading to the CMIR-SEREP method.

The SEREP method preserves the natural frequencies and mode shapes of the structural model during the condensation process. Let N be the number of DOFs and n the number of sensors used. Using the condensed model, the $n \times n$ mass and stiffness matrices can be obtained using the methods described in the previous chapter. An eigen-analysis on these condensed matrices yields the identified natural frequencies and mode shapes (termed the identified values), denoted as

$$\Lambda_{n \times n} = diag(\Omega_1^2 \cdots \Omega_n^2) \ \Phi_{\mathrm{p} n \times n} = [\varphi_{\mathrm{p}1} \cdots \varphi_{\mathrm{p}n}] \tag{55}$$

The modal matrix Φ_{p} in Eq. (55) is then expanded to include the secondary DOFs using

$$\varphi_{si} = -(\mathbf{K}_{\mathrm{ss}} - \Omega_i^2 \mathbf{M}_{\mathrm{ss}})^{-1} (\mathbf{K}_{\mathrm{sp}} - \Omega_i^2 \mathbf{M}_{\mathrm{sp}}) \varphi_{\mathrm{p}i}, \quad i = 1, \cdots, n \tag{56}$$

In principle, Eq. (56) can be used directly but the required inversion is costly. Miller's expansion can be employed to circumvent the inversion for every condensed frequency (Miller 1980):

$$\varphi_{si} = -(\mathbf{K}_{\mathrm{ss}}^{-1} \mathbf{K}_{\mathrm{sp}} + \Omega_i^2 [-\mathbf{K}_{\mathrm{ss}}^{-1} \mathbf{M}_{\mathrm{sp}} + \mathbf{K}_{\mathrm{ss}}^{-1} \mathbf{M}_{\mathrm{ss}} \mathbf{K}_{\mathrm{ss}}^{-1} \mathbf{K}_{\mathrm{sp}}]) \varphi_{\mathrm{p}i} \quad i = 1, \cdots, n \tag{57}$$

or the more exact Kidder expansion

$$\varphi_{si} = -(\mathbf{K}_{\mathrm{ss}}^{-1} + \Omega_i^2 \mathbf{K}_{\mathrm{ss}}^{-1} \mathbf{M}_{\mathrm{ss}} \mathbf{K}_{\mathrm{ss}}^{-1})(\mathbf{K}_{\mathrm{sp}} - \Omega_i^2 \mathbf{M}_{\mathrm{sp}}) \varphi_{\mathrm{p}i} \quad i = 1, \cdots, n \tag{58}$$

can be employed (Kidder 1973).

The expanded modal matrix can now be formed as

$$\Phi = \begin{bmatrix} \varphi_{p1} & \cdots & \varphi_{pn} \\ \varphi_{s1} & \cdots & \varphi_{sn} \end{bmatrix} \equiv \begin{bmatrix} \varphi_1 & \cdots & \varphi_n \end{bmatrix} \equiv \begin{bmatrix} \Phi_{\mathrm{p}} \\ \Phi_{\mathrm{s}} \end{bmatrix} \tag{59}$$

where Φ_{p} and Φ_{s} denote the $n \times N$ and $(N-n) \times N$ modal submatrices corresponding to the primary and secondary DOF partitions of Φ, respectively. To satisfy mass orthonormality, it is proposed to use the method used by Baruch (1982). That is, normalize Φ by

$$\bar{\varphi}_i = \varphi_i (\varphi_i^T \mathbf{M} \varphi_i)^{-1/2}, \quad i = 1, \cdots, n \tag{60}$$

where φ_i is the ith mode shape before normalization and given in Eq. (59). The expanded, mass normalized modal matrix is then finally computed from

$$\Theta = \Phi(\bar{\Phi}^T M \bar{\Phi})^{-1/2} \equiv \begin{bmatrix} \Theta_p \\ \Theta_s \end{bmatrix} \tag{61}$$

where $\bar{\Phi}$ is the modal matrix formed from the n normalized mode shapes in Eq. (60). The SEREP transformation now becomes

$$\mathbf{P}_{SE} = \begin{bmatrix} \Theta_p \\ \Theta_s \end{bmatrix} \Theta_p^{-1} \tag{62}$$

The stiffness and mass matrices of the condensed model refined by SEREP method are

$$\begin{aligned} \mathbf{K}_{SE} &= \mathbf{P}_{SE}^T \mathbf{K} \mathbf{P}_{SE} \\ \mathbf{M}_{SE} &= \mathbf{P}_{SE}^T \mathbf{M} \mathbf{P}_{SE} \end{aligned} \tag{63}$$

Note that the eigenvalues and eigenvectors obtained from Eq. (63) (for simplicity, termed as the analytically condensed values) should theoretically be identical to Eq. (55). Hence the procedure to obtain **K** and **M** is to assume an initial set of values for $\mathbf{K}_{sp}$, $\mathbf{K}_{ss}$, $\mathbf{M}_{sp}$ and $\mathbf{M}_{ss}$ to be used with Eq. (56) to be able to compute Eq. (63). The latter is then used to compute the eiqenvalues and eigenvectors and check whether they match Eq. (55).

The only underlying assumption is that the shape of each condensed mode resembles the full system mode, which inevitably is dependent on the choice of primary DOF. A number of authors have considered the selection of an appropriate primary set. Good results can only be obtained if there is an adequate number of well-spaced primary DOF.

The eigenvalues and eigenvectors from the identified and analytical condensed matrices are used in the objective function. Two different objective functions are investigated in this study. The first objective function, in terms of the differences between the identified eigenvalues and the analytical eigenvalues, is as follows

$$\varepsilon = \sum_{i=1}^{n} (1 - \frac{\Lambda_i^c}{\Lambda_i^I})^2 \tag{64}$$

where Λ_i^c and Λ_i^I are the analytical and identified i^{th} eigenvalue, respectively. The eigenvalues (natural frequencies) are usually not enough for checking the accuracy of the condensed model (Qu and Fu, 2000). Srinivasan and Kot (1992) noted that changes in mode shapes are a more sensitive indicator of damage than changes in natural frequencies. Therefore, the comparison of the mode shape or eigenvector becomes necessary. Hence, the second objective function makes use of both the eigenvalues and eigenvectors. To estimate the accuracy of the eigenvectors, a correlated coefficient for modal vector (CCFMV) value is defined as

$$CCFMV(\Phi_i^c, \Phi_i^I) = \frac{\left| \Phi_i^{Tc} \cdot \Phi_i^I \right|}{[(\Phi_i^{Tc} \cdot \Phi_i^c)(\Phi_i^{TI} \cdot \Phi_i^I)]^{1/2}} \tag{65}$$

Figure 2. Flowchart for identification of storey stiffness values using CMIR-DC

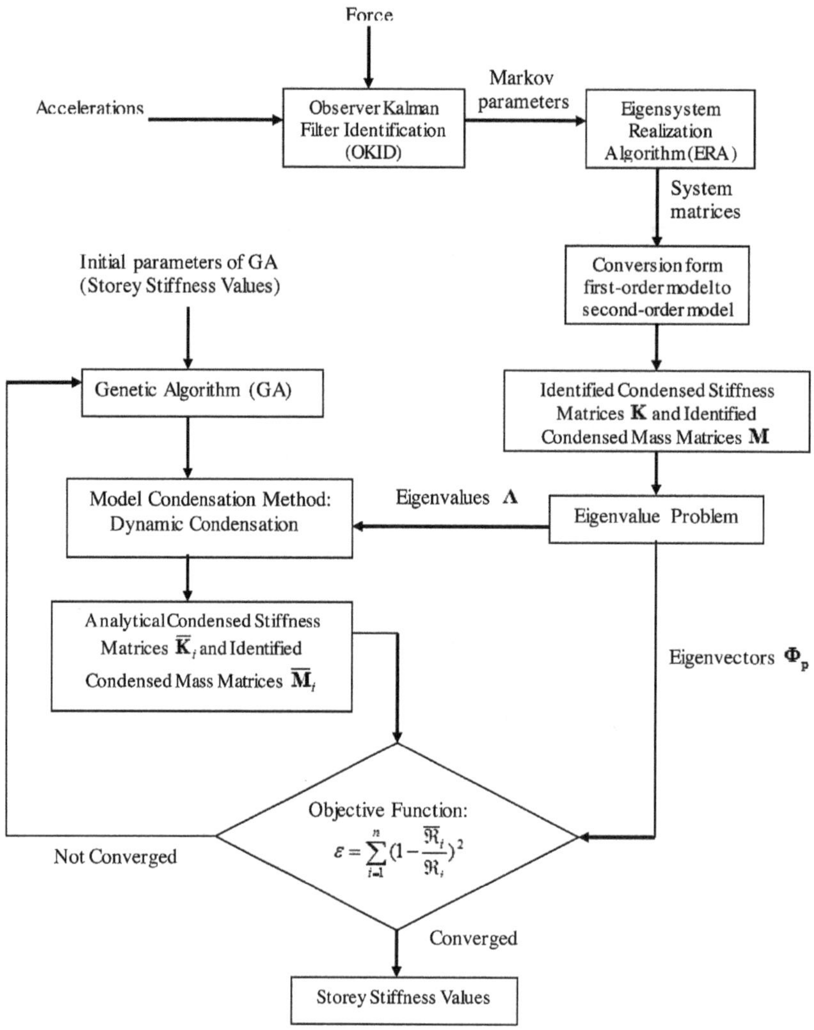

where Φ_i^c and Φ_i^I are the exact and the identified of the ith eigenvector, respectively. A CCFMV value close to 1 suggests that the two modes or vectors are well correlated and a value close to 0 indicates uncorrelated modes. Therefore, the second objective function is

$$\varepsilon = \sum_{i=1}^{n} [(1 - \frac{\Lambda_i^c}{\Lambda_i^I})^2 + (1 - CCFMV(\Phi_i^c, \Phi_i^I))^2] \tag{66}$$

The problem of recovering the stiffness values is transformed into minimization of Eqs. (64) and (66). Figure 3 shows the flowchart for identification of the stiffness of each storey using the CMIR-SEREP method.

FIXED AND NON-FIXED SENSOR APPROACHES

As mentioned earlier, a condensed system is identified corresponding to the number of sensors used, which does not necessarily give information on all unknown parameters. To this end, another contribution of this chapter is to extract sufficient information for structural identification. On the basis of the CMIR method, all stiffness parameters in the entire system can be recovered by extracting sufficient information using two different approaches, namely fixed sensor approach and non-fixed sensor approach. In some cases, it is possible to shift the sensors to maximize the information available for structural identification. Thus, non-fixed sensor approach is used in this circumstance. If it is not possible or convenient to shift the sensors, a novel way to obtain more information is to apply fixed sensor approach, and to deliberately ignore some sensors in different locations for identification of condensed stiffness matrices.

Approach 1: Incomplete Measurement With Fixed Sensors

Let N be the number of DOFs and n the number of sensors used. The computational procedure is as follows.

1. Identify the reduced stiffness matrix ($n \times n$) and mass matrix ($n \times n$) by OKID/ERA with conversion from first-order model to second-order model based on data from n sensors in the N -DOF system. The n eigenvalues and eigenvectors can then be computed. Apply model condensation to obtain the analytical condensed stiffness and mass matrix and solve the corresponding n eigenvalues and eigenvectors.
2. Repeat Step (a) as many times as possible by ignoring one sensor at a time. If necessary, repeat by ignoring more sensors until sufficient information is obtained. It can be shown that the number of ways to ignore sensors is ${}^{n}C_{n-m}$, where m is the number of sensor ignored.
3. Solve the objective function as described in previous section. Note that it is possible to obtain more information than required (i.e. N). This over-determined system can be solved using least square method or, in this study, by GA.

For example, 3 sensors are used to identify 6-DOF structural system. The first sensor is located at the first DOF, second sensor at the third DOF and third sensor at the sixth DOF. When one of the three sensors is ignored, there are ${}^{3}C_{2} = 3$ possible combinations. In a similar way, ${}^{3}C_{1} = 3$ combinations of the sensors can be obtained when any two of the three sensors are ignored.

Approach 2: Incomplete Measurement With Non-Fixed Sensors

Assuming all n sensors available are used in each shift, the number of ways to shift the sensors is ${}^{N}C_{n}$. The procedure is described below.

1. Identify the condensed stiffness matrix ($n \times n$) and mass matrix ($n \times n$) by OKID/ERA with conversion from first-order model to second-order model based on data from n sensors in the N -DOF system. The n eigenvalues and eigenvectors can then be computed. Apply model condensation to obtain the analytical condensed stiffness and mass matrix and solve the corresponding n eigenvalues and eigenvectors.

Figure 3. Flowchart for identification of storey stiffness values using CMIR-SEREP

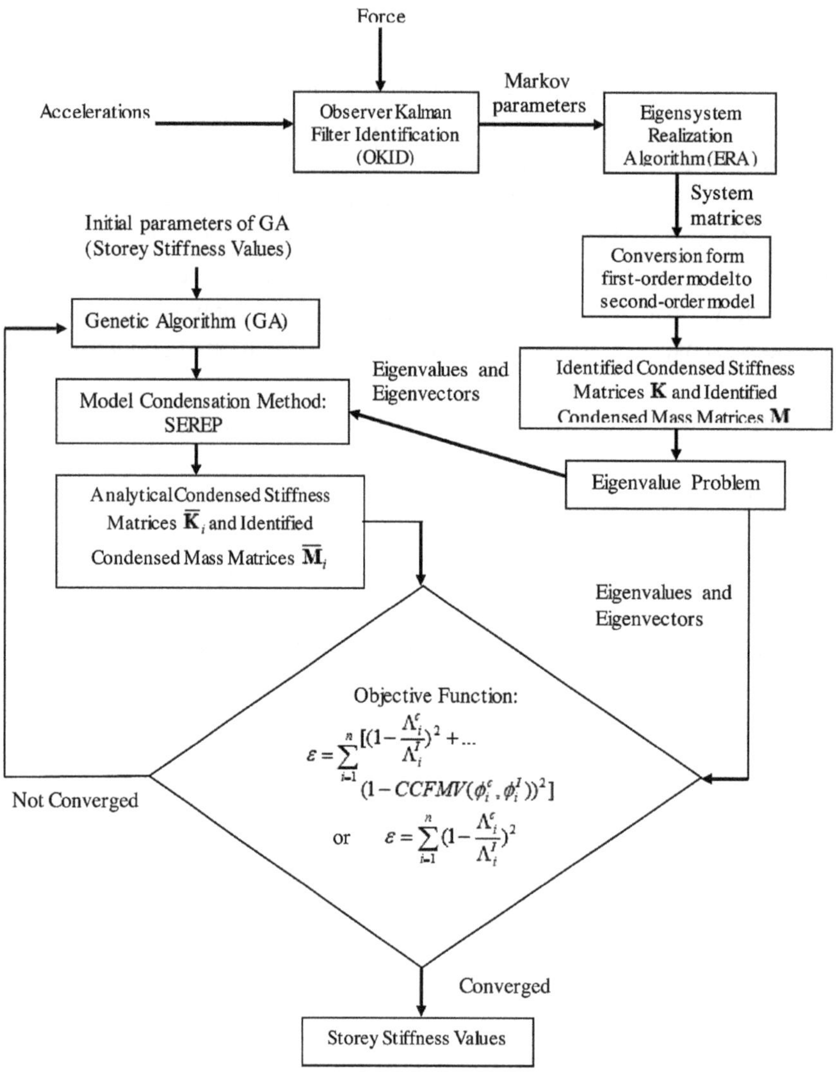

2. Repeat Step (a) by shifting the n sensors in different configurations, until sufficient information is obtained.
3. Recover storey stiffness values as in Step (c) of Approach 1.

Using the same previous example, if we shift the three sensors to any one of the six DOF, there are ${}^{6}C_{3} = 20$ combinations. In principle, the minimum number of sensors required in this approach is one. However, this would require N sensor configurations (i.e. re-arranged N-1 times in addition to the original configuration). This is not desirable in practice. On the other extreme, if N sensors are used, then only one sensor configuration is required (i.e. no further shifting needed), but this means complete measurement which defeats the purpose of identification based on incomplete measurement. There is, therefore, a trade-off between the number of sensors required and number of configurations required.

NUMERICAL RESULTS AND DISCUSSION

To validate the proposed damage identification approach with incomplete measurements, numerical simulation study is carried out to identify the stiffness and subsequently the location and extent of structural damage in the four storey shear building. In practice, the number of members (columns, beams, etc.) may be very large for a multi-storey building. It often suffices to detect stiffness changes (as indication of damage or deterioration) at some local level, e.g. for each storey of the building, rather than for each individual member. In this respect, it is fairly common to use a simple structural model such as lumped mass model or a shear building model for multi-storey buildings. The value for the mass, damping, and stiffness matrices are known and given in Table 1. The natural frequencies (12.2273, 26.2944, 39.3572, 50.8931 rad/sec) and damping factors (0.0150, 0.0150, 0.0185, 0.0223) of such a system will be used as benchmark values for comparing the various identification attempts. The DOFs are labeled in increasing order up to the structure. Here, the mass parameters are assumed as known. Damping is considered with 1.5% of critical damping ratio for both the first and second modes. The input is a Gaussian white noise excitation. The response measurements are generated by numerical simulation of the time response of linear time invariant models using MATLAB toolbox. The input is a random excitation whereas the response measurements are numerically simulated accelerations. Later, the signals are used in the identification procedure. A single excitation applied at the roof is considered. All input and output measurements are polluted by Gaussian white noise disturbances, and the standard deviation of the noise at each channel is adjusted such that it is equal to 10% of the root-mean-square of the unpolluted time history. In this study, no pre-processing of the data is considered so as to present a worst-case scenario. The CMIR-SC, CMIR-DC and CMIR-SEREP method are used for structural identification with incomplete measurements.

First, the undamaged case is considered here using incomplete measurements. The time histories are divided into 3,000 time steps with 0.001 second interval. Identification is carried out by using fixed sensor approach with 2 and 3 sensors. It is worthwhile to mention that there is an alternative to the non-fixed sensor approach, i.e. Markov parameter of each DOF can be found separately with the assumption that the input force must be the same in each case. For example, 3 sensors are used to identify 6-DOF structural system. The first sensor is located at the first DOF, second sensor at the third DOF and third sensor with the input force at the sixth DOF. Markov parameter of each DOF with sensor can be found at this stage. For the second stage, all the three sensors are shifted to the second, fourth and fifth DOFs with the same input force at the sixth DOF to identify Markov parameter of the second, fourth and fifth DOFs. Therefore, Markov parameter of all DOFs can be found. In other words, Markov parameters can be separately identified for each setup and combined together via a reference DOF (sixth DOF in this study) with the condition that excitation force is kept the same from setup to setup. The identified storey stiffness values are compared with the exact values in Tables 2-4 using three different proposed approaches.

Table 1. Mass, damping and stiffness matrices for the four-storey shear building considered

Mass (kg)				Stiffness (N/m x10^6)				Damping (Ns/m x10^3)			
2500	0	0	0	4.2	-1.8	0	0	3.897	-1.40	0	0
0	2000	0	0	-1.8	3.0	-1.2	0	-1.40	2.837	-0.93	0
0	0	1500	0	0	-1.2	1.8	-0.6	0	-0.93	1.777	-0.47
0	0	0	1000	0	0	-0.6	0.6	0	0	-0.47	0.717

Table 2. Identified storey stiffness values with CMIR-SC for four-storey shear building (undamaged case)

Storey	Exact Stiffness (kN/m)	Identified Stiffness in kN/m (% Error in Bracket)			
		No Noise		10% Noise	
		2 Sensors	3 Sensors	2 Sensors	3 Sensors
1	2400	2635.3 (9.8%)	2634.4 (9.8%)	2637.2 (9.9%)	2635.3 (9.8%)
2	1800	1970.5 (9.5%)	1977.5 (9.9%)	2134.6 (18.6%)	2054.7 (14.2%)
3	1200	1304.8 (8.7%)	1297.5 (8.1%)	1312.7 (9.4%)	1337.5 (11.5%)
4	600	656.6 (9.4%)	637.9 (6.3%)	652.6 (8.8%)	648.7 (8.1%)
Mean Absolute Error		9.4%	8.5%	11.7%	10.9%
Max. Absolute Error		9.8%	9.9%	18.6%	14.2%

Table 3. Identified storey stiffness values with CMIR-DC for four-storey shear building (undamaged case)

Storey	Exact Stiffness (kN/m)	Identified Stiffness in kN/m (% Error in Bracket)			
		No Noise		10% Noise	
		2 Sensors	3 Sensors	2 Sensors	3 Sensors
1	2400	2579.1 (7.5%)	2534.3 (5.6%)	2592.1 (8.0%)	2580.2 (7.5%)
2	1800	1925.2 (7.0%)	1883.4 (4.6%)	2013.1 (11.8%)	1972.3 (9.6%)
3	1200	1280.0 (6.67%)	1269.3 (5.8%)	1301.1 (8.4%)	1285.3 (7.1%)
4	600	612.3 (2.05%)	601.2 (0.2%)	613.3 (2.2)	610.1 (1.7%)
Mean Absolute Error		5.8%	4.1%	7.6%	6.5%
Max. Absolute Error		7.5%	5.8%	11.8%	9.6%

Table 4. Identified storey stiffness values with CMIR-SEREP for four-storey shear building (undamaged case)

Storey	Exact Stiffness (kN/m)	Identified Stiffness in kN/m (% Error in Bracket)			
		No Noise		10% Noise	
		2 Sensors	3 Sensors	2 Sensors	3 Sensors
1	2400	2565.3 (6.9%)	2496.1 (4.0%)	2522.1 (5.1%)	2588.1 (7.8%)
2	1800	1913.2 (6.3%)	1909.2 (6.1%)	1912.1 (6.2%)	1932.7 (7.4%)
3	1200	1250.3 (4.2%)	1213.2 (1.1%)	1232.2 (2.7%)	1287.8 (7.3%)
4	600	626.5 (4.4%)	620.1 (3.4%)	652.2 (8.7%)	630.1 (5.0%)
Mean Absolute Error		5.5%	3.6%	6.9%	5.7%
Max. Absolute Error		6.9%	6.1%	8.7%	7.8%

Effects of I/O Noise

In real world, measurements are inevitably contaminated by noise, such as ambient noise and the electrical noise of the data acquisition system, etc. It is therefore important to investigate the effects of I/O noise on the identification results in order to test the robustness of the identification strategy. Two different noise levels are considered: 0% and 10%. It is expected that the identification results with I/O noise should be worse than the results using clean data, i.e. without noise. It is seen that in Tables 2-4 the mean error of identified parameters is sensitive to the noise and the percentage error does show an increase with increasing I/O noise level. For example, using the CMIR-SEREP method, the mean errors are 5.5% and 6.9% for noise levels of 0% and 10%, respectively with 2 sensors.

Comparison of CMIR-SC, CMIR-DC and CMIR-SEREP

The three CMIR methods for incomplete measurement have been discussed. Generally, the CMIR-SC method gives worst results compared to the CMIR-DC and CMIR-SEREP method. Tables 2-4 present the identified stiffness values for the three proposed methods. In the CMIR-SC method, the GA is carried out with objective function shown in Eq. (37). The maximum error is 14.2% for the case of 3 sensors and 18.6% for the case of 2 sensors with 10% noise data with the CMIR-SC method. In the CMIR-DC method, the GA is carried out with objective function shown in Eq. (54). In the CMIR-SEREP method, the objective functions shown in Eqs. (64) and (66) are used. As a fair numerical comparison of CMIR-DC and CMIR-SEREP, the objective functions shown in Eq. (54) and Eq. (64) are used because both of them employ natural frequencies in the objective function. However, using the CMIR-DC method with 10% noise, the maximum errors are 11.8% and 9.6% for the cases of 2 and 3 sensors, respectively. Due to the fact that the inertia forces are accounted for in the CMIR-DC process, the identified storey stiffness values are generally more accurate than those identified by the CMIR-SC method. The maximum errors of the identified stiffness obtained from the CMIR-SEREP method are 8.7% for the case of 3 sensors and 7.8% for the case of 2 sensors with 10% noise data. The improvement over the CMIR-DC method is due to the fact that the CMIR-SEREP method preserves the eigenvalues (natural frequencies) and eigenvectors (mode shapes) during the condensation process.

Effects of Number of Sensors

Table 3 gives the identification results with the CMIR-DC method using different number of sensors. In the case where 2 sensors are used, the mean error of stiffness parameters is 7.6% and the maximum error is 11.8%. In the case where 3 sensors are used, the identification results are improved to give mean error of 6.5% and maximum error 9.6%. It can be seen that 3 sensors give generally more accurate results than 2 sensors as shown in Tables 2, 3 and 4 for the CMIR-SC, CMIR-DC and CMIR-SEREP methods respectively. More measurements (with good accuracy) provide better identification results. The identification accuracy is sensitive to the number of sensors, even though the system considered is relatively small (with 4-DOFs).

Damage Detection

Damage in vertical supporting members (such as columns) can be reflected by reduction in storey stiffness value. Two damage scenarios are studied: (1) single damage and (2) multiple damages. Damage Scenario 1 contains 30% damage in the fourth storey (i.e. the remaining stiffness is 70% of the original value). Damage Scenario 2 has two damage locations: 20% damage in the second storey and 40% in the third storey. Tables 5-6 and Figures 4-7 present the identified stiffness integrity index (as defined in Eq. 29) for Damage Scenarios 1 and 2, respectively. The three proposed CMIR methods are effective in identifying the damage locations and extents. The maximum error in the identified stiffness integrity index is 1.6% with the CMIR-SEREP method and 6.5% with the CMIR-DC method in Damage Scenario 1 under 10% I/O noise using 3 sensors. The maximum error for the CMIR-SC method is the largest among the three proposed CMIR methods -- 7.3% and 8.6% with both 3 and 2 sensors, respectively, under 10% noise in Damage Scenario 1. The results shown in Table 6 are encouraging with the mean absolute error of 3.3%, 3.0% and 1.7% for the CMIR-SC, CMIR-DC and CMIR-SEREP methods, respectively, with 2 sensors under 10% noise in Damage Scenario 2.

CONCLUSION

In general, structural identification based on numerical analysis of measured I/O signal is an inverse problem investigated for several reasons. If the aim is to obtain modal parameters, suitable methods such as the OKID/ERA presented in this chapter can be employed, and the solution is with respect to a first-order state space representation. If the aim is to create or update a model with appropriate physical parameters (e.g. stiffness and damping) of the structure, then the second-order representation is to be determined. Taking another step further, the solution of the inverse vibration problem can be used to

Table 5. Identified stiffness integrity indices under 10% noise with the fixed sensor approach for four-storey shear building (Damage Scenario 1)

Storey	Exact	Identified Stiffness Integrity Index (% Error in Bracket)					
		2 Sensors			3 Sensors		
		CMIR-SC	CMIR-DC	CMIR-SEREP	CMIR-SC	CMIR-DC	CMIR-SEREP
1	1.0	0.998 (-0.2%)	0.989 (-1.1%)	1.012 (1.2%)	1.008 (0.8%)	0.991 (-0.9%)	0.997 (-0.3%)
2	1.0	0.988 (-1.2%)	0.991 (-0.9%)	0.993 (-0.7%)	1.025 (2.5%)	1.022 (2.2%)	1.013 (1.3%)
3	1.0	1.065 (6.5%)	0.954 (-4.6%)	0.978 (-2.2%)	0.927 (-7.3%)	0.935 (-6.5%)	0.989 (-1.1%)
4	0.7	0.640 (-8.6%)	0.660 (-5.7%)	0.680 (-2.9%)	0.697 (-0.4%)	0.688 (-1.7%)	0.711 (1.6%)
Mean Absolute Error		4.1%	3.1%	1.7%	2.8%	2.8%	1.1%
Max. Absolute Error		8.6%	5.7%	2.9%	7.3%	6.5%	1.6%

Table 6. Identified stiffness integrity indices under 10% noise with the fixed sensor approach for four-storey shear building (Damage Scenario 2)

Storey	Exact	Identified Stiffness Integrity Index (% Error in Bracket)					
		2 Sensors			3 Sensors		
		CMIR-SC	CMIR-DC	CMIR-SEREP	CMIR-SC	CMIR-DC	CMIR-SEREP
1	1.0	0.998 (-0.2%)	0.991 (-0.9%)	1.018 (1.8%)	0.931 (-6.9%)	0.981 (-1.9%)	1.011 (1.1%)
2	0.8	0.833 (4.1%)	0.822 (2.75%)	0.792 (-1.0%)	0.799 (-0.1%)	0.813 (1.6%)	0.802 (0.3%)
3	0.6	0.553 (-7.8%)	0.621 (3.5%)	0.612 (2.0%)	0.611 (1.8%)	0.621 (3.5%)	0.611 (1.8%)
4	1.0	1.01 (1.0%)	0.950 (-5.0%)	0.980 (-2.0%)	0.988 (-1.2%)	1.010 (1.0%)	1.032 (3.2%)
Mean Absolute Error		3.3%	3.0%	1.7%	2.5%	2.0%	1.6%
Max. Absolute Error		7.8%	5.0%	2.0%	6.9%	3.5%	3.2%

Figure 4. Identified stiffness integrity indices for Damage Scenario 1 under 10% noise (2 sensors)

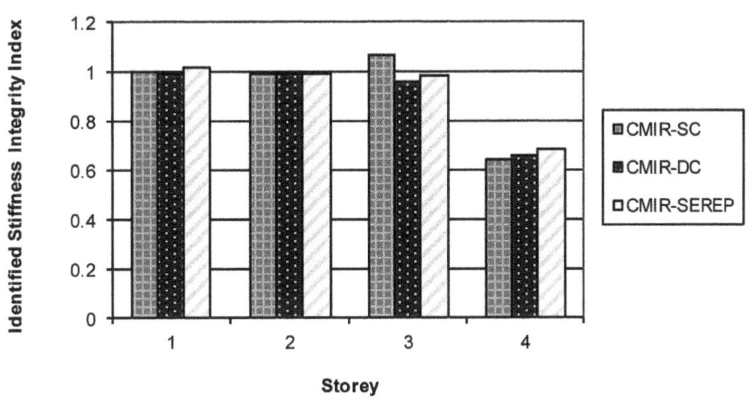

Figure 5. Identified stiffness integrity indices for Damage Scenario 1 under 10% noise (3 sensors)

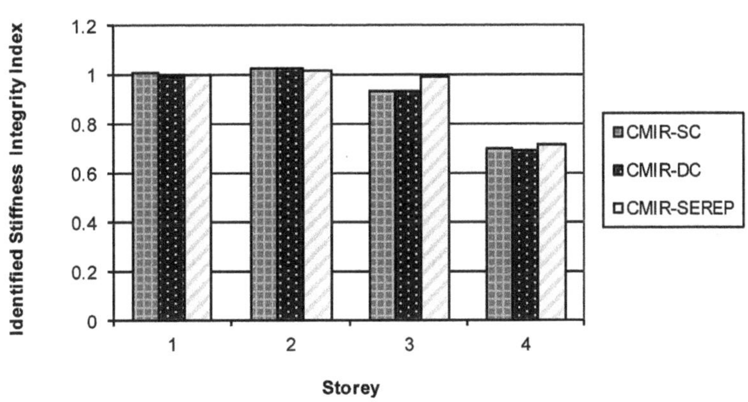

Figure 6. Identified stiffness integrity indices for Damage Scenario 2 under 10% noise (2 sensors)

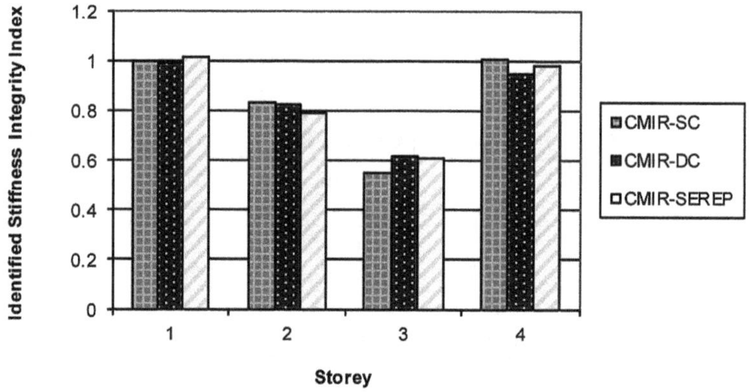

Figure 7. Identified stiffness integrity indices for Damage Scenario 2 under 10% noise (3 sensors)

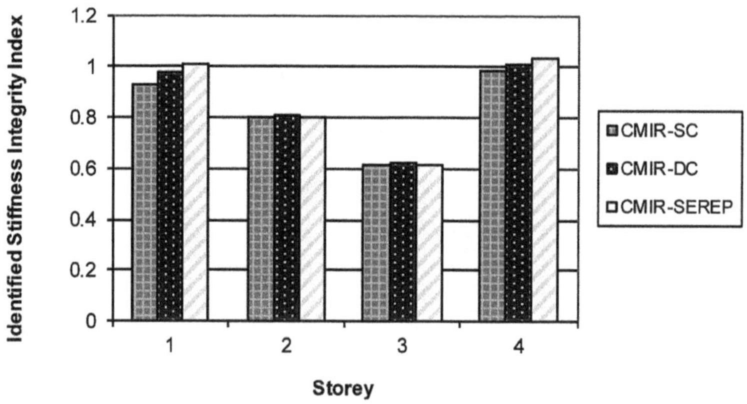

provide a non-destructive means to locate and even quantify structural damage by tracking the change in values of pertinent parameters such as stiffness. In this context, the main objectives of this chapter are to develop numerical strategies suitable for stiffness identification and damage assessment of structural systems. For incomplete measurement, condensed stiffness matrices can be identified in a similar manner. But to recover the full structural matrices, the CMIR method is formulated. The main significance of this method is that it is possible to obtain several condensed stiffness matrices, thereby enabling one to find individual stiffness coefficients for the structure based on incomplete measurement. GA is used as a robust optimization tool for recovering storey stiffness values from the condensed stiffness matrices. With regards to incomplete measurements, the CMIR method can be conducted with fixed sensors or with non-fixed (relocated) sensors. Three different model condensation methods, namely SC, DC and SEREP are employed to avoid the need of having complete measurement. The CMIR method developed in this chapter overcomes the necessity of having a full set of measurements, thereby allowing fewer sensors and actuators than the previous methods.

REFERENCES

Adeli, H. (2001). Special Issue: Health Monitoring of Structures. *Computer-Aided Civil and Infrastructure Engineering*, *16*(1).

Agbabian, M. S., Masri, S. F., Miller, R. K., & Caughey, T. K. (1991). System Identification Approach to Detection of Structural Changes. *J. Struct. Engrg. Mech.*, *117*(2), 370–390. doi:10.1061/(ASCE)0733-9399(1991)117:2(370)

Alvin, K. F., & Park, K. C. (1994). Second-Order Structural Identification Procedure via State-Space-Based System Identification. *AIAA Journal*, *32*(2), 397–406. doi:10.2514/3.11997

Baruch, M. (1982). Optimal Correction of Mass and Stiffness Matrices Using Measured Modes. *AIAA Journal*, *20*(11), 1623–1626. doi:10.2514/3.7995

Bernal, D., & Beck, J. (2004). Special Issue: Phase 1 of the IASC-ASCE Structural Health Monitoring Benchmark. *Journal of Engineering Mechanics*, *130*(1), 1–2. doi:10.1061/(ASCE)0733-9399(2004)130:1(1)

Chang, M., & Pakzad, S. N. (2014). Observer Kalman Filter Identification for Output-only Systems Using Interactive Structural Modal Identification Toolsuite. *Journal of Bridge Engineering*, *19*(5), 04014002. doi:10.1061/(ASCE)BE.1943-5592.0000530

Chen, H. P., & Tee, K. F. (2014). Structural Finite Element Model Updating Using Incomplete Ambient Vibration Modal Data. *Science China. Technological Sciences*, *57*(9), 1677–1688. doi:10.100711431-014-5619-9

Chen, H. P., Tee, K. F., & Ni, Y. Q. (2012). Mode Shape Expansion with Consideration of Analytical Modelling Errors and Modal Measurement Uncertainty. *Smart Structures and Systems*, *10*(4-5), 485–499. doi:10.12989ss.2012.10.4_5.485

DeAngelis, M., Lus, H., Betti, R., & Longman, R. W. (2002). Extraction Physical Parameters of Mechanical Models from Identified State Space Representations. *ASME Trans. J. Appl. Mech.*, *69*(5), 617–625. doi:10.1115/1.1483836

Doherty, J. E. (1987). Nondestructive Evaluation. In A. S. Kobayashi (Ed.), *Handbook on Experimental Mechanics*. Society for Experimental Mechanics, Inc.

Ghanem, R., & Shinozuka, M. (1995). Structural System Identification I: Theory. *Journal of Engineering Mechanics*, *121*(2), 255–264. doi:10.1061/(ASCE)0733-9399(1995)121:2(255)

Ghanem, R., & Sture, S. (2000). Special Issue: Structural Health Monitoring. *Journal of Engineering Mechanics*, *126*(7).

Guyan, R. J. (1965). Reduction of Stiffness and Mass Matrices. *AIAA Journal*, *3*(2), 380. doi:10.2514/3.2874

Hjelmstad, K. D., Banan, M. R., & Banan, M. R. (1995). Time Domain Parameter Estimation Algorithm for Structures. I: Computational Aspects. *Journal of Engineering Mechanics*, *121*(3), 424–434. doi:10.1061/(ASCE)0733-9399(1995)121:3(424)

Ho, B. L., & Kalman, R. E. (1965). Effective Construction of Linear State-Variable Models from Input/Output Functions. *Proc. 3rd Allerton Conf. Circuits and Systems Theory*, 449-459.

Houck. (1995). *A Genetic Algorithm for Function Optimization: A MATLAB Implementation*. North Carolina State University.

Juang, J.-N., Cooper, J. E., & Wright, J. R. (1988). An Eigensystem Realization Algorithm Using Data Correlations (ERA/DC) for Model Parameter Identification. *Control, Theory and Advanced Technology*, *4*(1), 5–14.

Juang, J.-N., & Pappa, R. S. (1985). An Eigensystem Realization Algorithm for Model Parameter Identification and Model Reduction. *Journal of Guidance, Control, and Dynamics*, *8*(5), 620–627. doi:10.2514/3.20031

Juang, J.-N., & Pappa, R. S. (1986). Effects of Noise on Modal Parameters Identified by the Eigensystem Realizations Algorithm. *Journal of Guidance, Control, and Dynamics*, *9*(3), 294–303. doi:10.2514/3.20106

Kidder, R. L. (1973). Reduction of Structural Frequency Equations. *AIAA Journal*, *11*(6), 892. doi:10.2514/3.6852

Koh, C. G., Quek, S. T., & Tee, K. F. (2002). Damage Identification of Structural Dynamic System. *Prof of the 2nd International Conference on Structural Stability and Dynamics*, 780-785.

Koh, C. G., Tee, K. F., & Quek, S. T. (2003). Stiffness and Damage Identification with Model Reduction Technique. *Prof of the 4th International Workshop on Structural Health Monitoring, Stanford University*, 525-532.

Koh, C. G., Tee, K. F., & Quek, S. T. (2006). Condensed Model Identification and Recovery for Structural Damage Assessment. *Journal of Structural Engineering*, *132*(12), 2018–2026. doi:10.1061/(ASCE)0733-9445(2006)132:12(2018)

Lin, C. B., Soong, T. T., & Natke, H. G. (1990). Real Time System Identification of Degrading Structures. *Journal of Engineering Mechanics*, *116*(10), 2258–2274. doi:10.1061/(ASCE)0733-9399(1990)116:10(2258)

Longman, R. W., & Juang, J.-N. (1987). A Variance Based Confidence Criterion for ERA Identified Modal Parameters. *Proc. of the AAS/AIAA Astrodynamics Conf.*

Lus, H. (2001). *Control Theory Based System Identification* (PhD thesis). Columbia University, New York, NY.

Lus, H., De Angelis, M., Betti, R., & Longman, R. W. (2003). Constructing Second-order Models of Mechanical Systems from Identified State Space Realizations. Part I: Theoretical Discussions. *Journal of Engineering Mechanics*, *129*(5), 477–488. doi:10.1061/(ASCE)0733-9399(2003)129:5(477)

Miller, C. A. (1980). Dynamic Reduction of Structural Models. *Journal of the Structural Division*, *106*(10), 2097–2108.

Moore, M. (2001). *Reliability of Visual Inspection for Highway Bridges, Volume 1.* Final Report, U.S. Department of Transportation report, FHWA-RD-01-020.

O'Callahan, J., Avitabile, P., & Riemer, R. (1989). System Equivalent Reduction Expansion Process (SEREP). *Proc. of the 7th International Modal Analysis Conference*, 29-37.

Papadopoulos, M., & Garcia, E. (1996). Improvement in Model Reduction Schemes Using the System Equivalent Reduction Expansion Process. *AIAA Journal*, *34*(10), 2217–2219. doi:10.2514/3.13383

Paz, M. (1985). *Structural Dynamics, Theory and Computation*. New York: Van Nostrand Reinhold.

Paz, M. (1989). Modified Dynamic Condensation Method. *Journal of Structural Engineering*, *115*(1), 234–238. doi:10.1061/(ASCE)0733-9445(1989)115:1(234)

Phan, M., Horta, L. G., Juang, J.-N., & Longman, R. W. (1993). Linear System Identification via an Asymptotically Stable Observer. *Journal of Optimization Theory and Applications*, *79*(1), 59–86. doi:10.1007/BF00941887

Phan, M., Horta, L. G., Juang, J.-N., & Longman, R. W. (1995). Improvement of Observer/Kalman Filter Identification (OKID) by Residual Whitening. *Journal of Vibration and Acoustics*, *117*(2), 232–238. doi:10.1115/1.2873927

Phan, M., Juang, J.-N., & Longman, R. W. (1992). Identification of Linear Multivariable Systems by Identification of Observers with Assigned Real Eigenvalues. *The Journal of the Astronautical Sciences*, *40*(2), 261–279.

Qin, S. J. (2006). An Overview of Subspace Identification. *Computers & Chemical Engineering*, *30*(10-12), 1502–1513. doi:10.1016/j.compchemeng.2006.05.045

Qu, Z.-Q., & Fu, Z.-F. (1998). New Structural Dynamic Condensation Method for Finite Element Models. *AIAA Journal*, *36*(7), 1320–1324. doi:10.2514/2.517

Qu, Z.-Q., & Fu, Z.-F. (2000). An Iterative Method for Dynamic Condensation of Structural Matrices. *J. Mechanical System and Signal Processing*, *14*(4), 667–678. doi:10.1006/mssp.1998.1302

Rivera, M. A., Singh, M. P., & Suarez, L. E. (1999). Dynamic Condensation Approach for Nonclassically Damped Structures. *AIAA Journal*, *37*(5), 564–571. doi:10.2514/2.774

Rytter, A. (1993). *Vibration Based Inspection of Civil Engineering Structures* (PhD Dissertation). Department of Building Technology and Structural Engineering, Aalborg University, Denmark.

Srinivasan, M. G., & Kot, C. A. (1992). Effects of Damage on the Modal Parameters of a Cylindrical Shell. *Proc. of the 10th International Modal Analysis Conference*, 529-535.

Tee, K. F. (2004). *Substructural Identification with Incomplete Measurement for Structural Damage Assessment* (Ph.D Dissertation). Department of Civil Engineering, National University of Singapore.

Tee, K. F., Cai, Y., & Chen, H. P. (2013). Structural Damage Detection Using Quantile Regression. *Journal of Civil Structural Health Monitoring*, *3*(1), 19–31. doi:10.100713349-012-0030-3

Tee, K. F., Koh, C. G., & Quek, S. T. (2003). System Identification and Damage Estimation via Substructural Approach. *Computational Structural Engineering*, *3*(1), 1–7.

Tee, K. F., Koh, C. G., & Quek, S. T. (2004). Substructural System Identification and Damage Estimation by OKID/ERA. *Proc. of the 3rd Asian-Pacific Symposium on Structural Reliability and Its Applications,* 637-647.

Tee, K. F., Koh, C. G., & Quek, S. T. (2005). Damage Assessment by Condensed Model Identification and Recovery with Substructural Approach. *Proc. of the International Conference on Experimental Vibration Analysis for Civil Engineering Structures*, 415-422.

Tee, K. F., Koh, C. G., & Quek, S. T. (2009). Numerical and Experimental Studies of a Substructural Identification Strategy. *Structural Health Monitoring*, *8*(5), 397–410. doi:10.1177/1475921709102089

Tseng, D.-H., Longman, R. W., & Juang, J.-N. (1994a). Identification of Gyroscopic and Nongyroscopic Second Order Mechanical Systems Including Repeated Problems. *Advances in the Astronautical Sciences*, *87*, 145–165.

Tseng, D.-H., Longman, R. W., & Juang, J.-N. (1994b). Identification of the Structure of the Damping Matrix in Second Order Mechanical Systems. *Advances in the Astronautical Sciences*, *87*, 166–190.

Wang, T., Zhang, L., & Tee, K. F. (2011). Extraction of Real Modes and Physical Matrices from Modal Testing. *Earthquake Engineering and Engineering Vibration*, *10*(2), 219–227. doi:10.100711803-011-0060-6

Yang, C. D., & Yeh, F. B. (1990). Identification, Reduction, and Refinement of Model Parameters by the Eigensystem Realization Algorithm. *Journal of Guidance, Control, and Dynamics*, *13*(6), 1051–1059. doi:10.2514/3.20578

ADDITIONAL READING

Balmes, E. (1997). New Results on the Identification of Normal Modes from Experimental Complex Modes. *Mechanical Systems and Signal Processing*, *11*(2), 229–243. doi:10.1006/mssp.1996.0058

Ewins, D. J. (1984). *Modal Testing: Theory and Practice*. Letchworth: Research Studies Press.

Hong, B. (2002). GA-based Identification of Structural Systems in Time Domain. *Ph.D Thesis*, National University of Singapore.

Horta, L. G., Juang, J.-N., & Longman, R. W. (1993). Discrete-Time Model Reduction in Limited Frequency Ranges. *Journal of Guidance, Control, and Dynamics*, *16*(6), 1125–1130. doi:10.2514/3.21136

Oreta, W. C., & Tanabe, T. A. (1994). Element Identification of Member Properties of Framed Structures. *Journal of Structural Engineering*, *120*(7), 1961–1976. doi:10.1061/(ASCE)0733-9445(1994)120:7(1961)

Tee, K. F., & Cai, Y. (2012). Statistical Structural Health Monitoring Using Quantile Regression, *Proc. of the 5th European Conference on Structural Control*, Genoa, Italy, June 18-20, Paper 096.

Tee, K. F., Koh, C. G., & Quek, S. T. (2003). Substructural Identification with Incomplete Measurement for Damage Assessment, *Proc. of the 1st International Conference on Structural Health Monitoring and Intelligent Infrastructure*, Tokyo, Japan, November 13-15, Vol.1, 379-386.

Tee, K. F., Koh, C. G., & Quek, S. T. (2005). Substructural First- and Second-Order Model Identification for Structural Damage Assessment. *Earthquake Engineering & Structural Dynamics*, *34*(15), 1755–1775. doi:10.1002/eqe.500

Udwadia, F. E., & Proskurowski, W. (1998). A Memory-matrix-based Identification Methodology for Structural and Mechanical System. *Earthquake Engineering & Structural Dynamics*, *27*(12), 1465–1481. doi:10.1002/(SICI)1096-9845(199812)27:12<1465::AID-EQE795>3.0.CO;2-7

Yun, C. B., & Bahng, E. Y. (2000). Substructural Identification Using Neural Networks. *Computers & Structures*, *77*(1), 41–52. doi:10.1016/S0045-7949(99)00199-6

Zhao, Q., Sawada, T., Hirao, K., & Nariyuki, Y. (1995). Localized Identification of MDOF Structures in the Frequency Domain. *Earthquake Engineering & Structural Dynamics*, *24*(3), 325–338. doi:10.1002/eqe.4290240303

KEY TERMS AND DEFINITIONS

Actuator: A device that converts energy into motion. It also can be used to apply a force.

Damage: Changes to the materials and/or geometric properties of structures, such as stiffness reduction.

Sensor: A device that is used to record acceleration, velocity, or displacement of the structures.

Structural Health Monitoring: Process of implementing a damage detection and characterization strategy for engineering structures.

System Identification: The process of constructing models from experimental data.

Chapter 7
Optimization of Soil Structure Effect by the Addition of Dashpots in Substratum Modelization

Souhaib Bougherra
Constantine University, Algeria

Mourad Belgasmia
Setif 1 University, Algeria

ABSTRACT

Soil structure interaction can significantly affect the behavior of buildings subjected to seismic attacks, wind excitation, and other dynamic loading types. Different researches were developed in the last decade demonstrating the importance of taking account of soil properties and its effect in changing the behavior of the structures. It is common practice to analyze the structures assuming a fixed base, but this approach is not appropriate for the reason that neglecting the soil parameters such as the stiffness and the damping affect the behavior of the structure. Therefore, the nonlinear static approach provided the nonlinear response behavior of a structure for different types of soil. In this chapter, the authors will discuss some proposed methods in taking account of soil-structure interaction that must be considered from the very beginning of the design process and its impact on the structural behavior optimization by adding springs and dashpots to reproduce the soil behavior.

INTRODUCTION

Soil-Structure Interaction (SSI) is a term used essentially to describe the forces generated that occur between different linked systems: the structure, the foundation and the soil surrounding the foundation. It is used also when these forces are able to change the dynamic response of the structure as well as the response of the soil when it is subjected to a free field ground motion.

DOI: 10.4018/978-1-5225-7059-2.ch007

Soil-structure interaction analysis is a special field of earthquake engineering that requires the intervention of different professionals (structural engineers, geotechnical engineers and engineering seismologists) who must collaborate between them to understand and to improve the technical approaches that should be considered in the design process by the structural engineers.

The purpose of this study is to evaluate the dynamic response of reinforced concrete building under lateral seismic loads and especially to demonstrate how the structure is going to behave when a changing in the soil parameters and an addition of springs and dashpots in the base of the structure is done.

This chapter summarizes some proposed methods in modeling soil-structure interaction effects on building structures, also, a brief history of the soil structure interaction and researches done in this field will be presented, and finally, the results of the case studied will be discussed.

BACKGROUND AND OBJECTIVES

One of the most important parameters affecting the linear and non-linear behavior and performance levels of structures is the soil–structure interaction (SSI) phenomenon, and for several decades, only soil flexibility was considered without soil inertia – springs modeled soil.

At that time, the key question of such a simple approach appeared to be damping. The material damping measured in the laboratories with soil samples proved to be considerably less than the damping measured in the dynamic field tests with rigid stamps resting on the soil surface.

It is clear that there is different types of time-varying loadings acting directly on the structure (periodic loads, impact loads and blast loads) but the most important loading and the most complicated one is the earthquake excitation that is applied to the soil surrounding the foundation of the structure.

The nature of the phenomenon of dynamic SSI began in 1930-40s [Eric Reissner, Karl Marguerre] that investigated the harmonically loaded soils and proved to be in inertial properties of the soil.

Although the effect of SSI have been the subject of numerous investigations in the past [Veletsos and Meek],[Aviles and Perez-Rocha],[Aviles and Suarez],[Gazetas and Mylonakis], [Jennings and Bielak], [Kausel], [,[Wolf, J. P], [Ciampoli, M. and Pinto, P. E], [Liping L, Wenjin G, Qiang X, Lili B, Yingmin L, Yuntian W]. However, there is still controversy regarding the role of SSI in the seismic performance of structures founded on soft soil.

In last decades, civil structures are gradually increasing in size and embedment. Effects like SSI from time to time are considered during the design procedures but in general, most designers define the base of the structure as a fixed base, that approach is not appropriate for the reason that neglecting the soil parameters such as the stiffness and the damping affects the behavior of the structure. In fact, SSI has been traditionally considered beneficial for seismic response of structures.

MODELIZATION OF SOIL STRUCTURE INTERACTION

Understanding of SSI principles varies widely across both the structural and geotechnical engineering disciplines. Most structural engineers can appreciate that SSI effects are more pronounced in soft soils, and many are aware that foundation input motions can differ from free-field ground motions.

Current methods of SSI analysis can be classified according to the choice of surface surrounding the basement in the soil that divides the soil-structure model into two parts: the external part and the internal part.

Nowadays, there exist two main methods for SSI problems: Direct and Substructure approaches, and to understand them, let us start with basic definitions.

Direct Method

In the "Direct" method, the model is analyzed as a complete system (the soil and the structure are usually discretized in a one finite element model with foundation and structural elements) Figure 1, and probably it is the easiest way to analyze soil-structure interaction for seismic excitation, from a conception point of view.

However, from a computational standpoint, it is difficult to obtain a direct solution of the SSI problem, because the number of dynamic degrees of freedom in the soil region is high, and systems that contains significant nonlinearities in the soil or the structural materials.

Substructure Method

The "Substructure" method, allows the complicated soil-structure system to be broken down into manageable parts that are combined in final to formulate the complete solution.

In this approach, two important steps are taken in consideration, the first one is to evaluate the free-field accelerogram is computed without considering the presence of the structure, then the dynamic stiffness coefficients of the soil that are interpreted by the spring-dashpot system are calculated. In the second step, the free-field accelerogram is used to analyze the structure having its base coupled with this discrete translational and rotational spring and dashpot elements.

The behavior of the soil is accounted for by simple mechanical elements such as springs and dashpots. Different configurations can be taken into account by connecting several springs and dashpots whose parameters are described in the next paragraphs.

Figure 1. Finite element model illustrating the direct method. [NIST]

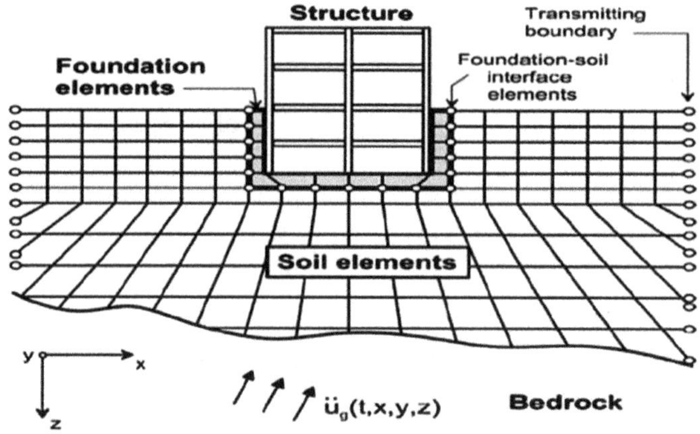

Figure 2. Substructure method in soil structure interaction [J.P WOLF]

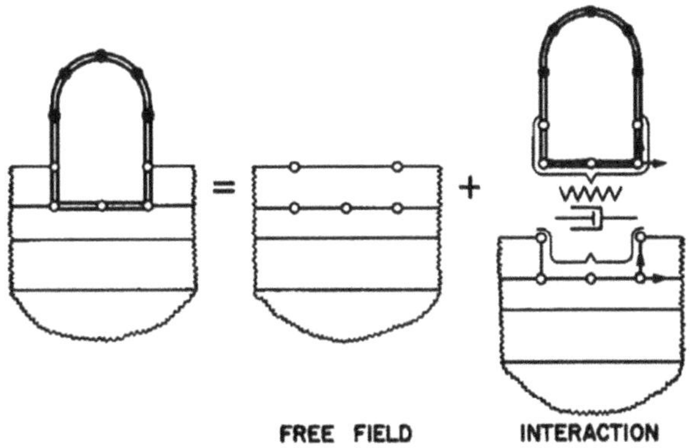

The stiffness and damping characteristics of the soil-foundation interaction are characterized using relatively simple impedance function models or a series of distributed springs and dashpots as schematized in figure 3 and figure 4.

Impedance

Historically, the first problems for impedances considering soil inertia were solved semi analytically for homogeneous half-space without the internal damping and for circular surface stamps. It turned out that horizontal impedances behaved more or less like pairs of springs and viscous dashpots. This fact created a base for using the above-mentioned "soil springs and dashpots" elements.

Figure 3. Model of springs [NIST]

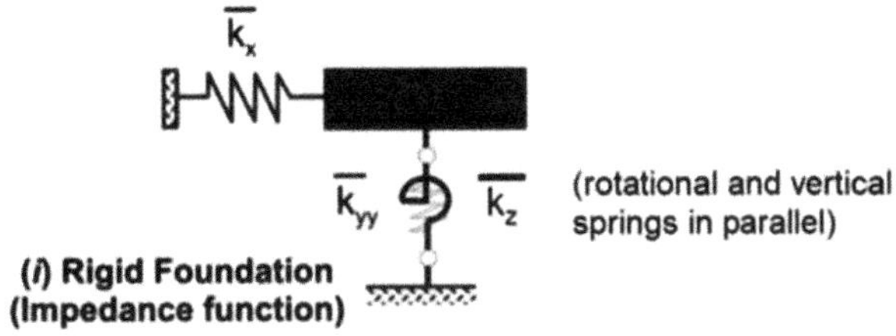

Figure 4. Model with springs and dashpots [NIST]

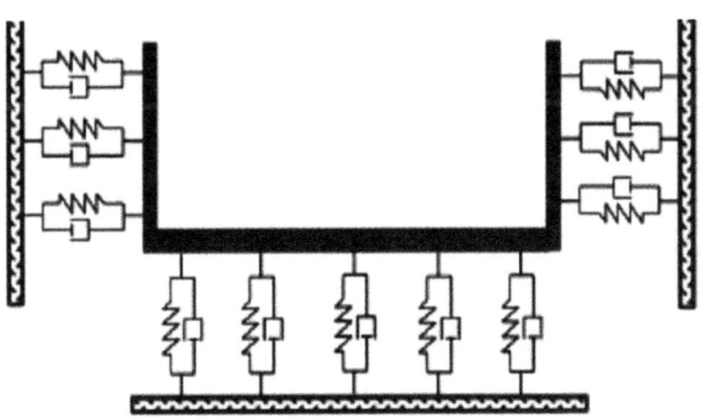

The foundation dashpot elements consider two different sources of damping; the hysteric(strain dependent) damping caused by the inelastic response of the soil supporting the foundation and the radiation (geometry dependent) damping caused by the stress waves that travel away from the foundation as the soil deforms. Radiation damping is much greater than material damping for most foundation configurations.

Many impedance function solutions are available for rigid circular or rectangular foundations located on the surface of, or embedded within, a uniform, elastic, or visco-elastic half-space and the two tables below were provided by Gazetas (1991).

Table 1 lists stiffness and damping coefficients named K and C respectively for the six degrees of freedom (three translational and three rotational) for an arbitrarily shaped foundation on surface of a homogeneous half-space, that include the approximate effects of radiation damping in the soil. Table 2 contains the same parameters but this time for a foundation embedded in a half space with an arbitrary basement shape.

Notes: D: embedded depth of foundation in soil; d: contact depth between foundation and soil; h=D-d/2; I_{bx}, I_{by}, I_{bz} : are Polar moment of inertia of the bottom surface of foundation about the x, y and

Table 1. Spring and dashpot coefficients for a foundation on surface of homogeneous half space

Movement		Spring constant K	Dashpot constant C
Vertical		$K_z = \frac{2GL}{1-v}\left(0.73 + 1.54\left(\frac{A_b}{4L^2}\right)^{0.75}\right)$	$C_z = \rho V_{La} A_b \tilde{c}_z$
Horizontal	Lateral direction	$K_y = \frac{2GL}{2-v}\left(2 + 2.50\left(\frac{A_b}{4L^2}\right)^{0.85}\right)$	$C_y = \rho V_s A_b \tilde{c}_y$
	Longitudinal direction	$K_y = K_y - \left[\frac{0.2}{0.75-v} GL\left(1 - \left(\frac{B}{L}\right)\right)\right]$	$C_x = \rho V_s A_b$
Rocking	Longitudinal x axis	$K_{rx} = \frac{G}{1-v} I_{bx}^{0.75}\left(\frac{L}{B}\right)^{0.25}\left(2.4 + \frac{B}{2L}\right)$	$C_{rx} = \rho V_{La} I_{bx} \tilde{c}_{rx}$
	Lateral y axis	$K_{ry} = \frac{3G}{1-v} I_{by}^{0.75}\left(\frac{L}{B}\right)^{0.15}$	$C_{ry} = \rho V_{La} I_{by} \tilde{c}_{ry}$
Torsion		$K_t = 3G I_{bz}^{0.75}\left(\frac{B}{L}\right)^{0.4}\left(\frac{I_{bz}}{B^4}\right)^{0.2}$	$C_t = \rho V_s I_{bz} \tilde{c}_t$

Table 2. Spring and dashpot coefficients for embedded foundations in a half space

Movement		Spring Constant	Dashpot Constant
Vertical		$K_{z,e} = K_z\left[1+\frac{1}{21}\left(\frac{D}{B}\right)\left(1+1.3\left(\frac{A_b}{4L^2}\right)\right)\right]\left[1+0.2\left(\frac{A_w}{A_b}\right)^{\frac{2}{3}}\right]$	$C_{z,e} = C_z + \rho V_{La} A_w$
Horizontal	Lateral direction	$K_{y,e} = K_y\left[1+0.15\left(\frac{D}{B}\right)^{0.5}\right]\left[1+0.52\left(\frac{hA_w}{BL^2}\right)^{0.4}\right]$	$C_{y,e} = C_y + 4\rho V_s Bd + 4\rho V_{La} Ld$
	Longitudinal direction	$K_{x,e} = K_x\left(\frac{K_{y,e}}{K_y}\right)$	$C_{x,e} = C_x + \rho V_{La} Bd + 4\rho V_s Ld$
Rocking	Longitudinal x axis	$K_{rx,e} = K_{rx}\left\{1+1.26\left(\frac{d}{B}\right)\left[1+\left(\frac{d}{B}\right)\left(\frac{d}{D}\right)^{-0.2}\left(\frac{B}{L}\right)^{0.5}\right]\right\}$	$C_{rx,e} = C_{rx} + \rho I_{bx}\left(\frac{d}{B}\right)$ $\left\{V_{La}\left(\frac{d^2}{B^2}\right)+3V_s+V_s\left(\frac{B}{L}\right)\left[1+\left(\frac{d^2}{B^2}\right)\right]\right\}\eta_r$ $\eta_{rx} = 0.25+0.65\sqrt{a_0}\left(d/D\right)^{-a_0/2}\left(D/B\right)^{-1/4}$
	Lateral y axis	$K_{ry,e} = K_{ry}\left\{1+0.92\left(\frac{d}{L}\right)^{0.6}\left[1.5+\left(\frac{d}{L}\right)^{1.9}\left(\frac{d}{L}\right)^{-0.6}\right]\right\}$	$C_{ry,e} = C_{ry}$ by replacing x by y and interchanging B by L with $a_0 = \omega B / V_s$

continued on following page

Table 2. Continued

Movement	Spring Constant	Dashpot Constant
Torsion	$K_{t,e} = K_t \Gamma_w \Gamma_{tre}$ $\Gamma_w = 1 + 0.4\left(\frac{D}{d}\right)^{0.5}\left(\frac{j_s}{j_r}\right)\left(\frac{B}{D}\right)^{0.6}$ $\Gamma_{tre} = 1 + 0.5\left(\frac{D}{B}\right)^{0.1}\left(\frac{B^4}{I_{bx}}\right)^{0.13}$ $j_s = \left(4/3\right)_{d\,(}$ $B^3 + L^3) + 4BLd\left(L + B\right)$ $j_r = \left(4/3\right)BL\left(B^2 + L^2\right)$	$C_{t,e} = C_t + 4\rho d\left[\left(\frac{1}{3}\right)V_{La}(B^3 + L^3\right] + V_s BL\left(L + B\right)\eta_t$ $\eta_t = \left(\frac{d}{D}\right)^{-0.5} a_0^2 / \left[a_0^2 + \left(\frac{1}{2}\right)\left(\frac{L}{B}\right)^{-1.5}\right]$

z axes; B, L: half-width and half-length of the foundation. G, Å Á Are the shear modulus, Poisson's ratio and mass density of the soil; A_w : area of foundation side contact with soil; A_b : actual area of bottom surface of foundation. $\tilde{c}_z, \tilde{c}_y, \tilde{c}_{rx}, \tilde{c}_{ry}, \tilde{c}_t$: Are the damping adjustment coefficients of surface foundation, which is determined by the table of $\left(a_0, L / B\right)$; V_s : shear wave velocity of the soil and finally V_{La} is Lymer's analog wave velocity that equals: $V_{La} = \dfrac{3.4}{\pi\left(1-\nu\right)} V_s$.

PUSHOVER ANALYSIS

Pushover analysis is a static nonlinear procedure who offered engineers a great opportunity to consider the inelastic behavior of structures and provided satisfactory predictions of structures seismic demands.

It is utilized generally to predict the behavior of structures under seismic loads. The work done by [Belgasmia and Moussaoui] demonstrates the importance of nonlinear analysis to prevent large deformations. The procedure is executed by applying a lateral forces distributed over the height of the structure and incrementally increasing until a collapse or a target displacement is reached (Figure 5). Moreover, it records the base shear-displacement relation. The pushover curve is based on force-deformation criteria for hinges developed by [ATC 40] and [FEMA 273] as shown in Figure 6.

A, B, C, D, and E are used to define the force deflection behavior of the hinge. Immediate Occupancy (IO), Life Safety (LS) and Collapse Prevention (CP), are used to define the acceptance criteria for the hinges. The results of the pushover analysis are approximated in a capacity curve in which a force versus displacement relationship represents the system's behavior (Figure 7).

Figure 5. Application of the distributed load on the structure modelled with springs [NIST]

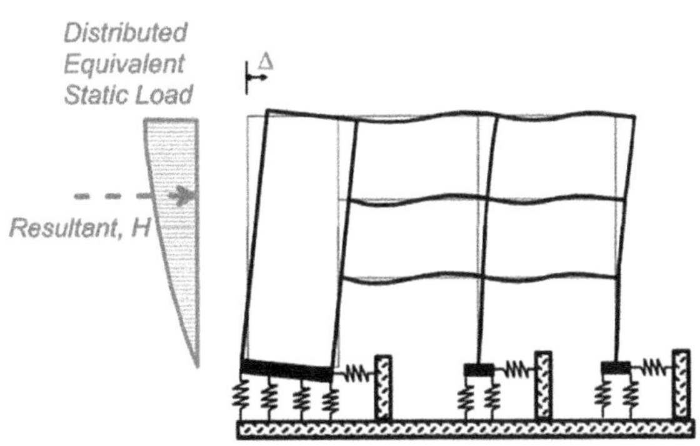

Figure 6. Capacity curve Hinges

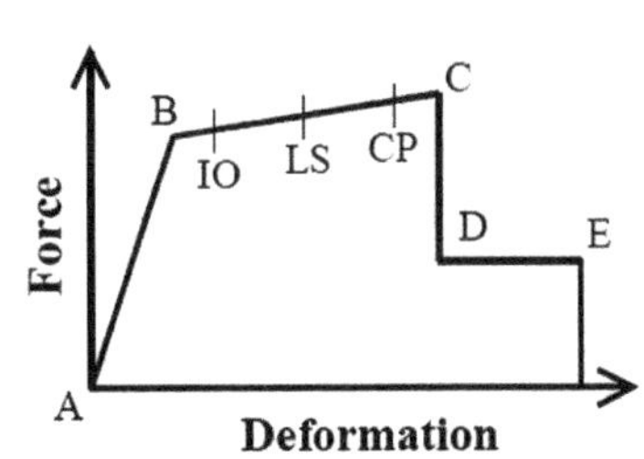

Figure 7. Typical Capacity curve

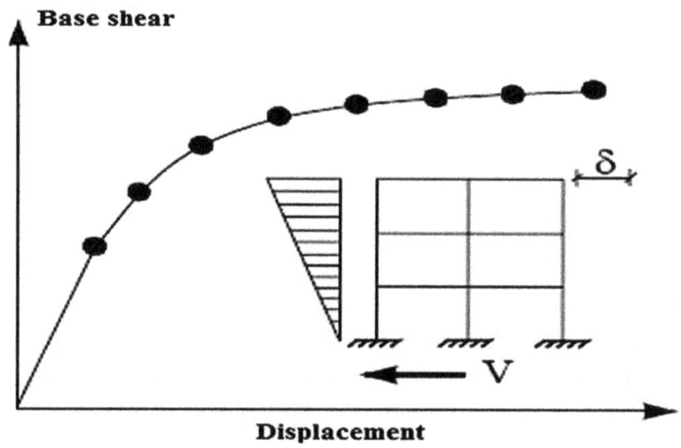

NUMERICAL MODELING

In this chapter a study of 3D nonlinear static analysis under seismic loads of a fourteen-story reinforced concrete building designed according to the Algerian code [C.B.A] and [R.P.A], with typical story height of 3.5m, the structure is a beam-column system with a flat slab at different floor levels. The strength of the concrete is 25 MPa and the steel reinforcement is 400 MPa. Plastic hinge property of the confined cross sections is determined by considering transversal and longitudinal reinforcement. In order to consider the influence of different ground soil conditions three wave shear(V_S) velocities, including 400 m/s,200 m/s,100 m/s that represents stiff, medium and soft soil are chosen, SAP2000 finite element software is used for three dimensional modeling and analyses of the building.(see figure 8).

In the current analysis, the authors will proceed as follows:

1. Establish the model of the RC high-rise building.
2. Apply the seismic load increasingly, the structural deformation is growing and plastic hinges are formed in some sections of the structure.
3. A collapse or a target displacement is reached when increasing the horizontal load.
4. Display the base shear-displacement curves and the floor displacement for two different cases:
    a. **Case 1:** The structure will be fixed at the base.
    b. **Case 2:** The interaction between the soil and the structure will be modeled with springs and dashpots for different types of soil: stiff, medium and soft soil.

ANALYSIS RESULTS

The results of the analysis given by SAP2000 software for different types of soil and for a fixed base are shown in Table 3 and Figures 9-11, which represents floor displacement, the Capacity curve, story drift plastic hinge formation respectively:

Table 3 shows that there is an augmentation in the lateral displacement between the model with fixed base and the one based on soft soil. The largest differences are obvious on the five top floors.

Figure 8. 3D models with fixed base and spring-dashpot elements

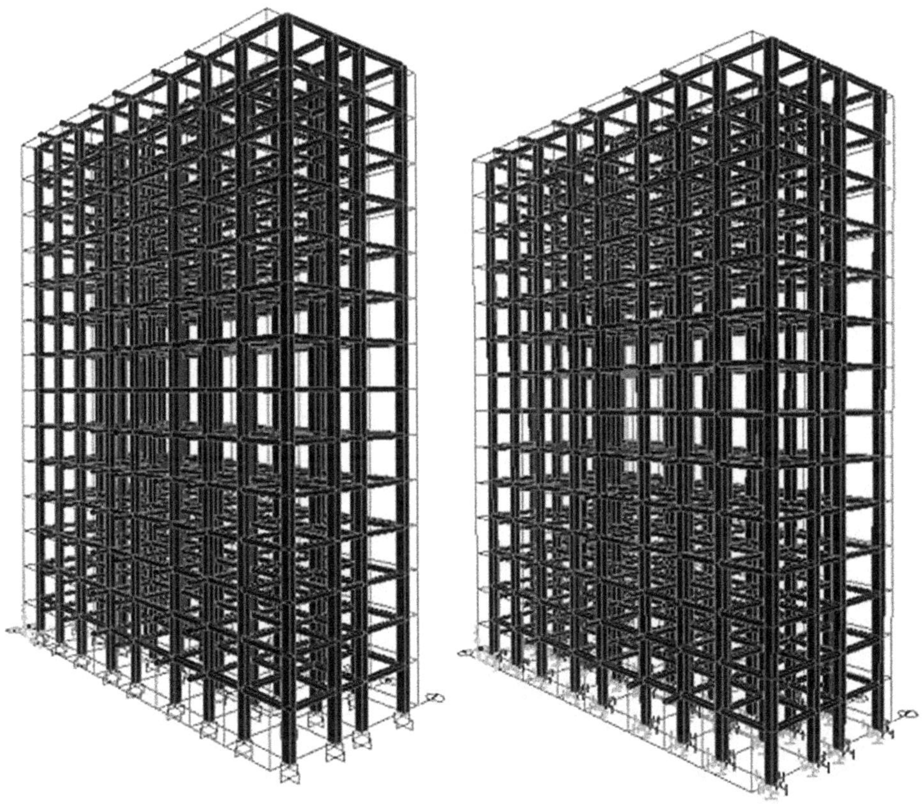

Table 3. Floor displacement for different types of soil

Story	Fixed base(m)	Stiff(m)	Medium(m)	Soft(m)
0	0	0.000015	0.000066	0.000244
1	0.030252	0.03048	0.041728	0.078495
2	0.090437	0.090767	0.114792	0.189983
3	0.170931	0.171317	0.20912	0.322467
4	0.264465	0.264871	0.317478	0.469256
5	0.364684	0.365081	0.433283	0.624119
6	0.466058	0.466426	0.550734	0.781454
7	0.564027	0.564355	0.664971	0.936449
8	0.655134	0.655419	0.772212	1.085215
9	0.737101	0.737344	0.869862	1.224898
10	0.808718	0.808921	0.956585	1.353759
11	0.869725	0.869892	1.032217	1.471218
12	0.920811	0.920946	1.097583	1.577875
13	0.963615	0.963721	1.154445	1.675493
14	1.000739	1.00082	1.205506	1.766946

Figure 9. Capacity curve for different types of soil

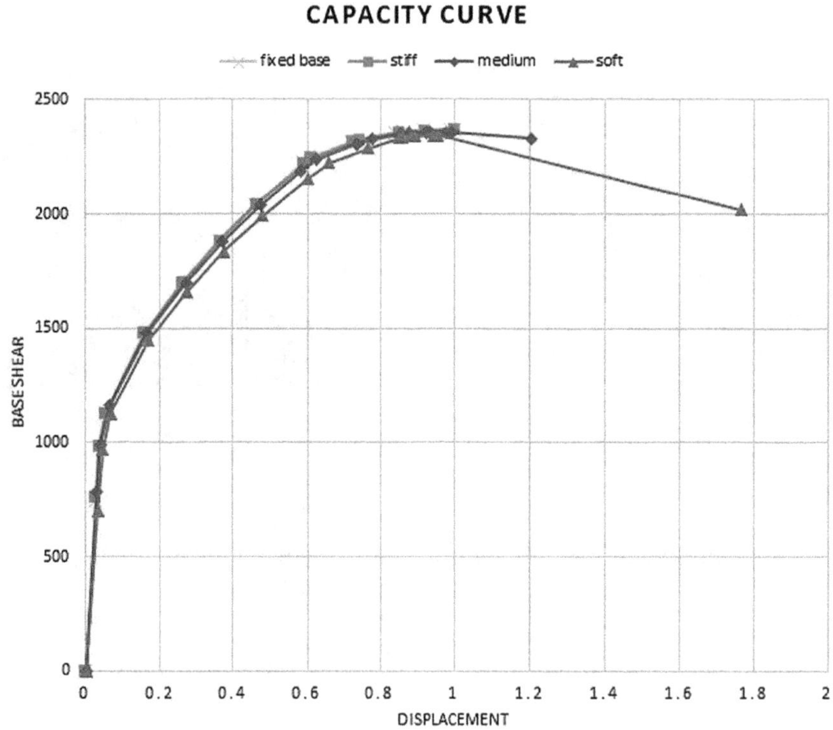

Figure 10. Inter story drift for different types of soil

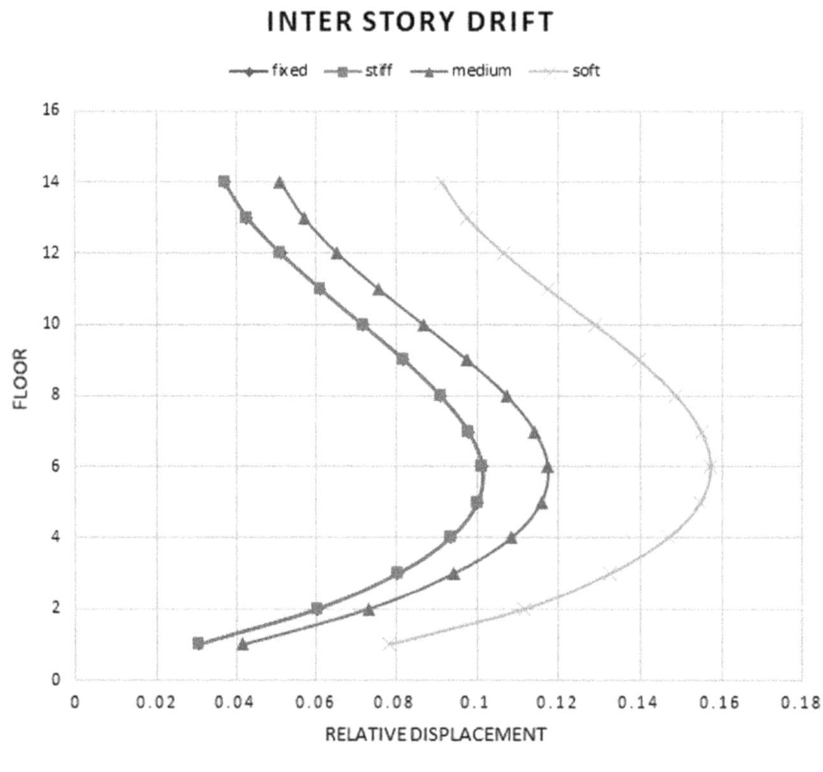

Figure 11. Plastic hinges distribution for fixed, stiff, medium and soft soil

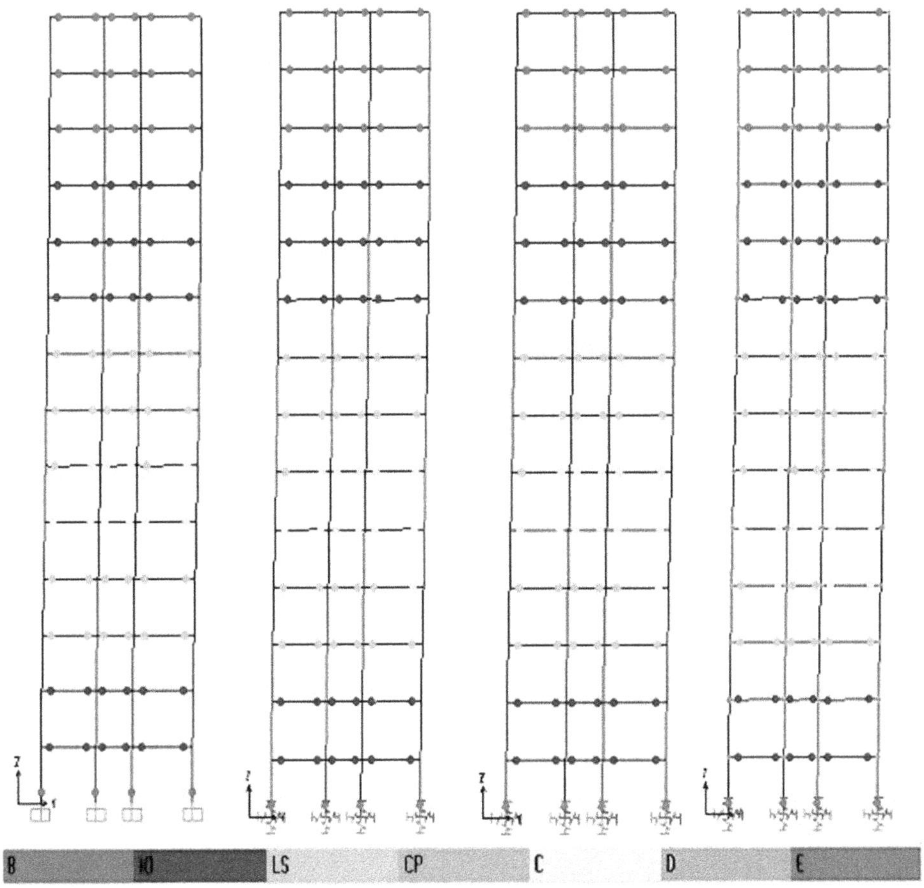

The capacity curves shown in figure 9 represents the relationship between the values of base shear and the top lateral displacement. As the curves show, effects of soil structure interaction are not obvious in the elastic stage of the curve but it is more apparent in the nonlinear stage. The authors deduce that with a given soft soil the structure tends to reflect a lower capacity than the structures with higher values of wave shear.

The curves of figure 10 represents the inter story drifts for different types of soil. It is obvious from this curves that SSI has an impact on structural behavior especially in the case of the inter story drift of the bottom floors. The authors deduce that the relative displacement increases with softer foundation soil.

Figure 11 shows that formation of plastic hinges changes when considering SSI effects, the yellow hinges represents the step that the structure collapsed, the position changes and the number of hinges increases with lower values of wave shear. In addition, the authors deduced that the hinges that represents the collapse statue in the model with a fixed base appear in the twelfth step, while in the models with horizontal and rocking springs, collapse hinges are formed in the eleventh step.

CONCLUSION

The study presented in this chapter was aimed to characterize the importance of taking account of soil-structure interaction and to show that structures are vulnerable to the changing of soil parameters. Introducing SSI by adding spring-dashpot elements in a 3D model has an effect on the behavior of the high-rise building, for a soft soil the building tend to have a lower capacity and collapses quickly than a building with a fixed base. As the authors discussed above, the lateral displacement and the inter story drift increases as the soil rigidity decreases. Neglecting the effect of SSI affects the structural behavior in an unconservative way. It is more relevant to introduce this effect and to do some changes in structural modelization procedures.

REFERENCES

ATC. (1996). 40, Seismic evaluation and retrofit of concrete buildings. Applied Technology Council, report ATC-40.

Avilés, J., & Pérez-Rocha, L. E. (2003). Soil–structure interaction in yielding systems. *Earthquake Engineering & Structural Dynamics*, *32*(11), 1749–1771. doi:10.1002/eqe.300

Avilés, J., & Suárez, M. (2002). Effective periods and dampings of building-foundation systems including seismic wave effects. *Engineering Structures*, *24*(5), 553–562. doi:10.1016/S0141-0296(01)00121-3

Belgasmia, M., & Moussaoui, S. (2013). Comparison of static pushover analysis in the case of small and large deformation with time history analysis using flexibility-based model for an existing structure. *International Journal of Current Engineering and Technologies*, *3*(2), 655–665.

Building Seismic Safety Council (US). (1997). *Federal Emergency Management Agency, & Applied Technology Council. NEHRP guidelines for the seismic rehabilitation of buildings* (Vol. 1). Federal Emergency Management Agency.

CBA 93. (1993). *Code du Béton Algérien*. CGS Alger.

Chopra, A. K., & Goel, R. K. (2002). A modal pushover analysis procedure for estimating seismic demands for buildings. *Earthquake Engineering & Structural Dynamics*, *31*(3), 561–582. doi:10.1002/eqe.144

Ciampoli, M., & Pinto, P. E. (1995). Effects of soil-structure interaction on inelastic seismic response of bridge piers. *Journal of Structural Engineering*, *121*(5), 806–814. doi:10.1061/(ASCE)0733-9445(1995)121:5(806)

Computers and Structures, Inc. (2000). Integrated structural analysis and design software, SAP2000 version 14.2.2. Berkeley, CA: Author.

Gazetas, G. (1990). Formulas and Charts for Impedances of Surface and Embedded Foundations. *Journal of Geotechnical Engineering*, *117*(9), 1363–1381. doi:10.1061/(ASCE)0733-9410(1991)117:9(1363)

Jennings, P. C., & Bielak, J. (1973). Dynamics of building-soil interaction. *Bulletin of the Seismological Society of America, 63*(1), 9–48.

Kausel, E. (2010). Early history of soil–structure interaction. *Soil Dynamics and Earthquake Engineering, 30*(9), 822–832. doi:10.1016/j.soildyn.2009.11.001

Liping, L., Wenjin, G., Qiang, X., Lili, B., Yingmin, L., & Yuntian, W. (2012). Analysis of Elasto-Plastic Soil-Structure Interaction System Using Pushover Method. *Proceedings of The Seminar on The 15th World Conference on Earthquake Engineering.*

Marguerre, K. (1931). Druckverteilung durch eine elastische Schicht auf starrer rauher Unterlage. *Ingenieur-Archiv, 2*(1), 108–117. doi:10.1007/BF02079817

Meek, J., & Veletsos, A. S. (1974, February). Simple models for foundations in lateral and rocking motion. In *Proceedings of the 5th World Conf. on Earthquake Engineering* (pp. 2610-2631). Academic Press.

Mylonakis, G., & Gazetas, G. (2000). Seismic soil-structure interaction: Beneficial or detrimental? *Journal of Earthquake Engineering, 4*(03), 277–301. doi:10.1080/13632460009350372

NIST. (2012). *GCR 12-917-21 (2012) Soil-structure interaction for building structures.* US Department of Commerce.

Reissner, E. (1936). Stationäre, axialsymmetrische, durch eine schüttelnde Masse erregte Schwingungen eines homogenen elastischen Halbraumes. *Ingenieur-Archiv, 7*(6), 381–396. doi:10.1007/BF02090427

R.P.A 99. (2003). *Règles Parasismiques Algériennes 1999.* Algiers: Centre National de Recherche Appliquée en Génie Parasismique.

Veletsos, A. S., & Meek, J. W. (1974). Dynamic behavior of building-foundation systems. *Earthquake Engineering & Structural Dynamics, 3*(2), 121–138. doi:10.1002/eqe.4290030203

Wolf, J. P., & Obernhuber, P. (1985). Non-linear soil-structure-interaction analysis using dynamic stiffness or flexibility of soil in the time domain. *Earthquake Engineering & Structural Dynamics, 13*(2), 195–212. doi:10.1002/eqe.4290130205

ADDITIONAL READING

Council, B. S. S. (2009). *NEHRP recommended seismic provisions for new buildings and other structures (FEMA P-750).* Washington, DC: Federal Emergency Management Agency.

Datta, T. K. (2010). *Seismic analysis of structures.* John Wiley & Sons. doi:10.1002/9780470824634

Deierlein, G. G., Reinhorn, A. M., & Willford, M. R. (2010). Nonlinear structural analysis for seismic design. *NEHRP seismic design technical brief, 4*, 1-36.

Filiatrault, A. (2013). *Elements of earthquake engineering and structural dynamics. Presses inter Polytechnique. GCR.* N. Applicability of Nonlinear Multiple-Degree-of-Freedom Modeling for Design.

Kausel, E., Roesset, J. M., & Christian, J. T. (1976). Nonlinear behavior in soil-structure interaction. *Journal of Geotechnical and Geoenvironmental Engineering, 102*(ASCE# 12579).

Vrettos, C. (1999). Vertical and rocking impedances for rigid rectangular foundations on soils with bounded non-homogeneity. *Earthquake Engineering & Structural Dynamics, 28*(12), 1525–1540. doi:10.1002/(SICI)1096-9845(199912)28:12<1525::AID-EQE879>3.0.CO;2-S

Wolf, J. (1985). *Dynamic soil-structure interaction (No. LCH-BOOK-2008-039).* Prentice Hall, Inc.

KEY TERMS AND DEFINITIONS

Capacity Curve: Curves generated by exposing a detailed structural model to one or more lateral load patterns and then increasing the magnitude of the total load to generate a nonlinear inelastic force-deformation relationship for the structure at a global level.

Dashpot: Is a mechanical device, a damper that resists motion via viscous friction.

Fixed Base: A combination of rigid foundation elements on a rigid base.

Free Field Ground Motion: Motions that are not affected by structural vibrations or the scattering of waves at and around the foundation.

Impedance Functions: Represents the frequency dependent stiffness and damping characteristics of soil foundation interaction.

Radiation Damping: Damping associated with wave propagation into the ground away from the foundation.

Soil-Structure Interaction: Is a term that describes the effect and the stress exchanging between the soil and the structure.

Chapter 8
Earthquake Resistant Design: Issues and Challenges

Md. Farrukh
Birla Institute of Technology, India

Nadeem Faisal
Birla Institute of Technology, India

Kaushik Kumar
Birla Institute of Technology, India

ABSTRACT

In the long history of mankind's existence, nature's forces have influenced human existence to a great extent. Of all natural disasters, the least understood and most destructive are earthquakes. Their claim of human lives and material losses constantly force people to search for better protection, still a great challenge for engineers and researchers worldwide. Although important progress has been done in understanding seismic activity and developing buildings technology, a better way of protecting buildings on large scale is still in search. The essential features of earthquake resistance structure are stable foundation design, regularity, ductility, adequate stiffness, redundancy, and ruggedness. The chapter focuses on increasing the knowledge dictum of earthquake resistant design and discusses the various sorts of issues and challenges. It also presents a wide view on optimization techniques that are required to be done in the latest technology currently in practice so as to achieve the optimum design techniques.

INTRODUCTION

Every year thousands or more than that earthquakes are felt by a human being worldwide out of which some range from very small intensity felt only by few persons to great intensities that destroy the whole cities (Federal Emergency Management Agency, 2010). Earthquakes have always been the major potential source of damage and casualties for humankind when compared with other natural hazards such as Tsunami, Cyclone, Floods, and Landslide. Although Places varies but the pattern with which earthquakes will strike remains same, since earthquake strikes without giving any sort of warnings, leaving

DOI: 10.4018/978-1-5225-7059-2.ch008

the whole cities in the wreckage and killing thousands of people (Gioncu & Mazzolani, 2011). The loss of human lives, economical losses depends upon the intensity with which grounds shakes, location and depth of earthquakes and the amount with which buildings get settled (Federal Emergency Management Agency, 2010).

People are killed not due to earthquakes but the buildings or the structure made by them does so. Till date, the number of earthquakes that have occurred yet and damages caused by it weather loss of human lives or financial losses are worse in areas which are densely populated. Where a large number of buildings have been constructed on a seismically active zone and also the due presence of older buildings that have not been properly refurbished (Henkel et. al., 2008). Earthquakes that occurs in isolated areas which are far away from dense populations have experienced rarely any damage so it is not necessary that earthquakes that are of strong magnitude will cause large destructions, even a small magnitude of earthquakes can create a similar scenario when it strikes a densely populated area (Gioncu & Mazzolani, 2011).

Earthquakes, are usually caused by number of factors, such as due to movement of magma within the volcanoes or due to expulsion in the earth crusts, but the earthquakes, which is initiated due to these factors are of very small intensity and causes very less damage, whereas major earthquakes are caused due to sudden movement of tectonic plates over the fault plane deep inside the earth's mantle because of which ground motion is experienced on the earth surface since transmission of energy waves from the bedrock takes place (Hamburger & Gumpertz, 2009),because of this ground motion earthquake forces are generated as structures dynamically respond to induced ground motion (King, 1998, Lindeburg & McMullin,2014). This action makes earthquake forces completely different from other imposed loads that act on any structures. Thus due the inelastic characteristics of the structures the earthquake forces can be influenced directly which give designer an opportunity to influence the earthquake forces to large extent by properly distributing lateral forces, by maintaining the regularity of building in plan and elevation, by limiting the developments of unwanted response mode and by selecting proper response mode (King, 1998).

MAIN PURPOSE OF DESIGN

Structure Design has always been one of the most interesting as well as creative segments in earthquake engineering for many decades. The main purpose of structure design is to create a structure, which is technical, as well as economically sound and it has the ability to resist and transmit the different forces acting on it, and the deformation induced by these forces should be within limits (Kargahi & Anderson, 2004, Rizk, 2010). Therefore, the major purpose of structure design is to construct the optimal structure. According to many authors optimisation not only means to consider the initial cost of buildings but it should focus on the benefits that have to come from structure in its service period, maintenance cost, damage cost and there is always been probabilities that during service period structure will suffer damage and failure (Rosenblueth, 1974, Dalton. et al. 2013). Gallagher pointed that even today in the field of modern analysis technology; optimum design technology, has not been properly accepted in practical design and it is difficult to determine the causes responsible for slow acceptance of design technologies in practical design fields (Gallagher, 1973).

In 1980, Esteva discussed the objectives and nature of earthquake resistant design as; *Engineering Design, which is rooted in our society, needs to be optimized.* It comprises of taking into consideration the alternate line of action, analyzing its consequences and making a foremost choice. In earthquake engineering, all alternate line of action comprises the adoption of both a seismic design criterion and a structural system, whereas analyzing its consequences involve measuring the structural response and estimating the cost of damage. The foremost choice is based on comparing initial, repair and maintenance cost. Esteva in its publications pointed that achievement in ongoing optimum earthquake resistant design technologies requires more than dimensioning structural members, studies of the mechanical behavior of structures and doing independent structure analysis. It requires a clear understanding of each of the above aspects and their relationship in the total design phase (Esteva, 1980). In civil engineering, design process usually has several phases (Beretro et. al., 1996, Beretro, 1996), out which four most common phases are;

1. Planning Phase
2. Preliminary Design Phase
3. Final Design Phase
4. Detailing Phase

The most difficult and the technical problems, which is associated with carrying out design process, is the formulation of design criteria (Bertero 1996). Design Criteria is basically a rule and guidelines, which must be, ensure so that the objectives of design are satisfied (Biggs, 1986). The major objectives are (1) Safety- it is the most important objectives since any damage to structure endangers human life. (2) Structure Performance- structural design must be satisfactorily well so that it can fulfill the purpose for which it is designed and is capable to transmit and resist forces acting on it. (3) Economy- after satisfying two objectives structure then design for minimum cost.

The modern Seismic design has its origin in late 1920's at that time seismic design mainly comprises of considering 10% to 15% of building weight as lateral forces on the buildings and it is applied uniformly on the buildings and the analysis are done (Henkel et. al., 2008, Priestley, 2013). In 1960's an instrument was developed, which is capable of recording the actual ground motion generated due to earthquake named as Accelerograph, they were installed within buildings at different levels due to which it becomes easier to understand the actual dynamic response of building when it is subjected to earthquakes (Hudson, 1979, Rupakhety, 2010). After knowing, the actual ground, motion as an input a new technology was developed, called inelastic integrated history analysis, with this it becomes clear that many buildings, which were designed, using earlier codes, have inadequate strength to resist the design level earthquakes without undergoing any severe damage. However, after doing an observation on the buildings it was found that this lack of strength did not usually result in buildings failure when they are subjected to earthquakes. Provided that due to development of inelastic deformation, strength can be maintained without any severe degradation. Conversely, the buildings, which have experienced severe strength loss, become unstable and then collapsed (Henkel et. al., 2008). With this knowledge, the designer's thought that *key to successful earthquake engineering design* lies on detailing of structural members so that preferable post-elastic mechanisms are properly identified and formation of undesirable global response mode is prevented. Therefore, a satisfactory earthquake engineering design is one where the designer, dominates the building response and took control over it by selecting favorable response mode, selecting acceptable zone for inelastic deformation and suppressing the formation of undesirable response modes (Henkel et. al., 2008, Booth, 2014).

ISSUES INVOLVED IN SEISMIC DESIGN

Bertero (1992) in its publication pointed out that, the principle issue that remains unsolved for the improvement of seismic design are related to these three elements (Bertero, 1992, Elnashi & Di, 2008):

1. Earthquake Input
2. Demands on the Structure

Earthquake Input

Earthquake input element consists of following interrelated issues: Design Earthquakes, Design Criteria, and Design Methodology. Design Earthquakes are depended on design criteria since design criteria are the key parameters, which define the general philosophy of earthquake design that, have been, accepted worldwide. Conceptually, design earthquakes are defined as those earthquake ground motions, out of all possible earthquake ground motion at the sites, which will drive any structure to its critical response (Bertero, 1992, Hamada, 2014). In actual practice, the application of these concepts is associated with many difficulties because, firstly, it is difficult to predict the actual dynamic characteristics of earthquake ground motions, which, is going to occur (Harris et al. 2011), and, secondly due to various performance levels (Limit States) that control the design, the critical response of structural system will vary (Bertero, 1992). Seismic codes have specified design earthquakes in terms of peak site accelerations, site intensity factors, and building code zone. However according to Applied Technology Council, trust on these indices is generally inadequate, therefore, methods using *Smoothed Linear Elastic Design Response Spectra* and *Ground Motion Spectra* (Trifunac, 2012) which is based on effective peak accelerations is recommended (Applied Technology Council, 1978).

Estimation of Potential Demands

The main uncertainties involved in the estimation of reliable demands are because of difficulties in predicting:

1. **Critical Seismic Excitation:** In the world, there are many seismically active site, designers, interested in designing any structure on these sites, however, sometimes it becomes difficult for them to decide for what ground motion design is to proceed because there are very few recorded ground accelerations available. Under this circumstances, he or she studies the somewhere else records having similar geographical features, and on the basis of that records he proceeds his design in a hope that the response spectrum established due to recorded ground acceleration will represent the excitation that is likely to happen at that site (Wang & Drenick, 1977, Katsanos et al. 2010). Therefore, this type of problems arises due to lack of properly, available earthquake design ground motions.
2. **Internal Forces and Deformations:** During earthquake, the estimations of internal forces like stresses generated cannot be estimated due to the fact that earthquake forces are generated by dynamic response of the structures to the earthquake induced ground motion and this make earthquake actions fundamentally different from any other imposed load. Due to this we are not able to estimate or analyse the stresses generated or for the mere fact we are not able to understand how

does the structure responses, how much internal forces are generated at each phase of structure like ground phase, middle phase or top phase and how much deflection or deformation is caused in the structure due to the earthquake.

3. The state of entire non-structural, superstructure, soil foundation when the critical earthquake ground motion occurs that is not properly selecting the mathematical models that are to be analyzed (Bertero, 1996, Kannan, 2014).

There, are some other issues also that need be to figure out, such as,

1. Even today as a matter of convenience, the basic of seismic design is still on an assumption that modified elastic or elastic acceleration response spectrum helps in establishing the best performance of a structure (Priestley, 1993, Priestley 2013). Even though the limitations of the above approach are clearly known but these approaches are still in used. The Limitations are:
    a. For any structure, the response at the moment of peak base shear is considered. Also, since durations effects are period dependent so from the above statement we find that the short period structure will suffer a great number of response cycles than long period structures.
    b. The peak displacement response of the inelastic and elastic system is very much variable and has a complex relationship than commonly accepted. Various rules, such as equal displacement and equal energy rules are employed without much logic and consistency.
2. In many, places it has been found that analysis and design are performed by two different specialists. The analyst focused on modeling the structures and doing lateral force analysis whereas designers are responsible for determining member sizes, properties of members, reinforcement quantities and detailing based on the results provided by analysist. Typically, the analysis is, more focused on analytical process rather than simulations of member characteristics. Therefore, due to the separation of analysis and design, it is appearing that analysis controls the design process rather than the reverse. In addition, there is room to examine the detailing practice and current design, using methods much of which, extrapolated, from static load considerations (Priestley, 1993)
3. Many researchers have focused on displacement based design rather than strength based (Priestley et al. 2007), but the processes which have been described are still strength based because of convincing and lack of design alternatives. Moehle (1992), in his research, pointed out the merits of displacement based design and ductility based design, but while doing comparison his starting point is still given stiffness and strength (Moehle, 1992).
4. In seismic design, enormous approximations are the involved but codified approach for seismic design perhaps limits that approximation to large extent, since sophisticated analytical techniques specified by codes are being accepted as a matter of routine in common design practices now a day. However, it appears that codified approach for seismic design bears a tenuous relationship to the expected performance of the structure (Priestley, 1993, Magenes & Penna, 2009).
5. Analysis, that is performed on the basis of elastic assumptions causes difficulties in finding out the values of structural parameters such as stiffness and damping, since, change in the properties of materials are not take into account during the progress of earthquake (Duggal, 2007).
6. Geological conditions and Soil-structure interaction have very great effects on structural performance, but still today, there is no precise method, that accurately incorporates these effects (Duggal, 2007).

CHALLENGES IN SEISMIC DESIGN

Safety is the fundamental right of communities leaving in build world (Universal Declaration of Human Rights, 1948). As per article 25 of the United Nations Universal Declaration of Human Rights 1948, "Everyone has the right to a standard of living adequate for the health and well-being of himself and of his family, including foods, clothing, and housing…" (Universal Declaration of Human Rights, 1948), but experiences from recent earthquakes in Haiti (2010), which kill about 222,570 people (The Statistical Portal, 2016), or in Japan (2011), that killed 15,894people (Oskin, 2017) and many such more clearly reveals that we are still far away from reaching this goal. In many cases, it seems that both design and detailing are done as per the code provisions but still damage occurred, that means something new has happened which was not predicted in the design practice. Therefore, the questions arise, what should be the exact reason for this unsatisfactory behavior:

- Imprecision in modeling ground motion?
- Imperfection in design concept?
- Not properly concerning ductility demand?

Today, the world of specialist has divided view on above question, some specialist consider that available design philosophy is proper only some improvement in construction detail is required whereas according to many another specialist, above mentioned questions is the real problem and some modification is required in design concept (Gioncu & Mazzolani, 2002). Therefore, in the light of this debate, the challenges in seismic design are presented below:

1. **Challenges in Concept**
    a. The important step toward the development of design and analysis methodologies is the study of structural response during an earthquake. Earlier due to less number of records in respect of a structural response, during a severe earthquake the design methods, which was, developed where based on easily understood hypothesis and having very little possibilities of verifying their reliability. Nowadays, due to the availability of a large number of instrumentations all over the world, it becomes possible to measure the ground motion for different site conditions. This, give the possibilities to emphasize an all-new aspect which was previously not considered in the design concept that is the difference in ground motion for far-field and near-field earthquakes (Gioncu &Mazzolani, 2002). The near-field regions are that area which extended to several kilometers from the fault rupture zone. Earlier, the majority of ground motions recorded is for far-field regions, therefore; the available concept refers to this earthquake type only. However, after the earthquake in Kobe (Japan) and Northridge (USA) it is, found that, these towns were, situated in the near-field regions, which gives an alarm that the earlier recorded ground motion for the far-field regions, cannot be used to study the actual earthquake response of structure in near-field regions. Such as (Gioncu et. al., 2000);
    b. In the near-field regions, the ground motion has a noticeable coherent pulse in displacement and velocity time history, whereas, acceleration time history has a distinct low frequency with short duration ground motion. While, in far-field regions, the records of velocity, displace-

ment, and accelerations have the cyclic movement characteristics with long duration ground motions (Davoodi et al. 2013).

c. During Kobe and Northridge earthquakes, it is found that the recorded velocities at the soil level were 150-200 cm/sec while for far-field regions the velocity value did not exceed 30 to 40 cm/sec. Therefore, it is found that in case of near-field regions, velocity is the key parameters in design concept whereas for far-field regions acceleration is the dominant parameters.

d. As compare to horizontal component, the vertical component is greater in near-field regions because of direct propagation of waves without any important modifications (Shrestha, 2015).

2. **Challenges in Design:** From the above-mentioned concept in respect of differences in ground motion in near and far-field regions, there are some major refinements in design concept;

   a. In near field regions, due to pulse characteristics of loads and a short period of ground motions, the case of high vibration mode increases as compared to far-field regions. When a structure is, subjected to pulse accelerations a wave is propagated from the structure, which causes large localized deformation in structure as well causes, inter-story drifts. Therefore, in this type of situations the design methodologies, which is, based on the single degree of freedom systems, is not sufficient to describe the exact behavior of structures (Gioncu &Mazzolani, 2002, Davoodi et al. 2012). Example can be understood from the earthquakes of Northridge (6.7 Magnitude Earthquake, Los Angeles, California, US, 1994) & Kobe (6.9 Magnitude Earthquake, Hashin, Japan, 1995) where great amount of damage occurred due to the fact that these towns are located in near field regions. So, the ground model adopted in design methodology on the basis of ground motions recorded in far field regions cannot be used to describe in proper manner the earthquake actions in near field regions.

   b. Due to impulse characteristics of loads and high velocity, the required ductility demand of structures may be very high. So, careful examination is needed to be done on the inelastic properties of the structure. At the same time, in near-field regions the short duration ground motion is most favorable factors, therefore a balance between the ductility demands and the effects of short duration of ground motion must be carefully analyzed (Davoodi et al. 2012).

   c. It is a pressing task for researchers to determine the ductility demand as a function of velocity actions since, due to impulse characteristics of load ductility demand increases but the same time due to great velocities there is a significant decrease in response.

   d. In case of steel structures, it becomes impossible sometimes to take advantages of plastic behavior of structures because of high earthquake velocity. Therefore, across the ductile fracture, the variation of energy dissipation is considered (Gioncu & Mazzolani, 2002).

3. **Challenges in Construction:** The various challenges associated with earthquake-resistant construction are;

   a. The project engineer whose is responsible for the construction site is unaware of the philosophy of earthquake resistant design since he/ she is fully dependent on the design provided by designers (Bertero, 1996).

   b. Many countries have revised their design codes and have developed new construction standard but the problem is, they are not strictly implemented in the execution stage (Chaphalkar, 2015).

c. Structures are constructed without giving propersetback distances and spaces between the buildings (Chaphalkar, 2015).
d. Contractors use cheap quality materials and give very less focus on seismic aspects.
e. The design and the detailing provided by designers, many of the time causes problems to engineers in the site, because of complex detailing (Gioncu & Mazollani, 2002).
f. Not having skilled labors and contractors, since they are not aware of seismic constructions.

RECENT ADVANCES IN EARTHQUAKE RESISTANT DESIGN

Recently there have been tremendous growth and developments of technologies which can be used improve earthquake resistant design. Few of them are discussed as below.

1. Shape Memory Alloys
2. Fiber Reinforced Polymer
3. Base Isolation
4. Vibrating Barriers

SHAPE MEMORY ALLOYS

Wiley and Buehler in 1960's developed a new material called Nickel Titanium, which exhibitsvery different properties from the available traditional materials. They found that a large deformed sample of alloys even reaching 8 to 15% of residual strain regain their original position (Fugazza, 2005).Because of this unusual effect, it is named as, shape memory effects and, due to the presence of alloy composition, it is called shape memory alloy (Fugazza, 2005). Later, it was found that it also exhibits the property of superelasticity at high temperature. Therefore, it is seen that shape memory alloys exhibit two unique characteristics (Chang & Araki, 2016);

1. Shape memory effects
2. Superelasticity

Shape memory effects are defined as the ability of alloys to regain its original (initial) shape when heated, and superelasticity is the ability of alloys to allow large deformations with no residual strain (Figure 1) (Chang & Araki, 2016). Because of its unique properties, it has gain lot of attention from the scientific community. Shape memory alloys have been successfully used in many fields such as in aerospace (Hartl & Laogudas, 2007), robotic (Kim et. al., 2006), automobile (Bellini et.al. 2009) and in biomedicals (Bansiddhi et.al. 2008). In recent years researchers have started focusing on using this material in earthquake engineering as an innovative material to protect structures from earthquake damage (Chang & Araki, 2016). There are many families of shape memory alloys having different applications in the field, depending on the phase transformation temperature, for example, iron-nickel-cobalt-aluminum and aluminum manganese shape memory alloys are used in seismic applications as their temperature varies in the range of -50 to 50° C(Chang & Araki, 2016). The first field application of shape memory alloys was in Michigan, where a bridge crack was repaired using this technology(Soroushian, 2001) later,

Figure 1. Shape memory alloys mechanism

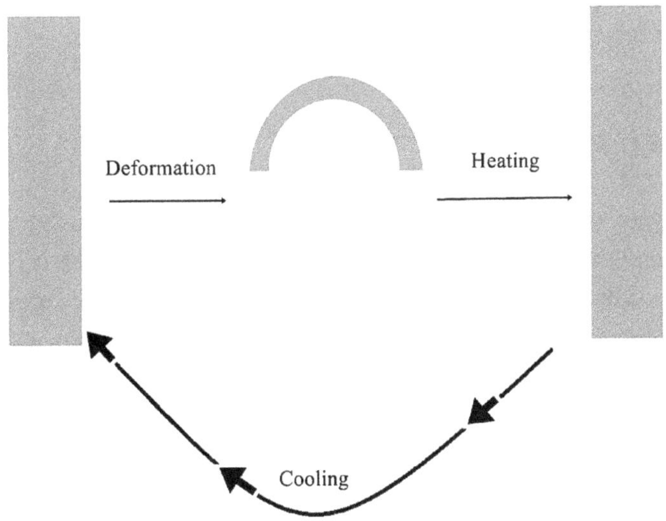

it was used of strengthening many old structures such as, bell tower of San Giorgio church Italy (Indirli & Castellano, 2008) and many Italian heritages which got damaged due to earthquakes (Chang & Araki, 2016). According to modern building codes, the structure needs to go significant non-structural and structural damages in order to dissipate energy during earthquakes. After the earthquakes, the structures that undergo large deformations are either, demolished or repaired; therefore, there are large economic losses in repairing these structures and rebuilding the demolished structures (Chang & Araki, 2016).

After the earthquake in New Zealand (2011) it was seen that approximately 40 to 50% of buildings were declared unusable because of severe damage and more than 500 buildings were demolished. Economic loss was estimated at $40 billion, which is equivalent to 20% of country GDP (Gross Domestic Product) (Chang & Araki, 2016). This draw a special attention to use a building system, which is capable of dissipating energy with minimum damage to structures and return to its original position after the earthquakes(Chancellor et.al., 2014), in this aspects shape memory alloys are the best building system which can be used in concrete and steel constructions to control large deformation caused by earthquakes. In several projects, for reducing permanent deformations, shape memory alloys, have been used in the critical regions of concrete (Saiidi & Wang, 2006). Such as in concrete beams (Abdulridha et.al., 2013); shape memory alloys have been coupled with main bars, in the region where large deformation will occur, shape memory alloys were also used in shear walls (Chang & Araki, 2016); the results, revealed thatthe use of elastic shape memory alloys, within shear walls, reduces large residual displacement. An attempt is also being done to use shape memory alloys in steel(Abolmaali et.al., 2006) and timber connections(Chang et.al., 2013). Traditional steel connection is often associated with local buckling after earthquakes and is costly to repair Ma et.al., (2007) improve the solution by modifying the connections with shape memory alloys bolt. Farhidzadeh and Roofeiin 2011 done analysis on 3-6-9 and 12-storey steel structures following connection as designed by Ma et.al., with 0, 5, 10 and 15%, eccentricities, the result which obtained after applying earthquake forces, shows that the base shear has considerably reduced as compared to normal connections(Rofooei & Farhidzadeh, 2011) Shape memory alloys, can also be used in bracing system since it helps in resolving the problems called pinching effects in the hysteretic loops within a structure(Araki, 2014) after earthquake. According to many research, if shape memory alloys are used in the bracing system, it helps in limiting residual drifts and inter-story within structures during an earthquake (Auricchio et.al, 2006).

FIBRE REINFORCED POLYMER

Many buildings, which are constructed in the seismic prone zone, have now become outdated and they are inadequate to withstand earthquake forces, if those structures are not properly retrofitted, they may get damaged in severe earthquakes (Yamamoto, 1992). Therefore, it is necessary to repair and retrofit these obsolete structures to reduce the risk of damage and casualties. Although there are no such things, which are fully earthquake proof by proper retrofitting, it is some way possible to improve the performance of the structures (Sarker, 2010). In this respect, a new material has been, developed which is capable of improving the seismic performance of structures with minimum cost, called Fibre reinforced polymer. Fibre reinforced polymer is a composite material, comprises of high strength fibre which areembedded in the resin matrix (Sarker, 2010). Fibre reinforced polymer are classified depending on the fibre used (Harichandran et.al, 2000), such as; (a) Glass fiber reinforced polymer (b) Carbon fibre reinforced polymer and (c) Aramid fibre reinforced polymer. Because of its, lightweight, high strength, the fibre reinforced polymer is being in used since 1980's (Harichandran et.al, 2000). In fibre reinforced polymer, fibre is the basic component responsible for carrying loads whereas the matrix material (plastic) transfers the shear (Sarker, 2010). In recent years, fibre reinforced polymer has become the best alternative of steel for the repair and retrofitting of seismic deficient structures, since fibre reinforced polymer has many advantages over steel such as(Sarker, 2010);

1. High stiffness
2. High Strength
3. High Corrosion Resistant
4. Easy to handle and install

Moreover, its high-temperature resistance and environmental resistance have made it the best choice for rehabilitation purpose.

Fibre reinforced polymer material has gained considerable popularity in RC structures, as various investigations (Benzoni et.al., 1996; Masukawa et.al., 1997;Seible et.al., 1997; Lavergne & Lobosssiere, 1997;Saadatmanesh et. al., 1997; Seible et.al., 1999; Mirmiran, &Shahawy, 1997; Fukuyama et.al., 1999;Pantelides et.al., 2000; Bousias et.al., 2004, Harajli et.al., 2006) have been done to study the, efficacy of fibre-reinforced polymer in strengthening of rectangular, square and circular columns; it was found that by confining the columns with fibre reinforced polymer there is a remarkable improvement in ductility and axial strength (Harajli et.al., 2006). Also, in reinforced concrete columns, fibre reinforced polymer sheet prevent longitudinal steel bar from buckling theirs by improving the performance of column under earthquake loading (Bousias et.al, 2004). In recent years, researchers have started focusing on using fibre reinforced polymer materials for retrofitting steel structures also (Sarker et.al., 2010), few types of research have been carried out, to find the potential of fibre reinforced polymer materials to repair as well as retrofit steel structures (Sarker et.al., 2010). In steel structures, fibre reinforced polymer is mainly used for strengthening steel girder, and for repairing corroded steel girders (Sarker et.al, 2010).

Many researches have shown that, by retrofitting damaged steel girder with fibre reinforced polymer materials, the elastic stiffness has increased from 10 to 35% (Gillespie et.al., 1996). Moreover, it also seems that, by repairing steel structure using a fibre-reinforced sheet, the load carrying capacity of steel structure and yield strength has considerably increased since it helps in regaining the lost capacity of the steel section (Sarker et.al, 2010).

BASE ISOLATION

Earthquake is not only a disaster; it is a natural phenomenon that occurs from ground movement. Earthquakes produce surface waves, which results in vibration of the structures and the ground on the top. Depending upon the intensities of these vibrations, it may cause ground buildings and structures to develop cracks, fissures, and settlements. The risk of loss of human life adds a serious dimension to the seismic design, putting an ethical responsibility to structural engineers. During past couple of decades, many new technologies and methods have been developed to reduce earthquakes forces that act on structures or absorbing a part of these seismic energies. One such system which is widely accepted and regarded as seismic protection systems is Base Isolation. It is also known as Seismic Base Isolation or Base Isolation System (Pressman, 2007). It is a popular way of protecting a structure against earthquake forces and ranks high in widely accepted systems of seismic protection system in earthquake-prone areas. It reduces the effect of an earthquake by isolating the structure from ground waves that occur due to surface waves resulting from earthquakes. Seismic isolation can be regarded as a part of design strategy since it uncouples the structure from negative effects of the ground motion. The word 'isolation' here means reduced interaction between ground and structure. When seismic isolation system is located under the structure, it is termed as "Base Isolation." Besides providing an isolation system to the structure, base isolation also acts as an energy dissipater, henceforth reducing the transmitted acceleration into the structure (Webster, 1994). The decoupling allows the structure to behave flexibly which improves its responses to an earthquake.

Concept of Base Isolation

Base Isolation concept could be explained through the example of a structure resting on frictionless rollers. If the ground shakes the rollers uninhibitedly roll, however, the structure does not move. Thus, no force is exchanged to the structure if there is a shaking in the ground, and essentially, the structure does not encounter the earthquake. Presently, if a similar structure is lying on flexible pads that give protection against lateral movements, a portion of the ground shakings will be moved to the structure above (Datta, 2010). These flexible pads are termed 'base isolators' and the structures secured through these devices are termed as Base-Isolated Buildings. The primary focus of the Base-Isolation is that it presents flexibility in the structures.

Challenges in Base Isolation

- A careful and thorough study is needed to distinguish most reasonable sort of device for a specific building or a structure.
- Also, base isolation may not be reasonable for every buildings and structures.

- The majority of the buildings which are suitable for Base isolations are low-to-medium-rise buildings which are resting on hard soil underneath.
- High rise building resting on soft soils are generally not appropriate for Base-Isolations.

Principle of Base Isolation

The basic principle of base isolation is to change the response of the building so the ground may move underneath the building without transmitting these responses and motions into the structure. A building which is rigid in nature will have zero periods. At the point when the ground will move, the acceleration incited in the building will be nearly equivalent to ground displacement and henceforth, there will be zero relative displacements between the structure and the ground. The ground and the structure will move about the similar amount. A structure that is perfectly flexible will have an infinite period. For such kind of structures, when the ground beneath the structure moves, there will be no acceleration or zero acceleration initiated in building and the relative displacement between the ground and the building will be equivalent to ground displacement (Reitherman, 2012). So, inflexible structures, the ground will move and buildings will not (Figure 2).

VIBRATING BARRIERS

Vibrating barriers (ViBa) are massive structures which are tuned to lessen the vibrations of existing structures in case of seismic activity (Figure 3). Structure-soil interaction (SSI) and structure-soil-structure (SSSI) govern functionality of ViBa. These processes involve the interaction between the ground and various buildings due to seismic activity. ViBa is designed to minimize these interactions, reducing the magnitude of vibrations transferred into infrastructure (Cacciola & Tombari, 2015)

Design and Implementation

Components

- Spring-like oscillator to absorb shockwaves
- Massive anchoring foundation made from are cycled waste material or excavated soil

Figure 2. Principle of base isolation: fixed buildings vs isolated buildings

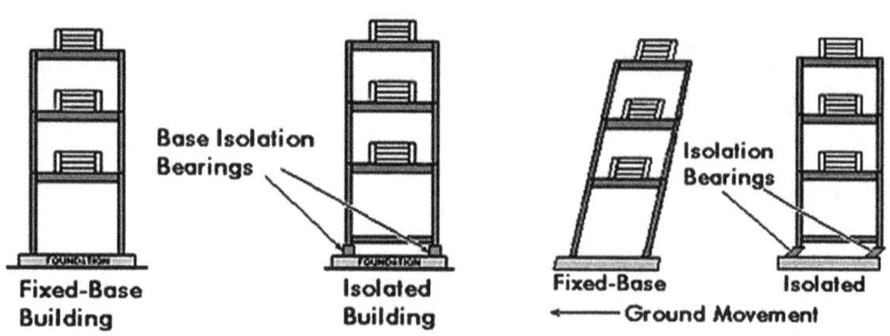

Figure 3. Concept of vibrating barriers

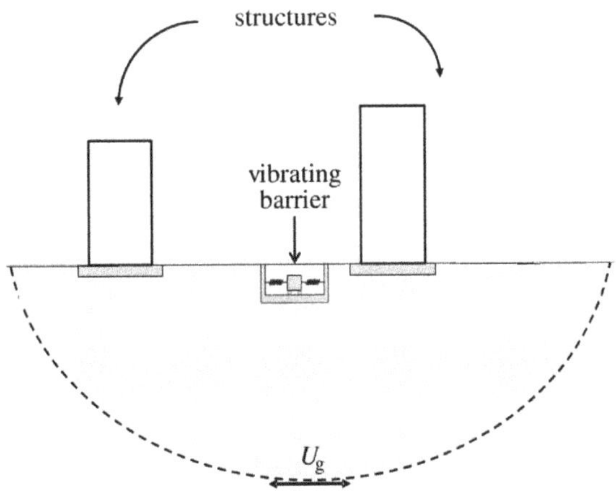

Placement

- Buried below ground
- Located in close proximity but detached from buildings rather than being integrated into the foundation

Function

- Incident vibrations from earthquakes cause the oscillator to move back and forth.
- This movement, anchored by the firm foundation, absorbs and disperses energy from seismic waves prior to impacting buildings
- Greatly reduces the magnitude of tremors experienced by protected structures
- Capable of simultaneously protecting multiple buildings rather than one individual structure due to its detached design

Benefits of ViBa

- ViBa reduces chronic damage on structures caused by regular human activities such as mining, transportation, and manufacturing
- The device's foundation can be produced from waste materials or excavated soil
- Minimizes production of air and water pollutants that would be generated if concrete or a less "green" material was used for constructing the foundation

Currently, only prototypes of ViBa have been constructed or tested. While it is not a faultless solution, it certainly has potential to revolutionize the field of earthquake engineering. Further development of the vibrating barrier will render it a viable solution for earthquake defense, minimizing the threats imposed by seismic activities more than ever before. Based on recent analysis of testing and research were done on this system, ViBa technology will greatly simplify the process of protecting cities against future earthquakes and other seismic events, without sacrificing utility, sustainability, and performance.

COMPARISON OF VARIOUS SEISMIC PROTECTION TECHNOLOGIES

Various kinds of seismic protection technologies being currently used and experimented are discussed below with their advantages and disadvantages (Table 1).

SOLUTIONS AND RECOMMENDATIONS

The solutions and recommendations which can provide beneficial to Earthquake resistant design are as follows: -

- The most important thing in creating earthquake resistant design and structures is the need to create awareness among people, designers, constructors, engineers, government, etc. to make them realize the importance of earthquake resistant structures that can prevent human life's as well as economical losses.
- Traditional design approach that is still being used in many underdeveloped countries needs to be upgraded bearing in mind to the latest challenges and consequences faced.
- Existing buildings which are not earthquake resistant needs special research so that technologies and inventions can be developed for them to prevent them from earthquakes bearing in mind these technologies need to be cost-effective.
- Prime importance needs to be given to earthquake resistant structures and design studies in colleges starting from undergraduate level to research level.
- Promotions to research work involving earthquake need to be given importance with proper findings by government and other agencies.
- Active participation of government at ground scale can contribute a significant role in improving earthquake resistant structures.
- Use of new researches like as mentioned above FRPs, Vibrating Barriers, Shape Memory Alloys, Base Isolation should be more and more used in construction of new buildings and structures.

Table 1. Comparison of various seismic protection technologies

Technology	Advantages	Disadvantages
Steel Reinforcement	Ductility; flexibility (able to bend rather than yield under stress)	Lacks sufficient strength to withstand great force
Base Isolation	Minimizes horizontal motion of structures	Must be installed in the original foundation; fails to sufficiently cushion vibrations
Damping Devices	Reduce magnitude of vibration transferred into buildings	Attached directly to structures; vibrations still directly affect buildings
Vibrating Barriers	Detached from buildings; able to significantly reduce seismic wave magnitude; capable of protecting network of multiple buildings	Must have extremely large mass to withstand forces from earthquakes

- Research to reduce cost of Fiber Reinforced Polymers (FRPs), Vibrating Barriers, Shape Memory Alloys, Base Isolation, should be given great emphasis as reduction in costs can lead to wider use of these technologies as presently they are being used in a very small scale due to high cost problem.
- Design codes need to be simplified and upgraded accordingly to suit the parameters.

FUTURE RESEARCH DIRECTIONS

The cost to reduce existing technologies holds the key to the future. Technologies like vibrating barriers, shape memory alloys, fibre reinforced polymers, base isolation is quite high cost driven and using them in existing structures can lead to great rise in cost of structures and henceforth, research which focuses on reducing the cost of these technologies can help in their wider application. A comparative analysis and study can enhance the use and application of these technologies and can provide us an insight to which technology is better in terms of earthquake resistant, easy to implementation, and cost effectiveness. Combining these technologies with one another can also pave a way to how much these technologies gel up with each other in terms of providing better and effective earthquake resistant structures.

CONCLUSION

In this chapter, a broad view of various challenges and issues and also various technologies which can further enhance and improve the earthquake resistant design has been presented. It should be noted that various design practices followed in various countries are unsafe and impractical. If this chapter has a message, it is to simplify the complexities of design and analysis of earthquake resistant design, also codified approach need to be modified according to suit the parameters and simplify the process.

The various technologies like base isolation, fiber-reinforced polymer, shape memory alloys and vibrating barriers need to be given more importance and testing, analysis, modification, design aspects of such technologies should be given prior importance. Technologies which are still in their prototype like vibrating barriers etc. need to be tested in real life scenarios to test and trust their dependency and reliability.

In the end, considering various assumptions and approximations inherent in the seismic design, we must keep the design and analysis process simple enough so that we might understand what we are doing.

ACKNOWLEDGMENT

The authors sincerely acknowledge the comments and suggestions of the reviewers that have been instrumental for improving and upgrading the chapter in its final form.

REFERENCES

Abdulridha, A., Palermo, D., Foo, S., & Vecchio, F. J. (2013). Behavior and modelling of super elastic shape-memory alloy reinforced concrete beams. *Engineering Structures*, *49*, 893–904. doi:10.1016/j.engstruct.2012.12.041

Abolmaali, A., Treadway, J., Aswath, P., Lu, F. K., & McCarthy, E. (2006). Hysteresis behaviour of t-stub connections with super elastic shape-memory fasteners. *Journal of Constructional Steel Research*, *62*(8), 831–838. doi:10.1016/j.jcsr.2005.11.017

Applied Technology Council 3.06 Amended. (1978). *Tentative provisions for the developments of seismic regulations for buildings*. Redwood City, CA: Applied Technology Council.

Araki, Y., Maekawa, N., Shrestha, K. C., Yamakawa, M., Koetaka, Y., Omori, T., & Kainuma, R. (2014). Feasibility of tension braces using Cu–Al–Mn superelastic alloy bars. *Structural Control and Health Monitoring*, *21*(10), 1304–1315. doi:10.1002tc.1644

Auricchio, F., Fugazza, D., & DesRoches, R. (2006). Earthquake performance of steel frames with nitinol braces. *Journal of Earthquake Engineering*, *10*(1), 45–66. doi:10.1080/13632460609350628

Bansiddhi, A., Sargeant, T., Stupp, S., & Dunand, D. (2008). Porous nickel titanium for bone implants: A review. *Acta Biomaterialia*, *4*(4), 773–782. doi:10.1016/j.actbio.2008.02.009 PMID:18348912

Bellini, A., Colli, M., & Dragoni, E. (2009). Mechatronic design of a shape memory alloy actuator for automotive tumble flaps: A case study. *IEEE Transactions on Industrial Electronics*, *56*(7), 2644–2656. doi:10.1109/TIE.2009.2019773

Benzoni, G., Ohtaki, T., & Priestley, M. J. N. (1996). *Seismic Performance of A Full Scale Bridge Column As Built and As Repaired*. Structural Systems Research Project, Report No. SSRP-96/07 University of California, San Diego.

Bertero, R. D., & Bertero, V. V., & TeranGilmore, A. (1996). Performance-based Earthquake Resistant Design Based on Comprehensive Philosophy and Energy Concepts. In *Proceeding 11th World Conference on Earthquake Engineering, 11WCEE 1996*. Acapulco, Mexico: Elsevier Science Ltd.

Bertero, V. V. (1992). Major issues and future directions in earthquake-resistant design. *Proceeding 10th World Conference on Earthquake Engineering 10WCEE,* 6407-6444.

Bertero, V. V. (1996). State of the art report on: Design Criteria. In *Proceeding 11th World Conference on Earthquake Engineering 11WCEE*. Acapulco, Mexico: Elsevier Science Ltd.

Biggs, J. M. (1986). *Introduction of structural engineering*. Englewood Cliffs, NJ: Prentice Hall.

Booth, E. (2014). *Earthquake Design Practice for Buildings* (Vol. 53). Earthquake Design Practice for Buildings. doi:10.1017/CBO9781107415324.004

Bousias, S. N., Triantafillou, T. C., Fardis, M. N., Spathis, L., & O'Regan, B. A. (2004). Fiber Reinforced Polymer Retrofitting of Rectangular Reinforced Concrete Columns with or without Corrosion. *ACI Structural Journal*, *101*(4), 512–520.

Cacciola, P., & Tombari, A. (2015). Vibrating barrier: a novel device for the passive control of structures underground motion. *Proceeding A of the Royal Society*, *471*(2179). 10.1098/rspa.2015.0075

Cacciola, P., & Tombari, A. (2015). Vibrating barrier: a novel device for the passive control of structures underground motion. *Proceeding A of the Royal Society*, *471*(2179). 10.1098/rspa.2015.0075

Chancellor, N. B., Eatherton, M. R., Roke, D. A., & Akbaş, T. (2014). Self-centering seismic lateral force resisting systems: High performance structures for the city of tomorrow. *Buildings*, *4*(3), 520–548. doi:10.3390/buildings4030520

Chang, W. S., & Araki, Y. (2016). Use of Shape Memory Alloys in Construction; A Critical Review. *Proceedings - Institution of Civil Engineers*, *169*(CE2).

Chang, W. S., Murakami, S., & Komatsu, K. (2013). Potential to use shape memory alloy in timber dowel-type connections. *Wood and Fiber Science*, *45*(3), 330–334.

Chaphalkar, M. (2015). *Earthquake resistant structures: Challenges Un-developed, Developing and Developed Countries.* Retrieved from https://www.linkedin.com/pulse/earthquake-resistant-structures-challenges-developing-chaphalkar-pe

Dalton, S. K., Atamturktur, S., Farajpour, I., & Juang, C. H. (2013). An optimization based approach for structural design considering safety, robustness, and cost. *Engineering Structures*, *57*, 356–363. doi:10.1016/j.engstruct.2013.09.040

Datta, T. K. (2010). *Seismic Analysis of Structures*. John Wiley and Sons. doi:10.1002/9780470824634.ch9

Davoodi, M., Jafari, M. K., & Hadiani, N. (2013). Seismic response of embankment dams under near-fault and far-field ground motion excitation. *Engineering Geology*, *158*, 66–76. doi:10.1016/j.enggeo.2013.02.008

Davoodi, M., Sadjadi, M., Goljahani, P., & Kamalian, M. (2012). Effects of Near-Field and Far -Field Earthquakes on seismic Response of SDOF System Considering Soil Structure Interaction. *Earthquake Engineering & Structural Dynamics*, *30*(12), 1769–1789. doi:10.1002/eqe.92

Duggal, S. K. (2007). *Earthquake Resistant Design of Structures*. Oxford University Press.

Elnashai, A. S., & Di Sarno, L. (2008). *Fundamentals of Earthquake Engineering. Fundamentals of Earthquake Engineering*. Academic Press. doi:10.1002/9780470024867

Esteva, L. (1980). Design: General. In *Design of Earthquake Resistant Structures* (pp. 54–97). New York: John Wiley & Sons.

Federal Emergency Management Agency. FEMA P-749. (2010). Earthquake resistant design concept. U.S. Department of Homeland Security, National Institute of Building Sciences Building Seismic Safety Council.

Fugazza, D. (2005). *Use of shape memory alloys device in earthquake engineering: Mechanical properties, Advanced Constitutive Modelling and Structural Applications* (Doctoral dissertation). University of Pavia.

Fukuyama, H., Suzuki, H., & Nakamura, H. (1999). Seismic retrofit of reinforced concrete columns by fiber sheet wrapping without removal of finishing mortar and side wall concrete. Fiber Reinforced Polymer Reinforcement for Reinforced Concrete Structures, ACI Report No SP-188.

Gallagher, R. H. (1973). *Introduction, Optimum structural design* (R. H. Gallagher & O. C. Zienkiewicz, Eds.). New York: John Wiley & Sons.

Gillespie, J. W., Mertz, D. R., Kasai, K., Edberg, W. M., Demitz, J. R., & Hodgson, I. (1996). Rehabilitation of Steel Bridge Girders: Large Scale Testing. *Proceeding of the American Society for Composites 11th Technical Conference on Composite Materials*, 231-240.

Gioncu, V., Mateescu, G., Tirca, L., & Anastasiadis, A. (2000). Influence of the type of seismic ground motions. In F. M. Mozzolani (Ed.), *Moment Resisting Connections of Steel Buildings Frames in Seismic Areas* (pp. 57–92). London: E& FN Spon.

Gioncu, V., & Mazzolani, F. (2011). Earthquake Engineering and for Structural Design. Spon Press, Taylor & Francis Group.

Gioncu, V., & Mazzolani, M. F. (2002). Ductility of Seismic Resistant Steel Structures, Spon Press, Taylor and Francis Group.

Hamada, M. (2014). Earthquake-resistant design and reinforcement. In Springer Series in Geomechanics and Geoengineering (pp. 75–124). Springer. doi:10.1007/978-4-431-54892-8_2

Hamburger, O., & Gumpertz, S. (2009). Facts for steel buildings; Earthquake and Seismic Design. American Institute of Steel Construction.

Harajli, M.H., Hantouche, E., & Soudki, K. (2006). Stress-Strain Model for Fiber-reinforced Polymer Jacketed Concrete Columns. *ACI Structural Journal*, *103*(5), 672–682.

Harichandran, R. S., & Baiyasi, M. I. (2000). Repair of corrosion damaged columns using FRP Wraps. Report No. RC-1386. Michigan State University.

Harris, R. A., Barall, M., Andrews, D. J., Duan, B., Ma, S., Dunham, E. M., ... Abrahamson, N. (2011). Verifying a Computational Method for Predicting Extreme Ground Motion. *Seismological Research Letters*, *82*(5), 638–644. doi:10.1785/gssrl.82.5.638

Hartl, D., & Lagoudas, D. C. (2007). Aerospace applications of shape-memory alloys. *Proceedings of the Institution of Mechanical Engineers. Part G, Journal of Aerospace Engineering*, *221*(4), 535–552.

Henkel, O., Holl, D., & Schalk, M. (2008). *Seismic design and Drywalling*. Knauf Gips KG.

Hudson, D. E. (1979). *Reading and interpreting, Strong Motion Acclerograms*. Earthquake Engineering Research Institute.

Indirli, M., & Castellano, M. G. (2008). Shape-memory alloy devices for the structural improvement of masonry heritage structures. *International Journal of Architectural Heritage*, *2*(2), 93–119. doi:10.1080/15583050701636258

Kannan, S. (2014). Innovative Mathematical Model for Earthquake Prediction. *Engineering Failure Analysis*, *41*, 890–895. doi:10.1016/j.engfailanal.2013.10.016

Kargahi, M., & Anderson, J. (2004). Structural Optimisation for Seismic Design. *13th World Conference on Earthquake Engineering, 13WCEE 2004.*

Katsanos, E. I., Sextos, A. G., & Manolis, G. D. (2010). Selection of earthquake ground motion records: A state-of-the-art review from a structural engineering perspective. *Soil Dynamics and Earthquake Engineering, 30*(4), 157–169. doi:10.1016/j.soildyn.2009.10.005

Kim, B., Lee, M. G., Lee, Y. P., Kim, Y., & Lee, G. (2006). An earthworm-like micro robot using shape-memory alloy actuator. *Sensors and Actuators. A, Physical, 125*(2), 429–437. doi:10.1016/j.sna.2005.05.004

King, A. (1998). Earthquake loads & Earthquake resistant design of buildings. In Building Research Association of New Zealand, BRANZ 1998. Porirua.

Lavergne, S., & Labossiere, P. (1997). Experimental Study of Concrete Columns Confined by a Composite Jacket under Combined Axial and Flexural Loads. *Proceedings CSCE Annual Conference*, 11-20.

Lindeburg, M. R., & McMullin, P. K. M. (2014). *Seismic Design of Building Structures* (11th ed.). Professional Publications, Inc.

Ma, H., Wilkinson, T., & Cho, C. (2007). Feasibility study on a self-centering beam to-column connection by using the superelastic behavior of shape-memory alloys. *Smart Materials and Structures, 16*(5), 1555–1563. doi:10.1088/0964-1726/16/5/008

Magenes, G., & Penna, A. (2009). Existing masonry buildings: general code issues and methods of analysis and assessment. *Eurocode 8 Perspectives from the Italian Standpoint, 3*, 185–198

Masukawa, J., Akiyama, H., & Saito, H. (1997). Retrofit of existing reinforced concrete piers by using carbon fiber sheet and aramid fiber sheet. *Proceedings of the 3rd Conference on Non-Metallic (FRP) Reinforcement for Concrete Structures*, 411-418.

Mirmiran, A., & Shahawy, M. (1997). Behavior of concrete columns confined by fiber composites. *Journal of Structural Engineering, 123*(5), 583–590. doi:10.1061/(ASCE)0733-9445(1997)123:5(583)

Moehle, J. P. (1992). Displacement-Based Design of RC Structures Subjected to Earthquakes. *Earthquake Spectra, 8*(3), 403–428. doi:10.1193/1.1585688

Oskin, B. (2017). *Japan Earthquake & Tsunami of 2011: Facts and Information, Live Science*. Retrieved from https://www.livescience.com/39110-japan-2011-earthquake-tsunami-facts.html

Pantelides, C. P., Gergely, J., Reaveley, L. D., & Volnyy, V. A. (2000b). Seismic Strengthening of Reinforced Concrete Bridge Pier with FRP Composites. *Proceedings of the 12th World Conference on Earthquake Engineering*, 127.

Pressman, A. (2007). *Architectural graphic standards*. John Wiley and Sons.

Priestley, M., Calvi, G. M., & Kowalsky, M. J. (2007). Displacement-based seismic design of structures. *Building, 23*(33), 1453–1460. doi:10.1016/S0141-0296(01)00048-7

Priestley, M. J. N. (1993). Myths and Fallacies in Earthquake Engineering - Conflicts between Design and Reality. New Zealand National Society for Earthquake Engineering Bulletin, 26, 329-335.

Priestley, M. J. N. (2013). Towards displacement-based design in seismic design codes. *Structural Engineering International: Journal of the International Association for Bridge and Structural Engineering, 23*(2), 111. doi:10.1080/10168664.2013.11985330

Reitherman, R. (2012). *Earthquakes and Engineers: An International History*. Reston, VA: ASCE Press. doi:10.1061/9780784410714

Rizk, A. S. S. (2010). Structural Design of Reinforced Concrete Tall Buildings. *CTBUH Journal*, (1), 34–41.

Rofooei, F., & Farhidzadeh, A. (2011). Investigation on the seismic behavior of steel mrf with shape-memory alloy equipped connections. *Procedia Engineering, 14*, 3325–3330. doi:10.1016/j.proeng.2011.07.420

Rosenblueth, E. (1974). Safety and structural design. In Reinforced Concrete Engineering. New York: John Wiley & Son.s

Rupakhety, R. (2010). *Contemporary issues in earthquake engineering research: processing of accelerometric data, modelling of inelastic structural response, and quantification of near-fault effects* (Doctoral dissertation). University of Iceland.

Saadatmanesh, H., Ehsani, M. R., & Jin, L. (1997). Repair of Earthquake-Damaged RC Columns with FRP Wraps. *ACI Structural Journal, 94*, 206–215.

Saiidi, M. S., & Wang, H. (2006). Exploratory study of seismic response of concrete columns with shape-memory alloys reinforcement. *ACI Structural Journal, 103*(3), 435–442.

Sarker, P., Begum, M., & Nasrin, S. (2011). Fibre reinforced polymers for structural retrofitting: A review. *Journal of Civil Engineering, 39*(1), 49-57.

Seible, F., Innamorato, D., Baumgartner, J., Karbhari, V., & Sheng, L. H. (1999). Seismic retrofit of flexural bridge spandrel columns using fiber reinforced polymer composite jackets. Fiber Reinforced Polymer Reinforcement for Reinforced Concrete Structures, ACI Report No-188.

Seible, F., Priestley, M. J. N., Hegemier, G. A., & Innamorato, D. (1997). Seismic Retrofit of RC Columns with Continuous Carbon Fiber Jackets. *Journal of Composites for Construction, 1*(2), 52–62. doi:10.1061/(ASCE)1090-0268(1997)1:2(52)

Shrestha, B. (2015). Seismic response of long span cable-stayed bridge to near-fault vertical ground motions. *KSCE Journal of Civil Engineering, 19*(1), 180–187. doi:10.100712205-014-0214-y

Soroushian, P., Ostowari, K., Nossoni, A., & Chowdhury, H. (2001). Repair andstrengthening of concrete structures through application of corrective posttensioning forces with shape-memory alloys. *Transportation Research Record: Journal of the Transportation Research Board, 1770*, 20–26. doi:10.3141/1770-03

The Statistics Portal. (2016). *Earthquakes with the highest death toll worldwide from 1900 to 2016*. Retrieved from www.statista.com/statistics/266325/death-toll-in-great-earthquakes/

Trifunac, M. D. (2012). Earthquake response spectra for performance based design-A critical review. *Soil Dynamics and Earthquake Engineering, 37*, 73–83. doi:10.1016/j.soildyn.2012.01.019

Universal Declaration of Human Rights. (1948). *Universal Declaration of Human Rights (Article-25).* Retrieved from http://www.un.org/en/universal-declaration-human-rights/

Wang, P. C., & Drenick, R. F. (1977). Critical seismic excitation and response of structures. Proceedings of 6WCEE, 2, 1040–1045.

Webster, A. C. (1994). *Technological advance in Japanese building design and construction.* ASCE Publications. doi:10.1061/9780872629325

Yamamoto, T. (1992). FRP Strengthening of RC columns for seismic retrofitting. In *Tenth World Conference*. Rotterdam: Balkema.

ADDITIONAL READING

Agrawal, P., & Shrikhande, M. (2006). *Earthquake Resistant Design of Structures.* Prentice Hall of India, Inc.

Chopra, A. K. (2007). *Dynamics Of Structures: Theory and Application to Earthquake Engineering* (2nd ed.). Prentice Hall of India.

Chowdhary, I., & Dasgupta, S. P. (2009). *Dynamics Of Structure and Foundation – A Unified Approach: 2 Applications.* CRC Press, Balkema.

Clough, R. W., & Penzien, J. (1993). *Dynamics of Structures.* New York: McGraw Hill, Inc.

Datta, T. K. (2010). *"Seismic Analysis of Structures", John Wiley & Sons (Asia).* Singapore: Pte Ltd.

Hart, G. C., & Wong, K. (2000). *Structural Dynamics for Structural Engineers.* New York: John Wiley & Sons, Inc.

Kramer, S. L. (1996), "Geotechnical Earthquake Engineering", Prentice Hall, 2007, ISBN 81-317-0718-0.

Lay, T., & Wallace, T. C. (1995) "Modern Global Seismology", Academic press, 1995, ISBN 0-12-732870-X.

Naeim, F., & Kelly, J. M. (1999). *Design of Seismic Isolated Structures: From Theory to Practice.* New York, USA: John Wiley and Sons, Inc. doi:10.1002/9780470172742

Reiter, L. (1989). *Earthquake Hazard Analysis: Issues and Insights.* Columbia University Press.

Wolf, J. P. (1985). *Dynamic Soil-Structure Interaction.* Englewood Cliffs, New Jersey: Prentice Hall.

Chapter 9
Optimized Foundation Design in Geotechnical Engineering

Mounir Bouassida
University of Tunis El Manar, Tunisia

Souhir Ellouze
University of Sfax, Tunisia

Wafy Bouassida
University of Tunis El Manar, Tunisia

ABSTRACT

The design of foundations constitutes a major step for each civil engineering structure. Indeed, the stability of those structures relies on cost-effective and adequately designed foundation solutions. To come up with an optimized design of a foundation, the geotechnical study passes several steps: the geotechnical survey including in situ and laboratory tests, the synthesis of geotechnical parameters to be considered for the design, and the suggestion of foundation solution avoiding over estimated cost and ensuring suitable method of execution. In this chapter, the three currently practiced categories of foundation are briefly introduced. Then, two illustrative Tunisian case histories are analyzed to explain, first, when the practiced foundation solution was inadequately chosen how a non-cost-effective solution can be avoided, and second, why an unsuitable foundation solution can lead to the stopping of the structure functioning and then how to proceed for the design of retrofit solution to be executed for restarting the functioning of the structure.

INTRODUCTION

Optimized foundation design (OFD) first passes by an adequately planned geotechnical survey which usually includes boreholes, in situ tests and laboratory tests. Such a program is decided on the basis of the structure dimensions (or area), load intensities and soil conditions. A well planned geotechnical survey (number, location and depth of boreholes and in situ investigation) is followed by a suitable synthesis

DOI: 10.4018/978-1-5225-7059-2.ch009

of geotechnical test results allowing the adoption of realistic geotechnical soil parameters. The latter constitutes the best starting point to think about an optimized design of foundations.

Three categories of foundations are currently practiced for civil engineering structures namely: shallow foundations, deep foundations and intermediate foundations related to reinforced or improved soils (Das, 2014). For each category the adequate type of foundation is decided on the basis of an optimized solution, e.g; cost effective and acceptable time of execution.

Shallow foundations include the following types: isolated footings; strip footings; crossed strip footing (either in one or in two directions) and rafted foundation.

Deep foundations also comprise a big variety of pile types (bored, driven, etc.), the optimized solution rather relies on the installation method of pile to warrant a reliable and cost-effective solution.

Ground improvement techniques represent the third category of foundation which can be considered in between the shallow and deep foundations. Several techniques can be adopted depending on the accorded priority for the project, i.e. to increase the bearing capacity and/or to reduce the settlement, or to accelerate the consolidation of compressible soils: stone columns, rigid inclusions, etc. (Indraratna et al, 2015).

In this chapter it is intended to highlight either the benefits or the disadvantages that can result from well planned or unsuitable geotechnical survey that can also lead to adequate or inadequate foundation solutions. Two Tunisian case histories are presented in detail to capture the learned lessons in regard to unsafe design in terms of non-cost effective or unsuitable foundation solution.

Foundation of Post Office in Tunis Centre: Case Study 1

The design of foundation of the post office in Tunis City, a ten-floor building (basement, ground floor, eight floors) is discussed. Due to the existence of deep soft clay layer the Tunisian Ministry of equipment decided the execution of 1 m diameter bored piles reaching 52 m depth. The foundation cost was approximately 40% of that of the whole project. Such an expensive solution was dictated by unacceptable long-term settlements compromising the stability of the structure. However, the applied load by the building is equivalent to quasi-uniform vertical stress of about 90 kPa. As such, in view of studying a foundation solution at a reasonable cost, the stone column and the sand compacted piles reinforcement techniques have revealed potential cost-effective solution (Datye and Nagaraj, 1981). Those techniques, however being adopted only for some oil tank projects in Tunis City, leads to a significant reduction in settlement to admissible limit. From the geotechnical survey which included boreholes and pressure meter tests up to 40 m depth it was concluded that crossed soft soil layers are under-consolidated silt-clay with limit pressure less than 1200 kPa.

Consideration of rafted foundation (Figure 1) leads to admissible bearing capacity equals to 100 kPa as estimated from the pressure meter method (Fascicule 62, titre V, 1994). It follows that the admissible bearing capacity of the oil tank complies with the applied load of 90kPa (Bouassida and Guetif, 2000).

The estimation of long-term settlement at the axis of rafted foundation resting on the unreinforced soil using the Terzaghi's method was of 45 cm, while at its border it equals 16.3 cm. Given these non-admissible values of settlement, the building tilting is with high risk such it has been observed for several other buildings in the same area of Tunis City like for six floors building located at the street Zaghloul which became out of service after about ten years (Bouassida & Klai, 2016). Therefore, before studying a piled foundation solution which is not cost-effective, it is obvious to analyze the reinforcement option using either stone columns or sand compacted piles. In fact, the reinforcement using by stone columns will

Figure 1. Geometry of rafted foundation (G' barycenter of applied loading)

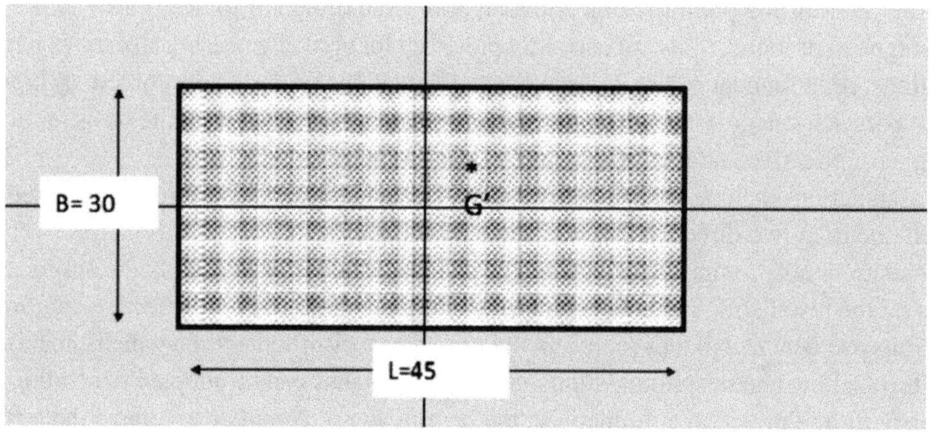

contribute in settlement reduction to comply with admissible values and also to accelerate the consolidation of reinforced soft layers (Bouassida, 2016). Accordingly, it is recommended to cover the surface of reinforced soil, before the execution of raft, by a blanket layer made up of a draining material preferably the same as that of columns, to evacuate the water which results from the consolidation of reinforced soil. Two models of reinforced soils are adopted to predict the ultimate vertical stress of reinforced soil. The ultimate vertical stress q_{ult} applied at the head of an isolated column is determined from Eq (1):

$$q_{ult} = \sigma_h tg^2\left(\frac{\pi}{4}+\frac{\varphi_c}{2}\right) \quad (1)$$

For the purpose of conservative design, it is recommended to consider the horizontal stress $\sigma_h = p_l^*$;

p_l^* = limit net pressure recorded from the pressure meter test.
σ_h= 230 kPa was considered.
φ_c=friction angle of the constitutive column material, its value is limited to 37° depending on the value of admissible stress of reinforced soil predicted by the French standard DTU 13.2, which are:

$$q_{adm}<2\sigma_h \quad (2a)$$

$$q_{adm}<q_{ult}/2 \quad (2b)$$

$$q_{adm}<800 \text{ kPa} \quad (2c)$$

Referring to the French method (2011) the admissible stress of the reinforced soil is equal to: $q_{adm} = q_r / 2 = 460$ kPa. It is noted that by this bearing capacity estimation does not take into account the contribution of initial soil. Using the group of columns model, the ultimate bearing capacity of reinforced soil is estimated as follows (Bouassida, 2016):

$q_r = \eta\sigma_s + (1-\eta)\sigma_c$

η= substitution factor
σ_c= ultimate vertical stress of the columns
σ_s= ultimate vertical stress of the initial soil

$$\sigma_c = \sigma_h K_p + 2C\sqrt{K_p} \tag{3}$$

$$\sigma_s = \sigma_h K_{ps} + 2C_s\sqrt{K_{ps}} \tag{4}$$

$$K_p = tg^2\left(\frac{\pi}{4} + \frac{\varphi}{2}\right) \tag{5}$$

$$K_{ps} = tg^2\left(\frac{\pi}{4} + \frac{\varphi_s}{2}\right) \tag{6}$$

$C_{c(s)}$ and $\varphi_{c(s)}$ denote the cohesion and friction angle of columns material and initial soil, respectively. For conservative design it is assumed C_c = 0.The value of undrained shear strength of soil C_s is estimated when the net limit pressure (p_l*) is less than 300 kPa, from the correlation proposed by Amar & Jézéquel, (1972):

$$C_u = \frac{\text{pl*}}{5,5} \tag{7}$$

The admissible stress of reinforced soil is:

$$\sigma_{adm} \leq \eta\frac{\sigma_c}{2} + (1-\eta)\frac{\sigma_s}{3} \tag{8}$$

From Eq (8) consider an allowable vertical stress equals 120 kPa the corresponding minimum substitution factor is $\eta_{\min} = 0.13$. The variation of admissible stress of reinforced soil versus the substitution factor is shown in Figure 2.

The suitable value the substitution factor will obviously be depending on the admissible settlement of reinforced soil. However, the columns' length should not exceed 15 m, based on considerations related to the calculation of bearing capacity of soil reinforced by floating columns (Bouassida et al, 2009). For settlement estimation, a column length equal to 10 m is adopted.

The thickness of the compressible layers (with pressure meter modulus E_M <3MPa) under the foundation of the building is about 40 m. As a result, settlement of the foundation should be calculated as

Figure 2. Variation of admissible stress of reinforced soil versus the substitution factor

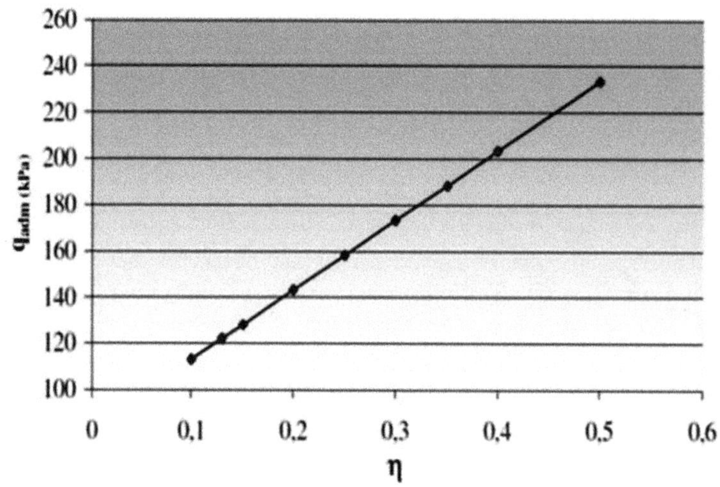

the sum of two terms: s_r: settlement of reinforced soil and s_{ur} settlement of unreinforced soil (Bouassida & Hazzar, 2012).

The settlement of rafted foundation is estimated by the pressure meter method, and other methods assuming linear elastic theory like the variational approach which rather uses group of columns modeling of reinforced soil.

Combining the verifications of bearing capacity and settlement a substitution factor equals 0.3 was adopted. The columns' installation of diameter of 1 min square mesh with an axis to axis columns spacing of 1.6 m was suggested.

The total settlement of unreinforced soil after the pressure meter method is equal to 16,3cm. Whereas the settlement of reinforced soil s_r is respectively 3,5cm, 2,9 cm and 2,8 cm as obtained by the French recommendations, the homogenized Young modulus and the variational method (Bouassida et al, 2003).

The executed piled foundation necessitated the installation of 69 bored piles of length 54 m with total cost of 1 Million US $. Whereas the installation of 486 columns would cost about 600,000.00 US $. In case the reinforcement using stone columns was agreed about 37% reduction in the cost of the foundation and 11% in the total cost of the post office building were affordable (Bouassida & Guetif, 2000).

Oil Tank Foundation at Rades: Case Study 2

It is a cylindrical steel tank of 33 m diameter built in the early nineties at the oil products storage area of the National Petroleum Company (SNDP) located in Rades in the Southern suburb of Tunis City. The tank foundation was improved by sand piles of 0.6 m diameter and length of 6m overlaid by a mattress layer of 1.2 m thickness. Few years after the commencement of tank operations, its cylindrical shell suffered severe buckling deformation. The settlement due to the primary consolidation of compressible layers was estimated at 20 cm after fifteen years of reservoir operation. The study of this pathological case makes it possible to draw some lessons about the design of foundation on ground reinforced by floating sand compacted columns.

Geometry and Foundation of the Tank

The working vertical load of the tank is q = 100 kPa. The tank is constituted by a steel shell resting on compacted blanket layer of thickness 1.5m overlying a series of compressible sands to highly compressible layers. The main role of blanket layer is to make the settlement of tank as much as uniform and, therefore, to minimize the risk of differential settlements. The execution of tank is preceded by a reinforcement of the ground, on the first six meters by sand compacted columns well known in Tunisia as sand piles.

Based on collected data from three boreholes executed up to 40m depth, and results of laboratory tests on several intact samples taken from 6m to 40m depth, the soil profile is subdivided into six layers (Figure 3). Geotechnical parameters of soil layers were adopted from existing geotechnical survey previously carried out for existing similar tank projects near by the studied tank herein.

Geotechnical investigations have revealed that the layer between 2m and 30m deep has low strength characteristics. Therefore, reinforcement of those weak layers revealed necessary. The executed reinforcement comprised the installation of 481 sand columns, of 6m length and 0.6m diameter, in non-regular pattern. The corresponding substitution factor was 11%. This solution essentially aimed to the acceleration of consolidation settlement in the upper layer of thickness 6 m.

Table 1 details the prediction of consolidation settlement using Eq (9) of unreinforced compressible layers, all assumed normally consolidated, by Terzaghi's method, up to 28 m depth:

$$s = H \frac{C_c}{1+e_0} log\left(1+\frac{\Delta\sigma'}{\sigma'_{v0}}\right) \tag{9}$$

Figure 3. Adopted geotechnical profile (Kanoun and Bouassida, 2008)

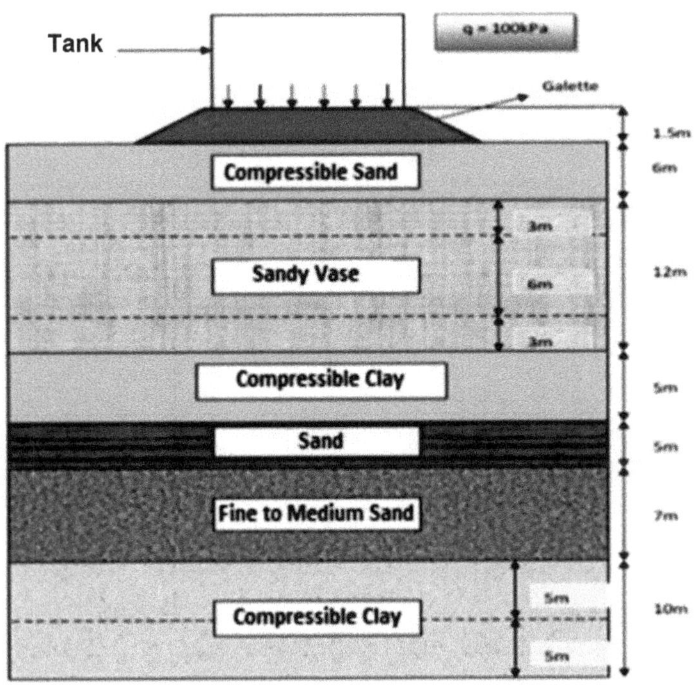

Table 1. Prediction of consolidation settlement at tank axis

Layer n°	depth(m)	C_c	e_0	σ'_0(kPa)	$\Delta\sigma$(kPa)	settlement (cm)
1	6	0.725	2.358	22.5	99.4	95.1
2	3	0.6	1	55.5	89.4	37.5
3	6	0.572	1.883	93	77	31.2
4	3	0.365	1.332	132.75	61	7.7
5	5	0.145	0.677	168	19.3	2.0
6	5	0.145	0.7	215.5	16	1.3
Total settlement (cm)						175

s = consolidation settlement
c_c = compression index
e_0 = initial void ratio
$\Delta\sigma'$ = excess of vertical stress due to the tank's load
σ'_{v0} = effective overburden stress

Table 1 indicates that within the first six meters depth (layer n°1) the long-term ground surface settlement is 95.1 cm which corresponds to approximately 50% of the total consolidation settlement. It should be emphasized that the long-term settlement of layers 2, 3 and 4 is approximately 75 cm, over a thickness of 12 m. Thus, the improvement by sand piles of 6m length will reduce the predicted settlement by about 44%, which is close to that of the upper layer. The total short-term settlement of unreinforced soil was estimated using a linear elastic calculation programmed in the software Columns 1.0 (Bouassida and Hazzar, 2012). Table 2 summarizes the geotechnical characteristics and the estimated settlement of soil layers.

From Table 2, it is noted that the short-term settlement, which occurs over the first six meters depth, is of the order of 30% of total predicted settlement. Assuming that short-term settlement occurs at the end of tank construction, the long-term residual settlement is approximately 35 cm, which is not acceptable for the tank stability.

Table 2. Geotechnical parameters of soil layers and short-term settlement estimation

N° layer	H (m)	C_u (kPa)	E (kPa)	ν	φ (°)	γ (kN/m³)	Settlement (cm)
1	6	10	2500	0.33	21	17.5	23.9
2	12	15	2000	0.4	0	17	47.8
3	5	30	7000	0.33	10	19	3.8
4	5	0	15000	0.25	37	18.5	1.4
5	7		4000	0.33	35	19	5.3
6	10	1	7000	0.33	17	20	3.0
	Total settlement (cm)						85.2

Reinforced Soil by Sand Columns

Verification of sand columns reinforcement was conducted by using software "Columns 1.01" (Bouassida & Hazzar, 2012). Consider the area ratio equals to 11% the admissible bearing capacity was verified for the executed reinforcement. The settlement of reinforced soil was estimated assuming linear elastic behavior; obtained results are given in Table 3.

The improvement by sand columns over the first 6 m depth, allowed limited settlement reduction, and its acceleration just in a few weeks. This settlement only represents about one-third of the total settlement, and about 50% of that occurring in the highly compressible silt layer located between 6m and 18m depth. Thus, the proposed design for settlement reduction (sand piles of 6m length) is largely underestimated.

Indeed, over fifteen years of tank service, the visual observed settlement is of the order of 15 to 20 cm. This post-construction settlement corresponds to that of highly compressible layer between 6 m and 18 m depth for which the degree of consolidation after fifteen years is of about 46%. While consolidation settlement of the reinforced layer was resorbed after the proof water test of the tank.

Retrofit Technique: Micropile Reinforcement

The appropriate solution is to minimize or even to eliminate the estimated residual primary consolidation settlement in the highly compressible silt layers (over 12m thickness) underlying the first reinforced layer by the sand piles. Two retrofit solutions were suggested to achieve this objective:

- Micropile reinforcement (MP) of length reaching the top side of sand layer of thickness 5 m (Figure 4).
- Reinforcement by inclined rigid inclusions (IRI).

This retrofit solution requires the disassembly of entire tank's shell for the installation of the micropiles in a mesh of increased spacing from the center to the border of tank. The micropile heads are connected by reinforced concrete beams are embedded ensuring uniform distribution of loads throughout the tank.

Table 3. Estimation of linear elastic settlement (Columns 1.01 software)

Layer n°	Settlement of unreinforced soil(cm)	Settlement of reinforced soil (cm)
1	23.9	3.0
2	47.8	47.8
3	3.8	3.8
4	1.4	1.4
5	5.3	5.3
6	3.0	3.0
Total settlement (cm)	**85.2**	**64.3**

Figure 4. Layout of reinforcing micropile embedded in reinforced concrete connecting beams (Simpro, 2010)

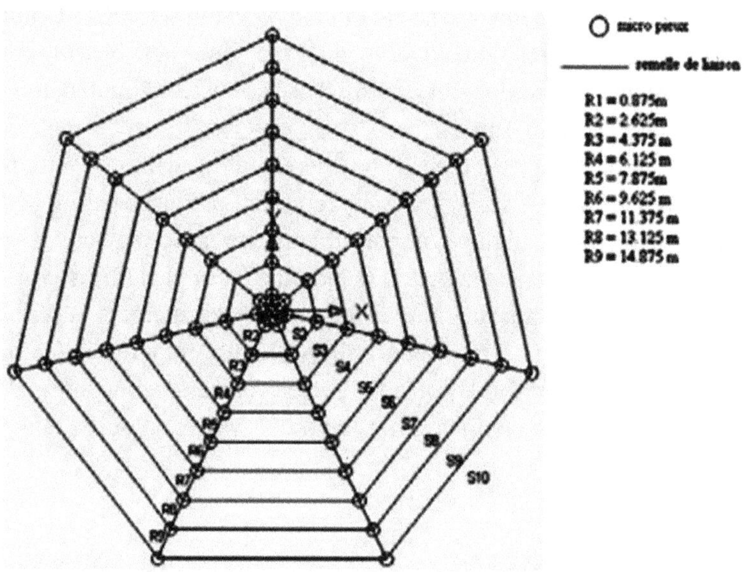

The micropiles of 30 cm diameter are assumed to react only by shaft resistance. A total of 64 micropiles was estimated of total length 1,600 lineal meter (Figure 4). This shaft capacity of micropiles was estimated based on pressure meter data selected from the project of Rades La Goulettebridge (Kanoun and Bouassida, 2008), due to the lack of specific geotechnical investigation for the oil tank. Although reinforcement using micropiles constitutes non-cost effective retrofit solution, it warrants the long-term tank stability without risk residual settlement.

Reinforcement by Inclined Rigid Inclusions

To avoid the entire disassembly of the tank and to proceed for repairing only the affected areas by buckling, installation of inclined rigid inclusions (IRI) embedded within the sand layer at 23 m depth can be designed (Figure 5). The carrying shaft load by IRI corresponds to limited proportion of total weight transmitted by the tank structure. Therefore, it will be necessary to estimate the total allowable shaft load to be balanced by the IRI and to deduce the required number of inclusions to the proportion of carried tank load. Consider this latter estimated as 67%, the remaining 33% of tank load will be balanced by the consolidated first layer which degree of consolidation increased by around 50% during the fifteen years of tank service. Since IRI are covered at the top by reinforced concrete raft, an enhanced load concentration is afforded, therefore the number of IRI is less than that of the classical micropiles. From an economical point of view, the reinforcement using inclined RI is then less expensive than that of micropiles connected by reinforced concrete beams.

Figure 5. Reinforcement using inclined rigid inclusions

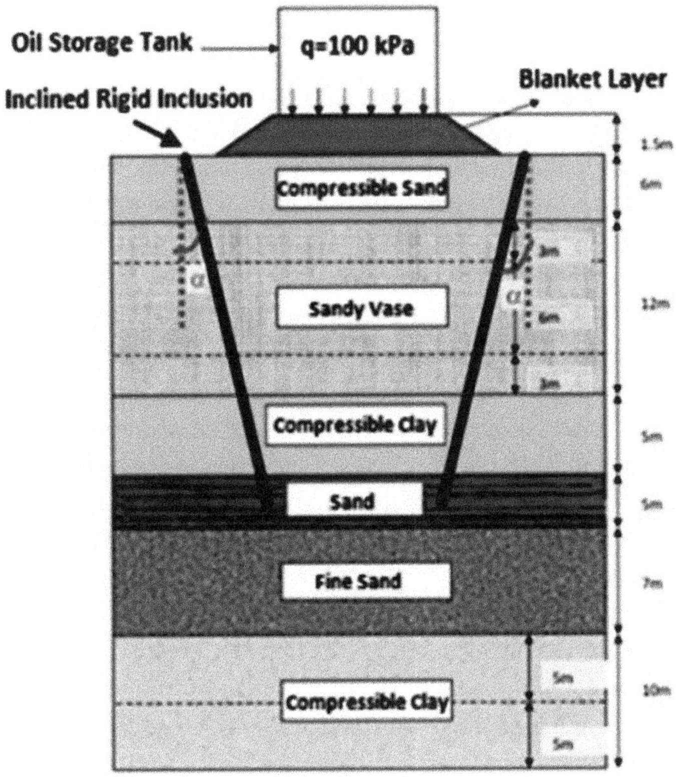

RECOMMENDATIONS

In order to repair the tank, a reinforcement using micropiles or rigid inclusions is required. Main objective of such reinforcement aims the neutralization of consolidation settlement of compressible layer located between 6 m and 18 m depth.

Due to the lack of pressure meter data for the project, and to assess the predictions, it is strongly recommended to carry out pressure meter tests up to 30 m depth. From recorded updated pressure meter data the partial consolidation during fifteen years of tank service can be estimated. On the basis of these data, an adequate design of reinforcement solution will permit a realistic determination of the number of vertical inclusions of total length from 27 m to 30 m with 2 m embedment within the dense sand layer located approximately from 23m to 28 m depth.

Last, it should be reminded that a specific geotechnical survey reveals necessary to avoid consideration of inadequate geotechnical data even if adopted for the design of neighboring projects.

CONCLUSION

This chapter discussed the optimization of design of foundations from analysis of two Tunisian case histories.

First, the execution of piled foundation for the post office building in the center of Tunis City revealed non-cost effective. It has been proven that feasibility and benefits of reinforcement using stone columns would be preferable to that of pile foundation which remains non-cost-effective solution. The effectiveness of reinforcement by stone columns relies both on significant reduction and accelerated long term settlement so that stability of post office building in post construction phase will not be affected by differential settlement, as confirmed by experiences in several countries.

Second, non-successful reinforcement with sand piles was experienced for an oil tank project due to the lack of geotechnical survey and well-designed reinforcement characterized by too short sand piles of length 6m.In fact, because the thickness of Tunis soft soil extends up to 25 m depth non-admissible consolidation settlement affected serviceability of the oil tank ceased after 15 years. Hence, reinforcement using micropiles of length 25 m depth was necessary to stop the remaining long-term settlement in initially unreinforced soft layers overlaid by the improved sand pile layer.

From the two analyzed Tunisian case studies an optimized foundation design is, first, never warranted without a well-planned geotechnical survey followed by a meaningful data synthesis for selecting the suitable geotechnical soil parameters. Second, cost-effective foundation alternatives can be thought using well designed ground improvement techniques. The latter requires careful analysis especially related to the long-term behavior of unreinforced soft soils which can be affected by the induced stress of the projected structure. As such, high costed retrofit solution can be avoided whenever serviceability of the structure is affected.

REFERENCES

Amar, S., & Jézéquel, J. (1972). Essais en place et en laboratoire sur sols cohérents,comparaison des résultats. *Bull. Liaison L.C.P., C*(58), 97–108.

Bouassida, M., & Guetif, Z. (2000). Etude comparative "pieux-colonnes" Cas du siège des chèques postaux à Tunis. *Actes du séminaire "Le renforcement des sols - état de l'art et perspectives en Tunisie*, 65-78.

Bouassida, M. (2016). Design of Column-Reinforced Foundations. J. Ross Publishing.

Bouassida, M., & Hazzar, L. (2015). Performance of soft clays reinforced by floating columns. Ground Improvement Cases Histories, Embankments with Special Reference to Consolidation and Other Physical Methods. In Part Two: Sands and Gravel Piles, Stone Columns and Other Rigid Inclusions. Butterworth Heinemann Publications. doi:10.1016/B978-0-08-100192-9.00016-8

Bouassida, M., & Klai, M. (2016). Caractérisation et étude du comportement de la vase de Tunis. In IVème Séminaire International "Innovation et Valorisation en Génie Civil et Matériaux de Construction. INVACO.

Bouassida, M., Porbaha, A., & Jellali, B. (2009). Limit Analysis of Rigid Foundations on Floating Columns. *International Journal of Geomechanics*, *9*(3), 89–101. doi:10.1061/(ASCE)1532-3641(2009)9:3(89)

Briaud, J. L., & Jordan, G. (1983). Pressure meter Design of Shallow Foundations. Texas A&M University System.

Das, M. B. (2017). *Principles of Geotechnical Engineering* (7th ed.). Mac Graw Hill.

Datye, K.R., & Nagaraju, S.S. (1981). Design approach and field control for stone columns». *Proc. 10th I.C.S.M.F.E.*, *3*, 637-640.

Documents Techniques Unifiés 13.2. (1983). *Travaux de fondations profondes pour le bâtiment*. CSTB.

Indraratna, B. (2015). Ground Improvement Cases Histories, Embankments with Special Reference to Consolidation and Other Physical Methods (vols. 1-2). Butterworth Heinemann Publications.

Kanoun, F., & Bouassida, M. (2008). Geotechnical aspects of Rades La Goulette project (Tunisia). *ISSMGE Bulletin*, *2*(3), 6–12.

L.C.P.C. (1993). *Règles de conception et de dimensionnement des fondations des ouvrages d'art*. Paris: Edit. LCPC.

Simpro. (2010). *Cas pathologique du réservoir n°16 de la SNDP (AGIL) de la zone d'hydrocarbures à Radès*. Retrieved from www.simpro-tn.com

ADDITIONAL READING

Ellouze, S., Bouassida, M., Hazzar, L., & Mroueh, H. (2010). On settlement of stone column foundation by Priebe's method. *Ground Improvement*, *163*(2), 101–107. doi:10.1680/grim.2010.163.2.101

Guetif, Z., Bouassida, M., & Debats, J. M. (2007). Improved Soft Clay Characteristics Due to Stone Column Installation. *Computers and Geotechnics*, *34*(2), 104–111. doi:10.1016/j.compgeo.2006.09.008

KEY TERMS AND DEFINITIONS

Area Ratio (Substitution Factor): Total cross-sectional area of reinforcing columns divided by the total loaded area (i.e., foundation).

Optimized Design: It combines the verifications, first of admissible bearing capacity and second of the admissible settlement, to determine the optimized area ratio.

Pressure Meter Test: It is an situ test performed within pre bore hole where the soil is laterally expanded subjected to horizontal stress, up to failure. From the change in volume versus applied horizontal stress two main parameters are determined: EM is the pressure meter modulus and, pl* is the limit net pressure. From those parameters and other correlated soil parameters the verification of bearing capacity and settlement are estimated for the short-term soil behavior.

APPENDIX

Nomenclature

B, L: breadth, length of foundation
E: Young modulus
H: thickness (m)
q_{ult}: ultimate vertical uniform stress
C_u: undrained cohesion (kPa)
s: settlement (cm)

Greek Symbols

γ: unit weight (kN/m^3)
σ_h: horizontal stress (kPa)
σ_s: ultimate vertical stress host soil (kPa)
σ_c: ultimate vertical stress on column (kPa)
φ_c: frictional angle of column material (°)
$\Delta\sigma$: excess if vertical stress (kPa)

Dimensionless Parameter

η: substitution factor (area ratio)
c_c: compression index
ν: Poisson's ratio
G': barycenter

Subscripts

ult: ultimate stress
c (s): column (soil)

Abbreviation

IRI: inclined rigid inclusion
SNDP: National Petroleum Company

Compilation of References

Abdulridha, A., Palermo, D., Foo, S., & Vecchio, F. J. (2013). Behavior and modelling of super elastic shape-memory alloy reinforced concrete beams. *Engineering Structures*, *49*, 893–904. doi:10.1016/j.engstruct.2012.12.041

Abolmaali, A., Treadway, J., Aswath, P., Lu, F. K., & McCarthy, E. (2006). Hysteresis behaviour of t-stub connections with super elastic shape-memory fasteners. *Journal of Constructional Steel Research*, *62*(8), 831–838. doi:10.1016/j.jcsr.2005.11.017

Adeli, H. (2001). Special Issue: Health Monitoring of Structures. *Computer-Aided Civil and Infrastructure Engineering*, *16*(1).

Agbabian, M. S., Masri, S. F., Miller, R. K., & Caughey, T. K. (1991). System Identification Approach to Detection of Structural Changes. *J. Struct. Engrg. Mech.*, *117*(2), 370–390. doi:10.1061/(ASCE)0733-9399(1991)117:2(370)

Aiken, I. D., Kelly, J. M., & Mahmoodi, P. (1990, May). The application of viscoelastic dampers to seismically resistant structures. In *Proceedings of Fourth US National Conference on Earthquake Engineering* (*Vol. 3*, pp. 459-468). Academic Press.

AISC. (2011). *Steel Construction* (14th ed.). AISC.

AkzoNobel. (2017, June 20). *Protective coatings*. Retrieved from AkzoNobel: http://www.international-pc.com/products/fire-protection/fire-protection-technical-information.aspx

Ali Mohamed, A. (1992). *Validité des cinétiques de séchage sous des conditions d'air variables* (PhD thesis). Université de Poitiers.

Ali, R. M. (2007). *Performance Based Design Of Offshore Structures Subjected To Blast Loading*. London: Imperial College.

Alvin, K. F., & Park, K. C. (1994). Second-Order Structural Identification Procedure via State-Space-Based System Identification. *AIAA Journal*, *32*(2), 397–406. doi:10.2514/3.11997

Amar, S., & Jézéquel, J. (1972). Essais en place et en laboratoire sur sols cohérents,comparaison des résultats. *Bull. Liaison L.C.P.*, *C*(58), 97–108.

Anatol Longinow, F. A. (2003). Blast resistant design with structural steel - common questions answered. *Modern Steel Construction*, 61 - 66.

André, J., Beale, R., & Baptista, A. M. (2015). New indices of structural robustness and structural fragility. *Structural Engineering and Mechanics*, *56*(6), 1063–1093. doi:10.12989em.2015.56.6.1063

API. (2000). *API recommended practice 2a-WSD (RP 2A-WSD)*. American Petroleum Institute.

Applied Technology Council 3.06 Amended. (1978). *Tentative provisions for the developments of seismic regulations for buildings*. Redwood City, CA: Applied Technology Council.

Araki, Y., Maekawa, N., Shrestha, K. C., Yamakawa, M., Koetaka, Y., Omori, T., & Kainuma, R. (2014). Feasibility of tension braces using Cu–Al–Mn superelastic alloy bars. *Structural Control and Health Monitoring*, *21*(10), 1304–1315. doi:10.1002tc.1644

ASFP. (2015). *Types of fire protection materials*. Retrieved October 14, 2017, from http://www.asfp.org.uk/Technical%20Services/types_of_fire_protection.php

ATC. (1996). 40, Seismic evaluation and retrofit of concrete buildings. Applied Technology Council, report ATC-40.

Auricchio, F., Fugazza, D., & DesRoches, R. (2006). Earthquake performance of steel frames with nitinol braces. *Journal of Earthquake Engineering*, *10*(1), 45–66. doi:10.1080/13632460609350628

Avilés, J., & Pérez-Rocha, L. E. (2003). Soil–structure interaction in yielding systems. *Earthquake Engineering & Structural Dynamics*, *32*(11), 1749–1771. doi:10.1002/eqe.300

Avilés, J., & Suárez, M. (2002). Effective periods and dampings of building-foundation systems including seismic wave effects. *Engineering Structures*, *24*(5), 553–562. doi:10.1016/S0141-0296(01)00121-3

Babalis, S. J., & Belessiotis, V. G. (2004). Influence of the drying conditions on the drying constants and moisture diffusivity during the thin-layer drying of figs. *Journal of Food Engineering*, *65*(3), 449–458. doi:10.1016/j.jfoodeng.2004.02.005

Baker, M. S. (2007). *Assessment of robustness*. Academic Press.

Bansiddhi, A., Sargeant, T., Stupp, S., & Dunand, D. (2008). Porous nickel titanium for bone implants: A review. *Acta Biomaterialia*, *4*(4), 773–782. doi:10.1016/j.actbio.2008.02.009 PMID:18348912

Baruch, M. (1982). Optimal Correction of Mass and Stiffness Matrices Using Measured Modes. *AIAA Journal*, *20*(11), 1623–1626. doi:10.2514/3.7995

Batoz, J. L., & Dhatt, G. (1990). *Modélisation des structures par élément finis* (Vol. 2). Hermès Paris.

Bea, R. W. (1991). *Design and characterize the offshore fire problem. Improved means of offshore platform fire resistance*. Berkeley, CA: University of California.

Belgasmia, M., & Moussaoui, S. (2013). Comparison of static pushover analysis in the case of small and large deformation with time history analysis using flexibility-based model for an existing structure. *International Journal of Current Engineering and Technologies*, *3*(2), 655–665.

Belhamri, A. (1992). *Etude des transferts de chaleur et de masse à l'intérieur d'un milieu poreux au cours du séchage* (PhD thesis). Université de Poitiers.

Belhamri, A. (2003). Characterization of the first falling rate period during drying of a porous material. *Drying Technology*, *21*(7), 1235–1252. doi:10.1081/DRT-120023178

Belhamri, A., & Fohr, J. P. (1996). Heat and mass transfer along a wetted porous plate in an air stream. *AIChE Journal. American Institute of Chemical Engineers*, *42*(7), 1833–1843. doi:10.1002/aic.690420705

Bellini, A., Colli, M., & Dragoni, E. (2009). Mechatronic design of a shape memory alloy actuator for automotive tumble flaps: A case study. *IEEE Transactions on Industrial Electronics*, *56*(7), 2644–2656. doi:10.1109/TIE.2009.2019773

Bennamoun, L., Arlabosse, P., & Leonard, A. (2013a). Review on fundamental aspect of application of drying process to wastewater sludge. *Renewable & Sustainable Energy Reviews*, *28*, 29–43. doi:10.1016/j.rser.2013.07.043

Bennamoun, L., & Belhamri, A. (2003). Design and simulation of a solar dryer for agriculture products. *Journal of Food Engineering*, *59*(2-3), 259–266. doi:10.1016/S0260-8774(02)00466-1

Bennamoun, L., & Belhamri, A. (2006). Numerical simulation of drying under variable external conditions: Application to solar drying of seedless grapes. *Journal of Food Engineering*, *76*(2), 179–187. doi:10.1016/j.jfoodeng.2005.05.005

Bennamoun, L., Belhamri, A., & Ali Mohamed, A. (2009). Application of a diffusion model to predict drying changes under variable conditions: Experimental and simulation study. *Fluid Dynamics and Materials Processing*, *5*(2), 177–191.

Bennamoun, L., Chen, Z., & Afzal, M. T. (2016). Microwave drying of wastewater sludge: Experimental and modeling stydy. *Drying Technology*, *34*(2), 235–243. doi:10.1080/07373937.2015.1040885

Bennamoun, L., Crine, M., & Leonard, A. (2013b). Convective drying of wastewater sludge: Introduction of shrinkage effect in mathematical modeling. *Drying Technology*, *31*(6), 643–654. doi:10.1080/07373937.2012.752743

Benzoni, G., Ohtaki, T., & Priestley, M. J. N. (1996). *Seismic Performance of A Full Scale Bridge Column As Built and As Repaired.* Structural Systems Research Project, Report No. SSRP-96/07 University of California, San Diego.

Bernal, D., & Beck, J. (2004). Special Issue: Phase 1 of the IASC-ASCE Structural Health Monitoring Benchmark. *Journal of Engineering Mechanics*, *130*(1), 1–2. doi:10.1061/(ASCE)0733-9399(2004)130:1(1)

Bertero, R. D., & Bertero, V. V., & TeranGilmore, A. (1996). Performance-based Earthquake Resistant Design Based on Comprehensive Philosophy and Energy Concepts. In *Proceeding 11th World Conference on Earthquake Engineering, 11WCEE 1996*. Acapulco, Mexico: Elsevier Science Ltd.

Bertero, V. V. (1992). Major issues and future directions in earthquake-resistant design. *Proceeding 10th World Conference on Earthquake Engineering 10WCEE,* 6407-6444.

Bertero, V. V. (1996). State of the art report on: Design Criteria. In *Proceeding 11th World Conference on Earthquake Engineering 11WCEE*. Acapulco, Mexico: Elsevier Science Ltd.

Biggs, J. M. (1964). *Introduction to structural dynamics*. New York: Mac Graw-Hill.

Biggs, J. M. (1986). *Introduction of structural engineering*. Englewood Cliffs, NJ: Prentice Hall.

Booth, E. (2014). *Earthquake Design Practice for Buildings* (Vol. 53). Earthquake Design Practice for Buildings. doi:10.1017/CBO9781107415324.004

Bouassida, M. (2016). Design of Column-Reinforced Foundations. J. Ross Publishing.

Bouassida, M., & Guetif, Z. (2000). Etude comparative "pieux-colonnes" Cas du siège des chèques postaux à Tunis. *Actes du séminaire "Le renforcement des sols - état de l'art et perspectives en Tunisie*, 65-78.

Bouassida, M., & Hazzar, L. (2015). Performance of soft clays reinforced by floating columns. Ground Improvement Cases Histories, Embankments with Special Reference to Consolidation and Other Physical Methods. In Part Two: Sands and Gravel Piles, Stone Columns and Other Rigid Inclusions. Butterworth Heinemann Publications. doi:10.1016/B978-0-08-100192-9.00016-8

Bouassida, M., & Klai, M. (2016). Caractérisation et étude du comportement de la vase de Tunis. In IVème Séminaire International "Innovation et Valorisation en Génie Civil et Matériaux de Construction. INVACO.

Bouassida, M., Porbaha, A., & Jellali, B. (2009). Limit Analysis of Rigid Foundations on Floating Columns. *International Journal of Geomechanics*, *9*(3), 89–101. doi:10.1061/(ASCE)1532-3641(2009)9:3(89)

Bousias, S. N., Triantafillou, T. C., Fardis, M. N., Spathis, L., & O'Regan, B. A. (2004). Fiber Reinforced Polymer Retrofitting of Rectangular Reinforced Concrete Columns with or without Corrosion. *ACI Structural Journal*, *101*(4), 512–520.

Bresler, B., & Iding, R. (n.d.). *Effect of Fire Exposure on Structural Response and Fireproofing Requirements of Structural Steel Frame Assemblies*. Wiss, Janney, Elstner Associates, Inc.

Briaud, J. L., & Jordan, G. (1983). Pressure meter Design of Shallow Foundations. Texas A&M University System.

Brode, H. L. (1955). Numerical solution of spherical blast waves. *Journal of Applied Physics, American Institute of Physics*.

Building Seismic Safety Council (US). (1997). *Federal Emergency Management Agency, & Applied Technology Council. NEHRP guidelines for the seismic rehabilitation of buildings* (Vol. 1). Federal Emergency Management Agency.

Cacciola, P., & Tombari, A. (2015). Vibrating barrier: a novel device for the passive control of structures underground motion. *Proceeding A of the Royal Society*, *471*(2179). 10.1098/rspa.2015.0075

CBA 93. (1993). *Code du Béton Algérien*. CGS Alger.

Chancellor, N. B., Eatherton, M. R., Roke, D. A., & Akbaş, T. (2014). Self-centering seismic lateral force resisting systems: High performance structures for the city of tomorrow. *Buildings*, *4*(3), 520–548. doi:10.3390/buildings4030520

Chang, K., Soong, T. T., Oh, S. T., & Lai, M. L. (1991). *Seismic response of a 2/5 scale steel structure with added viscoelastic dampers*. Academic Press.

Chang, M., & Pakzad, S. N. (2014). Observer Kalman Filter Identification for Output-only Systems Using Interactive Structural Modal Identification Toolsuite. *Journal of Bridge Engineering*, *19*(5), 04014002. doi:10.1061/(ASCE)BE.1943-5592.0000530

Chang, W. S., & Araki, Y. (2016). Use of Shape Memory Alloys in Construction; A Critical Review. *Proceedings - Institution of Civil Engineers*, *169*(CE2).

Chang, W. S., Murakami, S., & Komatsu, K. (2013). Potential to use shape memory alloy in timber dowel-type connections. *Wood and Fiber Science*, *45*(3), 330–334.

Chaphalkar, M. (2015). *Earthquake resistant structures: Challenges Un-developed, Developing and Developed Countries.* Retrieved from https://www.linkedin.com/pulse/earthquake-resistant-structures-challenges-developing-chaphalkar-pe

Chen, H. P., & Tee, K. F. (2014). Structural Finite Element Model Updating Using Incomplete Ambient Vibration Modal Data. *Science China. Technological Sciences*, *57*(9), 1677–1688. doi:10.100711431-014-5619-9

Chen, H. P., Tee, K. F., & Ni, Y. Q. (2012). Mode Shape Expansion with Consideration of Analytical Modelling Errors and Modal Measurement Uncertainty. *Smart Structures and Systems*, *10*(4-5), 485–499. doi:10.12989ss.2012.10.4_5.485

Chen, S.-H., Chen, M.-C., Chang, P.-C., Zhang, Q., & Chen, Y.-M. (2010). Guidelines for developing effective Estimation of Distribution Algorithms in solving single machine scheduling problems. *Expert Systems with Applications*, *37*(9), 6441–6451. doi:10.1016/j.eswa.2010.02.073

Chirife, J. (1983). Fundamentals of the drying mechanism during air dehydration of foods. In A.S. Mujumdar (Ed.), *Advances in drying II*. New York: Hemisphere Publication.

Chopra, A. (2007). Dynamic of structures: theory and applications to earthquake engineering (3rd ed.). Pearson Education, Inc.

Chopra, A. K., & Goel, R. K. (2002). A modal pushover analysis procedure for estimating seismic demands for buildings. *Earthquake Engineering & Structural Dynamics*, *31*(3), 561–582. doi:10.1002/eqe.144

Ciampoli, M., & Pinto, P. E. (1995). Effects of soil-structure interaction on inelastic seismic response of bridge piers. *Journal of Structural Engineering*, *121*(5), 806–814. doi:10.1061/(ASCE)0733-9445(1995)121:5(806)

Commend, S., & Zimmermann, T. (2001). Object-oriented nonlinear finite element programming: A primer. *Advances in Engineering Software*, *32*(8), 611–628. doi:10.1016/S0965-9978(01)00011-4

Computers and Structures, Inc. (2000). Integrated structural analysis and design software, SAP2000 version 14.2.2. Berkeley, CA: Author.

Constantinou, M. C., & Symans, M. D. (1992). *Experimental and analytical investigation of seismic response of structures with supplemental fluid viscous dampers*. Buffalo, NY: National Center for Earthquake Engineering Research.

Constantinou, M. C., Symans, M. D., Tsopelas, P., & Taylor, D. P. (1993). Fluid viscous dampers in applications of seismic energy dissipation and seismic isolation. *Proceedings ATC*, *17*(1), 581–592.

Corr, R. B., & Tam, V. H. Y. (1998). Gas explosion generated drag loads in offshore installations. *Journal of Loss Prevention in the Process Industries*, *11*(1), 43–48. doi:10.1016/S0950-4230(97)00054-5

Crank, J. (1975). The mathematics of diffusion (2nd ed.). Oxford, UK: Clarendon.

Daguenet, M. (1985). *Les séchoirs solaires: théorie et pratique*. UNESCO.

Dalton, S. K., Atamturktur, S., Farajpour, I., & Juang, C. H. (2013). An optimization based approach for structural design considering safety, robustness, and cost. *Engineering Structures*, *57*, 356–363. doi:10.1016/j.engstruct.2013.09.040

Das, N. K. (1988). *Safety analysis of steel building frames under dynamic wind loading* (Doctoral dissertation). Texas Tech University.

Das, M. B. (2017). *Principles of Geotechnical Engineering* (7th ed.). Mac Graw Hill.

Datta, D., Amaral, A. R. S., & Figueira, J. R. (2011). Single row facility layout problem using a permutation-based genetic algorithm. *European Journal of Operational Research*, *213*(2), 388–394. doi:10.1016/j.ejor.2011.03.034

Datta, T. K. (2010). *Seismic Analysis of Structures*. John Wiley and Sons. doi:10.1002/9780470824634.ch9

Datye, K.R., & Nagaraju, S.S. (1981). Design approach and field control for stone columns». *Proc. 10th I.C.S.M.F.E.*, *3*, 637-640.

Davenport, A. G. (1961). The application of statistical concepts to the wind loading of structures. *Proceedings - Institution of Civil Engineers*, *19*(4), 449–472. doi:10.1680/iicep.1961.11304

Davoodi, M., Jafari, M. K., & Hadiani, N. (2013). Seismic response of embankment dams under near-fault and far-field ground motion excitation. *Engineering Geology*, *158*, 66–76. doi:10.1016/j.enggeo.2013.02.008

Davoodi, M., Sadjadi, M., Goljahani, P., & Kamalian, M. (2012). Effects of Near-Field and Far -Field Earthquakes on seismic Response of SDOF System Considering Soil Structure Interaction. *Earthquake Engineering & Structural Dynamics*, *30*(12), 1769–1789. doi:10.1002/eqe.92

DeAngelis, M., Lus, H., Betti, R., & Longman, R. W. (2002). Extraction Physical Parameters of Mechanical Models from Identified State Space Representations. *ASME Trans. J. Appl. Mech.*, *69*(5), 617–625. doi:10.1115/1.1483836

Delannoy, C. (2002). *Programmer en langage c* (3rd ed.). Eyrolles Paris.

Denoël, V. (2005). *Application des méthodes d'analyse stochastique à l'étude des effets du vent sur les structures du génie civil* (Doctoral dissertation). University of Liège.

Denoël, V., & Degée, H. (2009). Asymptotic expansion of slightly coupled modal dynamic transfer functions. *Journal of Sound and Vibration*, *328*(1-2), 1–8. doi:10.1016/j.jsv.2009.08.014

Dhatt, G., & Touzot, G. (1984). Une présentation de la méthode des éléments finis (2nd ed.). Edition Maloine S.A.

Documents Techniques Unifiés 13.2. (1983). *Travaux de fondations profondes pour le bâtiment*. CSTB.

Doherty, J. E. (1987). Nondestructive Evaluation. In A. S. Kobayashi (Ed.), *Handbook on Experimental Mechanics*. Society for Experimental Mechanics, Inc.

Dong, W., Yuan, W., Zhou, X., & Wang, F. (2018). The fracture mechanism of circular/elliptical concrete rings under restrained shrinkage and drying from top and bottom surfaces. *Engineering Fracture Mechanics*, *189*, 148–163. doi:10.1016/j.engfracmech.2017.10.026

Dubois-pèlerin, Y., & Zimmermann, T. (1993). Object-oriented finite element programming III. An efficient implementation in C++. *Computer Methods in Applied Mechanics and Engineering*, *108*(1-2), 165–183. doi:10.1016/0045-7825(93)90159-U

Dubois-pèlerin, Y., Zimmermann, T., & Bomme, P. (1992). Object-oriented finite element programming II. A prototype programming in smaltalk. *Computer Methods in Applied Mechanics and Engineering*, *98*(3), 361–397. doi:10.1016/0045-7825(92)90004-4

Duflot, P., & Taylor, D. (2008). Experience and Practical Considerations in the Design of Viscous Dampers. *Third international conference*.

Duflot, P., Vigano, M. G., & Denoël, V. (2017). Method for Preliminary Design of a System of Viscous Dampers Applied to a Tall Building, *European African Conference on Wind Engineering*.

Duggal, S. K. (2007). *Earthquake Resistant Design of Structures*. Oxford University Press.

Eknes, M. L. T. M. (1994). Escalation of gas explosion event offshore. *Offshore Structural Design, Hazard, Safety and Engineering*. London: ERA Report No 94-0730, 3.1.1-3.1.16.

El-Baz, M. A. (2004). A genetic algorithm for facility layout problems of different manufacturing environments. *Computers & Industrial Engineering*, *47*(2–3), 233–246. doi:10.1016/j.cie.2004.07.001

Elnashai, A. S., & Di Sarno, L. (2008). *Fundamentals of Earthquake Engineering. Fundamentals of Earthquake Engineering*. Academic Press. doi:10.1002/9780470024867

ESDEP. (n.d.). *ESDEP 15A*. Retrieved July 5, 2017, from Dynamic analysis: http://fgg-web.fgg.uni-lj.si/%7E/pmoze/esdep/master/wg15a/l0100.htm

Esteva, L. (1980). Design: General. In *Design of Earthquake Resistant Structures* (pp. 54–97). New York: John Wiley & Sons.

Farid Alfawakhiri, L. A. (2003). *Blast resistant design with structural steel - common questions answered*. Modern Steel Construction.

Fealekari, M., & Chayjan, R. A. (2014). Optimization of convective drying process for Persian shallot using response surface method (RSM). *CIGR Journal*, *16*(2), 157–166.

Federal Emergency Management Agency. FEMA P-749. (2010). Earthquake resistant design concept. U.S. Department of Homeland Security, National Institute of Building Sciences Building Seismic Safety Council.

Filiatrault, A., & Cherry, S. (1990). Seismic design spectra for friction-damped structures. *Journal of Structural Engineering*, *116*(5), 1334–1355. doi:10.1061/(ASCE)0733-9445(1990)116:5(1334)

Fohr, J. P., Arnaud, G., Ali Mohamed, A., & Ben Moussa, H. (1990). Validity of drying kinetics. In A. S. Mujumdar & M. Roques (Eds.), *Drying 89* (pp. 269–275). New York: Hemisphere Publication.

Frey, F., & Jirousek, J. (2001). *Analyse des structures et milieu continus méthode des élément finis*. Suiss Federal institute of Technology.

Fugazza, D. (2005). *Use of shape memory alloys device in earthquake engineering: Mechanical properties, Advanced Constitutive Modelling and Structural Applications* (Doctoral dissertation). University of Pavia.

Fukuyama, H., Suzuki, H., & Nakamura, H. (1999). Seismic retrofit of reinforced concrete columns by fiber sheet wrapping without removal of finishing mortar and side wall concrete. Fiber Reinforced Polymer Reinforcement for Reinforced Concrete Structures, ACI Report No SP-188.

Gallagher, R. H. (1973). *Introduction, Optimum structural design* (R. H. Gallagher & O. C. Zienkiewicz, Eds.). New York: John Wiley & Sons.

Gazetas, G. (1990). Formulas and Charts for Impedances of Surface and Embedded Foundations. *Journal of Geotechnical Engineering*, *117*(9), 1363–1381. doi:10.1061/(ASCE)0733-9410(1991)117:9(1363)

Geoffroy, P. (1983). *Development and evaluation of a finite element for the static, dynamics nonlinear analysis of thin shells* (PhD thesis). University of Technology of Compiègne.

Gerald, C. F., & Wheatly, P. O. (1989). *Applied numerical Analysis* (4th ed.). Addison Wisley.

Ghanem, R., & Shinozuka, M. (1995). Structural System Identification I: Theory. *Journal of Engineering Mechanics*, *121*(2), 255–264. doi:10.1061/(ASCE)0733-9399(1995)121:2(255)

Ghanem, R., & Sture, S. (2000). Special Issue: Structural Health Monitoring. *Journal of Engineering Mechanics*, *126*(7).

Gillespie, J. W., Mertz, D. R., Kasai, K., Edberg, W. M., Demitz, J. R., & Hodgson, I. (1996). Rehabilitation of Steel Bridge Girders: Large Scale Testing. *Proceeding of the American Society for Composites 11th Technical Conference on Composite Materials*, 231-240.

Gioncu, V., & Mazzolani, F. (2011). Earthquake Engineering and for Structural Design. Spon Press, Taylor & Francis Group.

Gioncu, V., & Mazzolani, M. F. (2002). Ductility of Seismic Resistant Steel Structures, Spon Press, Taylor and Francis Group.

Gioncu, V., Mateescu, G., Tirca, L., & Anastasiadis, A. (2000). Influence of the type of seismic ground motions. In F. M. Mozzolani (Ed.), *Moment Resisting Connections of Steel Buildings Frames in Seismic Areas* (pp. 57–92). London: E& FN Spon.

Glover, F., Kelly, J. P., & Laguna, M. (1995). Genetic algorithms and tabu search: Hybrids for optimization. *Computers & Operations Research, 22*(1), 111–134. doi:10.1016/0305-0548(93)E0023-M

Goldberg, D. E. (1989). *Genetic Algorithms in Search, Optimization, and Machine Learning. Addison-Wesley Publishing Company*, Inc.

Guan, J., & Lin, G. (2016). Hybridizing variable neighborhood search with ant colony optimization for solving the single row facility layout problem. *European Journal of Operational Research, 248*(3), 899–909. doi:10.1016/j.ejor.2015.08.014

Guo, T., Xu, J., Xu, W., & Di, Z. (2014). Seismic Upgrade of Existing Buildings with Fluid Viscous Dampers: Design Methodologies and Case Study. *Journal of Performance of Constructed Facilities, 29*(6), 04014175. doi:10.1061/(ASCE)CF.1943-5509.0000671

Guyan, R. J. (1965). Reduction of Stiffness and Mass Matrices. *AIAA Journal, 3*(2), 380. doi:10.2514/3.2874

Hamada, M. (2014). Earthquake-resistant design and reinforcement. In Springer Series in Geomechanics and Geoengineering (pp. 75–124). Springer. doi:10.1007/978-4-431-54892-8_2

Hamburger, O., & Gumpertz, S. (2009). Facts for steel buildings; Earthquake and Seismic Design. American Institute of Steel Construction.

Hammouda, I., & Mihoubi, D. (2017). Influence of stationary and non-stationary conditions on drying time and mechanical properties of a porcelain slab. *Heat and Mass Transfer, 53*(12), 3571–3580. doi:10.100700231-017-2084-6

Harajli, M.H., Hantouche, E., & Soudki, K. (2006). Stress-Strain Model for Fiber-reinforced Polymer Jacketed Concrete Columns. *ACI Structural Journal, 103*(5), 672–682.

Harichandran, R. S., & Baiyasi, M. I. (2000). Repair of corrosion damaged columns using FRP Wraps. Report No. RC-1386. Michigan State University.

Harris, R. J. (1983). *The investigation and control of gas explosions in buildings and heating plant.* The University of Michigan.

Harris, R. A., Barall, M., Andrews, D. J., Duan, B., Ma, S., Dunham, E. M., ... Abrahamson, N. (2011). Verifying a Computational Method for Predicting Extreme Ground Motion. *Seismological Research Letters, 82*(5), 638–644. doi:10.1785/gssrl.82.5.638

Hartl, D., & Lagoudas, D. C. (2007). Aerospace applications of shape-memory alloys. *Proceedings of the Institution of Mechanical Engineers. Part G, Journal of Aerospace Engineering*, *221*(4), 535–552.

Haskell, G., & Lee, D. (1996). *Fluid viscous damping as an alternative to base isolation (No. CONF-960706)*. New York: American Society of Mechanical Engineers.

Haupt, R. L., & Haupt, S. E. (2004). *Practical Genetic Algorithms* (2nd ed.). Wiley Interscience.

Hauschild, M., & Pelikan, M. (2011). An introduction and survey of estimation of distribution algorithms. *Swarm and Evolutionary Computation*, *1*(3), 111–128. doi:10.1016/j.swevo.2011.08.003

Henkel, O., Holl, D., & Schalk, M. (2008). *Seismic design and Drywalling*. Knauf Gips KG.

Higgins, A. (2006). Scheduling of road vehicles in sugarcane transport: A case study at an Australian sugar mill. *European Journal of Operational Research*, *170*(3), 987–1000. doi:10.1016/j.ejor.2004.07.055

Hjelmstad, K. D., Banan, M. R., & Banan, M. R. (1995). Time Domain Parameter Estimation Algorithm for Structures. I: Computational Aspects. *Journal of Engineering Mechanics*, *121*(3), 424–434. doi:10.1061/(ASCE)0733-9399(1995)121:3(424)

Ho, B. L., & Kalman, R. E. (1965). Effective Construction of Linear State-Variable Models from Input/Output Functions. *Proc. 3rd Allerton Conf. Circuits and Systems Theory*, 449-459.

Houck. (1995). *A Genetic Algorithm for Function Optimization: A MATLAB Implementation*. North Carolina State University.

Hudson, D. E. (1979). *Reading and interpreting, Strong Motion Acclerograms*. Earthquake Engineering Research Institute.

Imbert, J. F. (1991). *Analyse des Structures par Eléments finis* (3rd ed.). Cépaduès Toulouse.

Indirli, M., & Castellano, M. G. (2008). Shape-memory alloy devices for the structural improvement of masonry heritage structures. *International Journal of Architectural Heritage*, *2*(2), 93–119. doi:10.1080/15583050701636258

Indraratna, B. (2015). Ground Improvement Cases Histories, Embankments with Special Reference to Consolidation and Other Physical Methods (vols. 1-2). Butterworth Heinemann Publications.

ISO19902. (2007). *Petroleum and natural gas industries — Fixed steel offshore structures*. International Standard Organization.

Izadifard, M. M. (2010). Ductility effects on the behaviour of steel structures under blast loading. *Iranian Journal of Science & Technology*, 49 - 62.

Izzuddin, D. L. (1997). Response of Offshore Structures to Explosion Loading. *International Journal of Offshore and Polar Engineering*, 212.

Jamayaha, S. E. G., Chou, S. K., & Wijeysundera, N. E. (1996). Drying of porous materials in a presence of solar radiation. *Drying Technology*, *14*(10), 2339–2369. doi:10.1080/07373939608917209

James, M. A. (1988). Analytical Methods for Determining Fire Resistance of Steel Members. In SFPE Handbook of fire engineering. Quincy, MA: Society of fire protection Engineers.

Jason, B. (2011). *Clever Algorithm* (1st ed.). Melbourne, Australia: LuLu.

Jennings, P. C., & Bielak, J. (1973). Dynamics of building-soil interaction. *Bulletin of the Seismological Society of America*, *63*(1), 9–48.

Jodin, P. G. P. (1980). Experimental and theoritical study of cracks in mixed mode conditions. In J. C. Radon (Eds.), Fracture and Fatigue: Elasto-plasticity, thin sheet and micromechanisims problems (pp. 350 - 352). London: Imperial College, London.

Juang, J.-N., Cooper, J. E., & Wright, J. R. (1988). An Eigensystem Realization Algorithm Using Data Correlations (ERA/DC) for Model Parameter Identification. *Control, Theory and Advanced Technology*, *4*(1), 5–14.

Juang, J.-N., & Pappa, R. S. (1985). An Eigensystem Realization Algorithm for Model Parameter Identification and Model Reduction. *Journal of Guidance, Control, and Dynamics*, *8*(5), 620–627. doi:10.2514/3.20031

Juang, J.-N., & Pappa, R. S. (1986). Effects of Noise on Modal Parameters Identified by the Eigensystem Realizations Algorithm. *Journal of Guidance, Control, and Dynamics*, *9*(3), 294–303. doi:10.2514/3.20106

Kandemir-Mazanoglu, E. C., & Mazanoglu, K. (2017). An optimization study for viscous dampers between adjacent buildings. *Mechanical Systems and Signal Processing*, *89*, 88–96. doi:10.1016/j.ymssp.2016.06.001

Kannan, S. (2014). Innovative Mathematical Model for Earthquake Prediction. *Engineering Failure Analysis*, *41*, 890–895. doi:10.1016/j.engfailanal.2013.10.016

Kanoun, F., & Bouassida, M. (2008). Geotechnical aspects of Rades La Goulette project (Tunisia). *ISSMGE Bulletin*, *2*(3), 6–12.

Kareem, A. (2008). Numerical simulation of wind effects: A probabilistic perspective. *Journal of Wind Engineering and Industrial Aerodynamics*, *96*(10-11), 1472–1497. doi:10.1016/j.jweia.2008.02.048

Kargahi, M., & Anderson, J. (2004). Structural Optimisation for Seismic Design. *13th World Conference on Earthquake Engineering, 13WCEE 2004*.

Katsanos, E. I., Sextos, A. G., & Manolis, G. D. (2010). Selection of earthquake ground motion records: A state-of-the-art review from a structural engineering perspective. *Soil Dynamics and Earthquake Engineering*, *30*(4), 157–169. doi:10.1016/j.soildyn.2009.10.005

Kausel, E. (2010). Early history of soil–structure interaction. *Soil Dynamics and Earthquake Engineering*, *30*(9), 822–832. doi:10.1016/j.soildyn.2009.11.001

Khama, R., Aissani, F., & Alkama, R. (2016). Design and performance testing of an industrial-scale indirect solar dryer. *Journal of Engineering Science and Technology*, *11*(9), 1263–1281.

Khilwani, N., Shankar, R., & Tiwari, M. K. (2012). *Facility layout problem: an approach based on a group decision-making system and psychoclonal algorithm*. Academic Press.

Kidder, R. L. (1973). Reduction of Structural Frequency Equations. *AIAA Journal*, *11*(6), 892. doi:10.2514/3.6852

Kim, B., Lee, M. G., Lee, Y. P., Kim, Y., & Lee, G. (2006). An earthworm-like micro robot using shape-memory alloy actuator. *Sensors and Actuators. A, Physical*, *125*(2), 429–437. doi:10.1016/j.sna.2005.05.004

King, A. (1998). Earthquake loads & Earthquake resistant design of buildings. In Building Research Association of New Zealand, BRANZ 1998. Porirua.

Ki-Yeob Kang, K.-H. C.-M. (2016). Dynamic response of structural models according to characteristics. *Ocean Engineering*, 174 - 190.

Kočí, J., Maděra, J., Jerman, M., & Černý, R. (2012). Optimization methods for determination of moisture diffusivity of building materials in the drying phase. *WIT Transactions on Ecology and The Environment*, *165*, 323–333. doi:10.2495/ARC120291

Koh, C. G., Quek, S. T., & Tee, K. F. (2002). Damage Identification of Structural Dynamic System. *Prof of the 2nd International Conference on Structural Stability and Dynamics*, 780-785.

Koh, C. G., Tee, K. F., & Quek, S. T. (2003). Stiffness and Damage Identification with Model Reduction Technique. *Prof of the 4th International Workshop on Structural Health Monitoring, Stanford University*, 525-532.

Koh, C. G., Tee, K. F., & Quek, S. T. (2006). Condensed Model Identification and Recovery for Structural Damage Assessment. *Journal of Structural Engineering*, *132*(12), 2018–2026. doi:10.1061/(ASCE)0733-9445(2006)132:12(2018)

Korkida, M. K., Maroulis, Z. B., & Marinos-Kouris, D. (2002). Heat and mass transfer coefficients in drying: Compilation of Literature Data. *Drying Technology*, *20*(1), 1–18. doi:10.1081/DRT-120001363

Kris, J., & Lars, K. (1999). *La bible du programmeur c/c*. Editions Reynauld Goulet.

Kumar, K. R., Hadjinicola, G. C., & Lin, T. (1995). A heuristic procedure for the single-row facility layout problem. *European Journal of Operational Research*, 87.

L.C.P.C. (1993). *Règles de conception et de dimensionnement des fondations des ouvrages d'art*. Paris: Edit. LCPC.

Lampinen, M. J., & Ojala, K. T. (1993). Mathematical modeling of web drying. In R. A. Mashelkar (Ed.), *Advances transport process IX* (pp. 271–347). Mujumdar: Elsevier. doi:10.1016/B978-0-444-89737-4.50011-9

Lavergne, S., & Labossiere, P. (1997). Experimental Study of Concrete Columns Confined by a Composite Jacket under Combined Axial and Flexural Loads. *Proceedings CSCE Annual Conference*, 11-20.

Lin, C. B., Soong, T. T., & Natke, H. G. (1990). Real Time System Identification of Degrading Structures. *Journal of Engineering Mechanics, 116*(10), 2258–2274. doi:10.1061/(ASCE)0733-9399(1990)116:10(2258)

Lindeburg, M. R., & McMullin, P. K. M. (2014). *Seismic Design of Building Structures* (11th ed.). Professional Publications, Inc.

Liping, L., Wenjin, G., Qiang, X., Lili, B., Yingmin, L., & Yuntian, W. (2012). Analysis of Elasto-Plastic Soil-Structure Interaction System Using Pushover Method. *Proceedings of The Seminar on The 15th World Conference on Earthquake Engineering*.

Liu, X. D., Wang, X. Z., Pan, Y. K., Cao, C. W., & Liu, D. Y. (2002). R & D of drying technology in China. In A. S. Mujumdar (Ed.), Drying 2002 (pp. 49–63). Academic Press.

Lloyds. (2014). *Guidance notes for risk based analysis: fire loads and protection*. Lloyd's Register.

Longman, R. W., & Juang, J.-N. (1987). A Variance Based Confidence Criterion for ERA Identified Modal Parameters. *Proc. of the AAS/AIAA Astrodynamics Conf.*

Lus, H. (2001). *Control Theory Based System Identification* (PhD thesis). Columbia University, New York, NY.

Lus, H., De Angelis, M., Betti, R., & Longman, R. W. (2003). Constructing Second-order Models of Mechanical Systems from Identified State Space Realizations. Part I: Theoretical Discussions. *Journal of Engineering Mechanics, 129*(5), 477–488. doi:10.1061/(ASCE)0733-9399(2003)129:5(477)

M, B. J. (1964). *Introduction of Structural Dynamics*. New York: McGraw-Hill.

Maake, W., Eckert, H. J., & Cauchepin, J. L. (1993). *Manuel technique du froid: bases-composant-calcul*. PYC.

Magenes, G., & Penna, A. (2009). Existing masonry buildings: general code issues and methods of analysis and assessment. *Eurocode 8 Perspectives from the Italian Standpoint, 3*, 185–198

Ma, H., Wilkinson, T., & Cho, C. (2007). Feasibility study on a self-centering beam to-column connection by using the superelastic behavior of shape-memory alloys. *Smart Materials and Structures, 16*(5), 1555–1563. doi:10.1088/0964-1726/16/5/008

Marguerre, K. (1931). Druckverteilung durch eine elastische Schicht auf starrer rauher Unterlage. *Ingenieur-Archiv, 2*(1), 108–117. doi:10.1007/BF02079817

Marko, J., Thambiratnam, D., & Perera, N. (2004). Influence of damping systems on building structures subject to seismic effects. *Engineering Structures, 26*(13), 1939–1956. doi:10.1016/j.engstruct.2004.07.008

Martinez-Rodrigo, M., & Romero, M. L. (2003). An optimum retrofit strategy for moment resisting frames with nonlinear viscous dampers for seismic applications. *Engineering Structures, 25*(7), 913–925. doi:10.1016/S0141-0296(03)00025-7

Masukawa, J., Akiyama, H., & Saito, H. (1997). Retrofit of existing reinforced concrete piers by using carbon fiber sheet and aramid fiber sheet. *Proceedings of the 3rd Conference on Non-Metallic (FRP) Reinforcement for Concrete Structures*, 411-418.

Meek, J., & Veletsos, A. S. (1974, February). Simple models for foundations in lateral and rocking motion. In *Proceedings of the 5th World Conf. on Earthquake Engineering* (pp. 2610-2631). Academic Press.

Miller, C. A. (1980). Dynamic Reduction of Structural Models. *Journal of the Structural Division, 106*(10), 2097–2108.

Mills, C. A. (1987). The design of concrete structure to resist explosions and weapon effects. In *Proceedings of the first Int. Conference on concrete hazard protections*, (pp. 61 - 73). Edinburgh, UK: Academic Press.

Mirmiran, A., & Shahawy, M. (1997). Behavior of concrete columns confined by fiber composites. *Journal of Structural Engineering, 123*(5), 583–590. doi:10.1061/(ASCE)0733-9445(1997)123:5(583)

Moehle, J. P. (1992). Displacement-Based Design of RC Structures Subjected to Earthquakes. *Earthquake Spectra, 8*(3), 403–428. doi:10.1193/1.1585688

Moore, D. B. F. W. (2003). Design of structural connections to Eurocode 3. Watford: Leonardo da Vinci.

Moore, M. (2001). *Reliability of Visual Inspection for Highway Bridges, Volume 1.* Final Report, U.S. Department of Transportation report, FHWA-RD-01-020.

Moreschi, L. M. (2000). *Seismic design of energy dissipation systems for optimal structural perfromance* (Doctoral dissertation). Virginia Tech.

Moyne C., & Roques M. (1986). Réalités et perspectives du séchage. *Revue Générale du Thermique*, 292.

Mylonakis, G., & Gazetas, G. (2000). Seismic soil-structure interaction: Beneficial or detrimental? *Journal of Earthquake Engineering, 4*(03), 277–301. doi:10.1080/13632460009350372

Newland, D. E. (2012). *An introduction to random vibrations, spectral & wavelet analysis.* Courier Corporation.

Ngo, P. M. (2007). Blast Loading and Blast Effects on Structures – An Overview. *EJSE*, 79-80.

NIST. (2012). *GCR 12-917-21 (2012) Soil-structure interaction for building structures.* US Department of Commerce.

Nutec, F. (2008). *Basic Safety Offshore.* Falck Nutec.

O'Callahan, J., Avitabile, P., & Riemer, R. (1989). System Equivalent Reduction Expansion Process (SEREP). *Proc. of the 7th International Modal Analysis Conference*, 29-37.

Oskin, B. (2017). *Japan Earthquake & Tsunami of 2011: Facts and Information, Live Science.* Retrieved from https://www.livescience.com/39110-japan-2011-earthquake-tsunami-facts.html

Ou-Yang, C., & Utamima, A. (2013). Hybrid Estimation of Distribution Algorithm for Solving Single Row Facility Layout Problem. *Computers & Industrial Engineering*, *66*(1), 95–103. doi:10.1016/j.cie.2013.05.018

Paik, J. K. (2011). Quantitative assessment of hydrocarbon explosion and fire risks in offshore installations. Elsevier.

Pantelides, C. P., Gergely, J., Reaveley, L. D., & Volnyy, V. A. (2000b). Seismic Strengthening of Reinforced Concrete Bridge Pier with FRP Composites. *Proceedings of the 12th World Conference on Earthquake Engineering*, 127.

Papadopoulos, M., & Garcia, E. (1996). Improvement in Model Reduction Schemes Using the System Equivalent Reduction Expansion Process. *AIAA Journal*, *34*(10), 2217–2219. doi:10.2514/3.13383

Paz, M. (1985). *Structural Dynamics, Theory and Computation*. New York: Van Nostrand Reinhold.

Paz, M. (1989). Modified Dynamic Condensation Method. *Journal of Structural Engineering*, *115*(1), 234–238. doi:10.1061/(ASCE)0733-9445(1989)115:1(234)

Pel, L., Landman, K. A., & Kaasschieter, E. F. (2002). Analytic solution for the non-linear drying problem. *International Journal of Heat and Mass Transfer*, *45*(15), 3173–3180. doi:10.1016/S0017-9310(02)00025-X

Phan, M., Horta, L. G., Juang, J.-N., & Longman, R. W. (1993). Linear System Identification via an Asymptotically Stable Observer. *Journal of Optimization Theory and Applications*, *79*(1), 59–86. doi:10.1007/BF00941887

Phan, M., Horta, L. G., Juang, J.-N., & Longman, R. W. (1995). Improvement of Observer/Kalman Filter Identification (OKID) by Residual Whitening. *Journal of Vibration and Acoustics*, *117*(2), 232–238. doi:10.1115/1.2873927

Phan, M., Juang, J.-N., & Longman, R. W. (1992). Identification of Linear Multivariable Systems by Identification of Observers with Assigned Real Eigenvalues. *The Journal of the Astronautical Sciences*, *40*(2), 261–279.

Pitaluga. (1992 - 1995). *Optimized Fire Safety of Offshore Structures (OFSOS)*. Brite-Euram Project.

Pressman, A. (2007). *Architectural graphic standards*. John Wiley and Sons.

Priestley, M. J. N. (1993). Myths and Fallacies in Earthquake Engineering - Conflicts between Design and Reality. New Zealand National Society for Earthquake Engineering Bulletin, 26, 329-335.

Priestley, M. J. N. (2013). Towards displacement-based design in seismic design codes. *Structural Engineering International: Journal of the International Association for Bridge and Structural Engineering*, *23*(2), 111. doi:10.1080/10168664.2013.11985330

Priestley, M., Calvi, G. M., & Kowalsky, M. J. (2007). Displacement-based seismic design of structures. *Building*, *23*(33), 1453–1460. doi:10.1016/S0141-0296(01)00048-7

Profire. (2014). *Section factors*. Retrieved June 20, 2017, from W/D, A/P, M/D Calculation Method: http://www.profire.com.tr/eng/222-page-wd-ap-md-calculation-method.aspx

Qiao, S. Z. (2010). Advanced CFD modeling on vapour dispenser and vapour cloud explosion. *Journal of Loss Prevention*, 843-848.

Qin, S. J. (2006). An Overview of Subspace Identification. *Computers & Chemical Engineering*, *30*(10-12), 1502–1513. doi:10.1016/j.compchemeng.2006.05.045

Qu, Z.-Q., & Fu, Z.-F. (1998). New Structural Dynamic Condensation Method for Finite Element Models. *AIAA Journal*, *36*(7), 1320–1324. doi:10.2514/2.517

Qu, Z.-Q., & Fu, Z.-F. (2000). An Iterative Method for Dynamic Condensation of Structural Matrices. *J. Mechanical System and Signal Processing*, *14*(4), 667–678. doi:10.1006/mssp.1998.1302

R.P.A 99. (2003). *Règles Parasismiques Algériennes 1999*. Algiers: Centre National de Recherche Appliquée en Génie Parasismique.

Rabha, D. K., Muthukumar, P., & Somayaji, C. (2017). Experimental investigation of thin layer kinetics of ghost chilli pepper (Capsicum Chinense Jacq.) dried in a forced convection solar tunnel dryer. *Renewable Energy*, *105*, 583–589. doi:10.1016/j.renene.2016.12.091

Ras, A., & Boumechra, N. (2016). Seismic energy dissipation study of linear fluid viscous dampers in steel structure design. *Alexandria Engineering Journal*, *55*(3), 2821–2832. doi:10.1016/j.aej.2016.07.012

Reissner, E. (1936). Stationäre, axialsymmetrische, durch eine schüttelnde Masse erregte Schwingungen eines homogenen elastischen Halbraumes. *Ingenieur-Archiv*, *7*(6), 381–396. doi:10.1007/BF02090427

Reitherman, R. (2012). *Earthquakes and Engineers: An International History*. Reston, VA: ASCE Press. doi:10.1061/9780784410714

Ripon, S. K. N., Glette, K., Mirmotahari, O., Høvin, M., & Tørresen, J. (2009). *Pareto Optimal Based Evolutionary Approach for Solving Multi-Objective Facility Layout Problem*. Academic Press.

Ripon, K. S. N., Glette, K., Khan, K. N., Hovin, M., & Torresen, J. (2013). Adaptive variable neighborhood search for solving multi-objective facility layout problems with unequal area facilities. *Swarm and Evolutionary Computation*, *8*, 1–12. doi:10.1016/j.swevo.2012.07.003

Rivera, M. A., Singh, M. P., & Suarez, L. E. (1999). Dynamic Condensation Approach for Nonclassically Damped Structures. *AIAA Journal*, *37*(5), 564–571. doi:10.2514/2.774

Rizk, A. S. S. (2010). Structural Design of Reinforced Concrete Tall Buildings. *CTBUH Journal*, (1), 34–41.

Robert, D. C., David, S., & Malkus, M. E. P. (1988). *Concepts and Applications of Finite Element analysis*. University of Wisconsin Madison.

Roberts, J. B., & Spanos, P. D. (2003). *Random vibration and statistical linearization*. Courier Corporation.

Robertson. (1959). Proposed criteria for defining load failure of beam floors and roof constructions during fire test. *Journal of Research of the National Bureau of Standards, 63C*, 121.

Rofooei, F., & Farhidzadeh, A. (2011). Investigation on the seismic behavior of steel mrf with shape-memory alloy equipped connections. *Procedia Engineering, 14*, 3325–3330. doi:10.1016/j.proeng.2011.07.420

Rosenblueth, E. (1974). Safety and structural design. In Reinforced Concrete Engineering. New York: John Wiley & Son.s

Rupakhety, R. (2010). *Contemporary issues in earthquake engineering research: processing of accelerometric data, modelling of inelastic structural response, and quantification of near-fault effects* (Doctoral dissertation). University of Iceland.

Rytter, A. (1993). *Vibration Based Inspection of Civil Engineering Structures* (PhD Dissertation). Department of Building Technology and Structural Engineering, Aalborg University, Denmark.

Saadatmanesh, H., Ehsani, M. R., & Jin, L. (1997). Repair of Earthquake-Damaged RC Columns with FRP Wraps. *ACI Structural Journal, 94*, 206–215.

Saaed, T. E., Nikolakopoulos, G., Jonasson, J. E., & Hedlund, H. (2015). A state-of-the-art review of structural control systems. *Journal of Vibration and Control, 21*(5), 919–937. doi:10.1177/1077546313478294

Said, H., & El-Rayes, K. (2013). Performance of global optimization models for dynamic site layout planning of construction projects. *Automation in Construction, 36*, 71–78. doi:10.1016/j.autcon.2013.08.008

Saiidi, M. S., & Wang, H. (2006). Exploratory study of seismic response of concrete columns with shape-memory alloys reinforcement. *ACI Structural Journal, 103*(3), 435–442.

Samarghandi, H., & Eshghi, K. (2010). An efficient tabu algorithm for the single row facility layout problem. *European Journal of Operational Research, 205*(1), 98–105. doi:10.1016/j.ejor.2009.11.034

Samarghandi, H., Taabayan, P., & Jahantigh, F. F. (2010). A particle swarm optimization for the single row facility layout problem. *Computers & Industrial Engineering, 58*(4), 529–534. doi:10.1016/j.cie.2009.11.015

Sarker, P., Begum, M., & Nasrin, S. (2011). Fibre reinforced polymers for structural retrofitting: A review. *Journal of Civil Engineering, 39*(1), 49-57.

Seible, F., Innamorato, D., Baumgartner, J., Karbhari, V., & Sheng, L. H. (1999). Seismic retrofit of flexural bridge spandrel columns using fiber reinforced polymer composite jackets. Fiber Reinforced Polymer Reinforcement for Reinforced Concrete Structures, ACI Report No-188.

Seible, F., Priestley, M. J. N., Hegemier, G. A., & Innamorato, D. (1997). Seismic Retrofit of RC Columns with Continuous Carbon Fiber Jackets. *Journal of Composites for Construction, 1*(2), 52–62. doi:10.1061/(ASCE)1090-0268(1997)1:2(52)

Seyyedhasani, H., & Dvorak, J. S. (2017). Using the Vehicle Routing Problem to reduce field completion times with multiple machines. *Computers and Electronics in Agriculture*, *134*, 142–150. doi:10.1016/j.compag.2016.11.010

Shinozuka, M. (1971). Simulation of multivariate and multidimensional random processes. *The Journal of the Acoustical Society of America*, *49*(1B), 357–368. doi:10.1121/1.1912338

Shrestha, B. (2015). Seismic response of long span cable-stayed bridge to near-fault vertical ground motions. *KSCE Journal of Civil Engineering*, *19*(1), 180–187. doi:10.100712205-014-0214-y

Simiu, E., & Scanlan, R. H. (1986). *Wind effects on structures: an introduction to wind engineering*. John Wiley.

Simpro. (2010). *Cas pathologique du réservoir n°16 de la SNDP (AGIL) de la zone d'hydrocarbures à Radès*. Retrieved from www.simpro-tn.com

Solimanpur, M., Vrat, P., & Shankar, R. (2005). An ant algorithm for the single row layout problem in flexible manufacturing systems. *Computers & Operations Research*, *32*(3), 583–598. doi:10.1016/j.cor.2003.08.005

Soong, T. T., & Costantinou, M. C. (Eds.). (2014). *Passive and active structural vibration control in civil engineering* (Vol. 345). Springer.

Soong, T. T., & Spencer, B. F. Jr. (2002). Supplemental energy dissipation: State-of-the-art and state-of-the-practice. *Engineering Structures*, *24*(3), 243–259. doi:10.1016/S0141-0296(01)00092-X

Soroushian, P., Ostowari, K., Nossoni, A., & Chowdhury, H. (2001). Repair andstrengthening of concrete structures through application of corrective posttensioning forces with shape-memory alloys. *Transportation Research Record: Journal of the Transportation Research Board*, *1770*, 20–26. doi:10.3141/1770-03

Srinivasan, M. G., & Kot, C. A. (1992). Effects of Damage on the Modal Parameters of a Cylindrical Shell. *Proc. of the 10th International Modal Analysis Conference*, 529-535.

Starossek, U. M. H. (2008). Approaches to measures of structural robustness. Seoul, South Korea: Academic Press.

Stolz, A. K. (2016). A large blast simulator for the experimental investigation of explosively loaded building components. *Chemical Engineering Transactions*, 151–156. doi:10.3303/CET1648026

Symans, M. D., Charney, F. A., Whittaker, A. S., Constantinou, M. C., Kircher, C. A., Johnson, M. W., & McNamara, R. J. (2008). Energy dissipation systems for seismic applications: Current practice and recent developments. *Journal of Structural Engineering*, *134*(1), 3–21. doi:10.1061/(ASCE)0733-9445(2008)134:1(3)

Szarka, I. (2015). *Structural Integrity Management ensuring robustness and barriers*. Stavanger: University of Stavanger.

Tam, V. H. Y. (1990). Modelling of missile energy from gas explosions offshore. *International conference on the management and engineering of fire safety and loss prevention onshore and offshore.*

Tee, K. F. (2004). *Substructural Identification with Incomplete Measurement for Structural Damage Assessment* (Ph.D Dissertation). Department of Civil Engineering, National University of Singapore.

Tee, K. F., Koh, C. G., & Quek, S. T. (2004). Substructural System Identification and Damage Estimation by OKID/ERA. *Proc. of the 3rd Asian-Pacific Symposium on Structural Reliability and Its Applications,* 637-647.

Tee, K. F., Cai, Y., & Chen, H. P. (2013). Structural Damage Detection Using Quantile Regression. *Journal of Civil Structural Health Monitoring*, *3*(1), 19–31. doi:10.100713349-012-0030-3

Tee, K. F., Koh, C. G., & Quek, S. T. (2003). System Identification and Damage Estimation via Substructural Approach. *Computational Structural Engineering*, *3*(1), 1–7.

Tee, K. F., Koh, C. G., & Quek, S. T. (2005). Damage Assessment by Condensed Model Identification and Recovery with Substructural Approach. *Proc. of the International Conference on Experimental Vibration Analysis for Civil Engineering Structures*, 415-422.

Tee, K. F., Koh, C. G., & Quek, S. T. (2009). Numerical and Experimental Studies of a Substructural Identification Strategy. *Structural Health Monitoring*, *8*(5), 397–410. doi:10.1177/1475921709102089

The Statistics Portal. (2016). *Earthquakes with the highest death toll worldwide from 1900 to 2016.* Retrieved from www.statista.com/statistics/266325/death-toll-in-great-earthquakes/

TM5. (1990). *Design of structures to resist the effects of accidental explosions.* Department of Army - TM5-1300.

TMR. (2009). *Resistance to accidental and catastrophic fires: General principles.* MTS.

Toğrul, I. T., & Pehlivan, D. (2003). Modelling of drying kinetics of single apricot. *Journal of Food Engineering*, *58*(1), 23–32. doi:10.1016/S0260-8774(02)00329-1

Trifunac, M. D. (2012). Earthquake response spectra for performance based design-A critical review. *Soil Dynamics and Earthquake Engineering*, *37*, 73–83. doi:10.1016/j.soildyn.2012.01.019

Tseng, D.-H., Longman, R. W., & Juang, J.-N. (1994a). Identification of Gyroscopic and Nongyroscopic Second Order Mechanical Systems Including Repeated Problems. *Advances in the Astronautical Sciences*, *87*, 145–165.

Tseng, D.-H., Longman, R. W., & Juang, J.-N. (1994b). Identification of the Structure of the Damping Matrix in Second Order Mechanical Systems. *Advances in the Astronautical Sciences*, *87*, 166–190.

Tyler, R. G. (1985). Test on a Brake Lining Damper for Structures. *Bulletin of the New Zealand National Society for Earthquake Engineering*, *18*(3), 280–284.

Universal Declaration of Human Rights. (1948). *Universal Declaration of Human Rights (Article-25).* Retrieved from http://www.un.org/en/universal-declaration-human-rights/

Uno, Y., & Hirabayashi, N. (2009). *Facility Layout Method Using Evolution Strategies with Correlated Mutations*. Academic Press.

Urdan, T. C. (2011). *Statistics in Plain English. Routledge: Taylor & Francis Group* (3rd ed.). New York: Routledge.

van Brakel, J. (1980). Mass transfer in convective drying. In Advances in drying I (pp. 217-265). New York: Hemisphere Publication.

Vasilis, K. (2013). Calculation of blast loads for application to structural components. European Commission - Joint Research Centre.

Veletsos, A. S., & Meek, J. W. (1974). Dynamic behavior of building-foundation systems. *Earthquake Engineering & Structural Dynamics*, *3*(2), 121–138. doi:10.1002/eqe.4290030203

Wang, P. C., & Drenick, R. F. (1977). Critical seismic excitation and response of structures. Proceedings of 6WCEE, 2, 1040–1045.

Wang, K., Choi, S. H., & Lu, H. (2015). A hybrid estimation of distribution algorithm for simulation-based scheduling in a stochastic permutation flowshop. *Computers & Industrial Engineering*, *90*, 186–196. doi:10.1016/j.cie.2015.09.007

Wang, T., Zhang, L., & Tee, K. F. (2011). Extraction of Real Modes and Physical Matrices from Modal Testing. *Earthquake Engineering and Engineering Vibration*, *10*(2), 219–227. doi:10.100711803-011-0060-6

Webster, A. C. (1994). *Technological advance in Japanese building design and construction*. ASCE Publications. doi:10.1061/9780872629325

Whittaker, A. S., Bertero, V. V., Thompson, C. L., & Alonso, L. J. (1991). Seismic testing of steel plate energy dissipation devices. *Earthquake Spectra*, *7*(4), 563–604. doi:10.1193/1.1585644

Wolf, J. P., & Obernhuber, P. (1985). Non-linear soil-structure-interaction analysis using dynamic stiffness or flexibility of soil in the time domain. *Earthquake Engineering & Structural Dynamics*, *13*(2), 195–212. doi:10.1002/eqe.4290130205

Yamamoto, T. (1992). FRP Strengthening of RC columns for seismic retrofitting. In *Tenth World Conference*. Rotterdam: Balkema.

Yang, C. D., & Yeh, F. B. (1990). Identification, Reduction, and Refinement of Model Parameters by the Eigensystem Realization Algorithm. *Journal of Guidance, Control, and Dynamics*, *13*(6), 1051–1059. doi:10.2514/3.20578

Zhamalov, A. Z. (1989). Use of solar energy for drying construction materials. *Applied Solar Energy*, *25*(4), 68–70.

Zhang, Y., & Li, X. (2011). Estimation of distribution algorithm for permutation flow shops with total flow-time minimization. *Computers & Industrial Engineering*, *60*(4), 706–718. doi:10.1016/j.cie.2011.01.005

Zimmermann, T., Dubois-pèlerin, Y., & Bomme, P. (1992). Object-oriented finite element programming I. Governing principles. *Computer Methods in Applied Mechanics and Engineering*, *98*, 291–303.

Related References

To continue our tradition of advancing academic research, we have compiled a list of recommended IGI Global readings. These references will provide additional information and guidance to further enrich your knowledge and assist you with your own research and future publications.

Abed, S., Khir, T., & Ben Brahim, A. (2016). Thermodynamic and Energy Study of a Regenerator in Gas Turbine Cycle and Optimization of Performances. *International Journal of Energy Optimization and Engineering*, *5*(2), 25–44. doi:10.4018/IJEOE.2016040102

Abu Bakar, W. A., Abdullah, W. N., Ali, R., & Mokhtar, W. N. (2016). Polymolybdate Supported Nano Catalyst for Desulfurization of Diesel. In T. Saleh (Ed.), *Applying Nanotechnology to the Desulfurization Process in Petroleum Engineering* (pp. 263–280). Hershey, PA: IGI Global. doi:10.4018/978-1-4666-9545-0.ch009

Addo-Tenkorang, R., Helo, P., & Kantola, J. (2016). Engineer-To-Order Product Development: A Communication Network Analysis for Supply-Chain's Sustainable Competitive Advantage. In R. Addo-Tenkorang, J. Kantola, P. Helo, & A. Shamsuzzoha (Eds.), *Supply Chain Strategies and the Engineer-to-Order Approach* (pp. 43–59). Hershey, PA: IGI Global. doi:10.4018/978-1-5225-0021-6.ch003

Adebiyi, I. D., Popoola, P. A., & Pityana, S. (2016). Mitigation of Wear Damage by Laser Surface Alloying Technique. In E. Akinlabi, R. Mahamood, & S. Akinlabi (Eds.), *Advanced Manufacturing Techniques Using Laser Material Processing* (pp. 172–196). Hershey, PA: IGI Global. doi:10.4018/978-1-5225-0329-3.ch007

Ahmad, W. (2016). Sulfur in Petroleum: Petroleum Desulfurization Techniques. In T. Saleh (Ed.), *Applying Nanotechnology to the Desulfurization Process in Petroleum Engineering* (pp. 1–52). Hershey, PA: IGI Global. doi:10.4018/978-1-4666-9545-0.ch001

Ahmed, I., Ahmad, N., Mehmood, I., Haq, I. U., Hassan, M., & Khan, M. U. (2016). Applications of Nanotechnology in Transportation Engineering. In A. Khitab & W. Anwar (Eds.), *Advanced Research on Nanotechnology for Civil Engineering Applications* (pp. 180–207). Hershey, PA: IGI Global. doi:10.4018/978-1-5225-0344-6.ch006

Aikhuele, D. (2018). A Study of Product Development Engineering and Design Reliability Concerns. *International Journal of Applied Industrial Engineering*, *5*(1), 79–89. doi:10.4018/IJAIE.2018010105

Al-Najar, B. T., & Bououdina, M. (2016). Bioinspired Nanoparticles for Efficient Drug Delivery System. In M. Bououdina (Ed.), *Emerging Research on Bioinspired Materials Engineering* (pp. 69–103). Hershey, PA: IGI Global. doi:10.4018/978-1-4666-9811-6.ch003

Al-Shebeeb, O. A., Rangaswamy, S., Gopalakrishan, B., & Devaru, D. G. (2017). Evaluation and Indexing of Process Plans Based on Electrical Demand and Energy Consumption. *International Journal of Manufacturing, Materials, and Mechanical Engineering*, *7*(3), 1–19. doi:10.4018/IJMMME.2017070101

Alexakis, H., & Makris, N. (2016). Validation of the Discrete Element Method for the Limit Stability Analysis of Masonry Arches. In V. Sarhosis, K. Bagi, J. Lemos, & G. Milani (Eds.), *Computational Modeling of Masonry Structures Using the Discrete Element Method* (pp. 292–325). Hershey, PA: IGI Global. doi:10.4018/978-1-5225-0231-9.ch012

AlMegren, H. A., Gonzalez-Cortes, S., Huang, Y., Chen, H., Qian, Y., Alkinany, M., ... Xiao, T. (2016). Preparation of Deep Hydrodesulfurzation Catalysts for Diesel Fuel using Organic Matrix Decomposition Method. In H. Al-Megren & T. Xiao (Eds.), *Petrochemical Catalyst Materials, Processes, and Emerging Technologies* (pp. 216–253). Hershey, PA: IGI Global. doi:10.4018/978-1-4666-9975-5.ch009

Alshammari, A., Kalevaru, V. N., Bagabas, A., & Martin, A. (2016). Production of Ethylene and its Commercial Importance in the Global Market. In H. Al-Megren & T. Xiao (Eds.), *Petrochemical Catalyst Materials, Processes, and Emerging Technologies* (pp. 82–115). Hershey, PA: IGI Global. doi:10.4018/978-1-4666-9975-5.ch004

Amel, M. (2016). Synthesis, Characterizations, and Biological Effects Study of Some Quinoline Family. In M. Bououdina (Ed.), *Emerging Research on Bioinspired Materials Engineering* (pp. 160–196). Hershey, PA: IGI Global. doi:10.4018/978-1-4666-9811-6.ch006

Amna, T., Haasan, M. S., Khil, M., & Hwang, I. (2016). Impact of Electrospun Biomimetic Extracellular Environment on Proliferation and Intercellular Communication of Muscle Precursor Cells: An Overview – Intercellular Communication of Muscle Precursor Cells with Extracellular Environment. In M. Bououdina (Ed.), *Emerging Research on Bioinspired Materials Engineering* (pp. 247–265). Hershey, PA: IGI Global. doi:10.4018/978-1-4666-9811-6.ch009

Amuda, M. O., Lawal, T. F., & Akinlabi, E. T. (2017). Research Progress on Rheological Behavior of AA7075 Aluminum Alloy During Hot Deformation. *International Journal of Materials Forming and Machining Processes*, *4*(1), 53–96. doi:10.4018/IJMFMP.2017010104

An, M., & Qin, Y. (2016). Challenges of Railway Safety Risk Assessment and Maintenance Decision Making. In B. Rai (Ed.), *Handbook of Research on Emerging Innovations in Rail Transportation Engineering* (pp. 173–211). Hershey, PA: IGI Global. doi:10.4018/978-1-5225-0084-1.ch009

Anil, M., Ayyildiz-Tamis, D., Tasdemir, S., Sendemir-Urkmez, A., & Gulce-Iz, S. (2016). Bioinspired Materials and Biocompatibility. In M. Bououdina (Ed.), *Emerging Research on Bioinspired Materials Engineering* (pp. 294–322). Hershey, PA: IGI Global. doi:10.4018/978-1-4666-9811-6.ch011

Armutlu, H. (2018). Intelligent Biomedical Engineering Operations by Cloud Computing Technologies. In U. Kose, G. Guraksin, & O. Deperlioglu (Eds.), *Nature-Inspired Intelligent Techniques for Solving Biomedical Engineering Problems* (pp. 297–317). Hershey, PA: IGI Global. doi:10.4018/978-1-5225-4769-3.ch015

Arokiyaraj, S., Saravanan, M., Bharanidharan, R., Islam, V. I., Bououdina, M., & Vincent, S. (2016). Green Synthesis of Metallic Nanoparticles Using Plant Compounds and Their Applications: Metallic Nanoparticles Synthesis Using Plants. In M. Bououdina (Ed.), *Emerging Research on Bioinspired Materials Engineering* (pp. 1–34). Hershey, PA: IGI Global. doi:10.4018/978-1-4666-9811-6.ch001

Atik, M., Sadek, M., & Shahrour, I. (2017). Single-Run Adaptive Pushover Procedure for Shear Wall Structures. In V. Plevris, G. Kremmyda, & Y. Fahjan (Eds.), *Performance-Based Seismic Design of Concrete Structures and Infrastructures* (pp. 59–83). Hershey, PA: IGI Global. doi:10.4018/978-1-5225-2089-4.ch003

Aydin, A., Akyol, E., Gungor, M., Kaya, A., & Tasdelen, S. (2018). Geophysical Surveys in Engineering Geology Investigations With Field Examples. In N. Ceryan (Ed.), *Handbook of Research on Trends and Digital Advances in Engineering Geology* (pp. 257–280). Hershey, PA: IGI Global. doi:10.4018/978-1-5225-2709-1.ch007

Azevedo, N. M., Lemos, J. V., & Rocha de Almeida, J. (2016). Discrete Element Particle Modelling of Stone Masonry. In V. Sarhosis, K. Bagi, J. Lemos, & G. Milani (Eds.), *Computational Modeling of Masonry Structures Using the Discrete Element Method* (pp. 146–170). Hershey, PA: IGI Global. doi:10.4018/978-1-5225-0231-9.ch007

Bamufleh, H. S., Noureldin, M. M., & El-Halwagi, M. M. (2016). Sustainable Process Integration in the Petrochemical Industries. In H. Al-Megren & T. Xiao (Eds.), *Petrochemical Catalyst Materials, Processes, and Emerging Technologies* (pp. 150–163). Hershey, PA: IGI Global. doi:10.4018/978-1-4666-9975-5.ch006

Banerjee, S., Gautam, R. K., Gautam, P. K., Jaiswal, A., & Chattopadhyaya, M. C. (2016). Recent Trends and Advancement in Nanotechnology for Water and Wastewater Treatment: Nanotechnological Approach for Water Purification. In A. Khitab & W. Anwar (Eds.), *Advanced Research on Nanotechnology for Civil Engineering Applications* (pp. 208–252). Hershey, PA: IGI Global. doi:10.4018/978-1-5225-0344-6.ch007

Bas, T. G. (2017). Nutraceutical Industry with the Collaboration of Biotechnology and Nutrigenomics Engineering: The Significance of Intellectual Property in the Entrepreneurship and Scientific Research Ecosystems. In T. Bas & J. Zhao (Eds.), *Comparative Approaches to Biotechnology Development and Use in Developed and Emerging Nations* (pp. 1–17). Hershey, PA: IGI Global. doi:10.4018/978-1-5225-1040-6.ch001

Beale, R., & André, J. (2017). *Design Solutions and Innovations in Temporary Structures*. Hershey, PA: IGI Global. doi:10.4018/978-1-5225-2199-0

Behnam, B. (2017). Simulating Post-Earthquake Fire Loading in Conventional RC Structures. In P. Samui, S. Chakraborty, & D. Kim (Eds.), *Modeling and Simulation Techniques in Structural Engineering* (pp. 425–444). Hershey, PA: IGI Global. doi:10.4018/978-1-5225-0588-4.ch015

Ben Hamida, I., Salah, S. B., Msahli, F., & Mimouni, M. F. (2018). Distribution Network Reconfiguration Using SPEA2 for Power Loss Minimization and Reliability Improvement. *International Journal of Energy Optimization and Engineering*, *7*(1), 50–65. doi:10.4018/IJEOE.2018010103

Benjamin, S. R., de Lima, F., & Rathoure, A. K. (2016). Genetically Engineered Microorganisms for Bioremediation Processes: GEMs for Bioremediaton. In A. Rathoure & V. Dhatwalia (Eds.), *Toxicity and Waste Management Using Bioremediation* (pp. 113–140). Hershey, PA: IGI Global. doi:10.4018/978-1-4666-9734-8.ch006

Bhaskar, S. V., & Kudal, H. N. (2017). Effect of TiCN and AlCrN Coating on Tribological Behaviour of Plasma-nitrided AISI 4140 Steel. *International Journal of Surface Engineering and Interdisciplinary Materials Science*, *5*(2), 1–17. doi:10.4018/IJSEIMS.2017070101

Bhowmik, S., Sahoo, P., Acharyya, S. K., Dhar, S., & Chattopadhyay, J. (2016). Effect of Microstructure Degradation on Fracture Toughness of 20MnMoNi55 Steel in DBT Region. *International Journal of Manufacturing, Materials, and Mechanical Engineering*, *6*(3), 11–27. doi:10.4018/IJMMME.2016070102

Bhutto, A. W., Abro, R., Abbas, T., Yu, G., & Chen, X. (2016). Desulphurization of Fuel Oils Using Ionic Liquids. In H. Al-Megren & T. Xiao (Eds.), *Petrochemical Catalyst Materials, Processes, and Emerging Technologies* (pp. 254–284). Hershey, PA: IGI Global. doi:10.4018/978-1-4666-9975-5.ch010

Bhuyan, D. (2018). Designing of a Twin Tube Shock Absorber: A Study in Reverse Engineering. In K. Kumar & J. Davim (Eds.), *Design and Optimization of Mechanical Engineering Products* (pp. 83–104). Hershey, PA: IGI Global. doi:10.4018/978-1-5225-3401-3.ch005

Bouloudenine, M., & Bououdina, M. (2016). Toxic Effects of Engineered Nanoparticles on Living Cells. In M. Bououdina (Ed.), *Emerging Research on Bioinspired Materials Engineering* (pp. 35–68). Hershey, PA: IGI Global. doi:10.4018/978-1-4666-9811-6.ch002

Brunetti, A., Sellaro, M., Drioli, E., & Barbieri, G. (2016). Membrane Engineering and its Role in Oil Refining and Petrochemical Industry. In H. Al-Megren & T. Xiao (Eds.), *Petrochemical Catalyst Materials, Processes, and Emerging Technologies* (pp. 116–149). Hershey, PA: IGI Global. doi:10.4018/978-1-4666-9975-5.ch005

Bügler, M., & Borrmann, A. (2016). Simulation Based Construction Project Schedule Optimization: An Overview on the State-of-the-Art. In F. Miranda & C. Abreu (Eds.), *Handbook of Research on Computational Simulation and Modeling in Engineering* (pp. 482–507). Hershey, PA: IGI Global. doi:10.4018/978-1-4666-8823-0.ch016

Calderon, F. A., Giolo, E. G., Frau, C. D., Rengel, M. G., Rodriguez, H., Tornello, M., ... Gallucci, R. (2018). Seismic Microzonation and Site Effects Detection Through Microtremors Measures: A Review. In N. Ceryan (Ed.), *Handbook of Research on Trends and Digital Advances in Engineering Geology* (pp. 326–349). Hershey, PA: IGI Global. doi:10.4018/978-1-5225-2709-1.ch009

Carmona-Murillo, J., & Valenzuela-Valdés, J. F. (2016). Motivation on Problem Based Learning. In D. Fonseca & E. Redondo (Eds.), *Handbook of Research on Applied E-Learning in Engineering and Architecture Education* (pp. 179–203). Hershey, PA: IGI Global. doi:10.4018/978-1-4666-8803-2.ch009

Ceryan, N. (2016). A Review of Soft Computing Methods Application in Rock Mechanic Engineering. In P. Samui (Ed.), *Handbook of Research on Advanced Computational Techniques for Simulation-Based Engineering* (pp. 1–70). Hershey, PA: IGI Global. doi:10.4018/978-1-4666-9479-8.ch001

Ceryan, N., & Can, N. K. (2018). Prediction of The Uniaxial Compressive Strength of Rocks Materials. In N. Ceryan (Ed.), *Handbook of Research on Trends and Digital Advances in Engineering Geology* (pp. 31–96). Hershey, PA: IGI Global. doi:10.4018/978-1-5225-2709-1.ch002

Ceryan, S. (2018). Weathering Indices Used in Evaluation of the Weathering State of Rock Material. In N. Ceryan (Ed.), *Handbook of Research on Trends and Digital Advances in Engineering Geology* (pp. 132–186). Hershey, PA: IGI Global. doi:10.4018/978-1-5225-2709-1.ch004

Chandrasekaran, S., Silva, B., Patil, A., Oo, A. M., & Campbell, M. (2016). Evaluating Engineering Students' Perceptions: The Impact of Team-Based Learning Practices in Engineering Education. *International Journal of Quality Assurance in Engineering and Technology Education, 5*(4), 42–59. doi:10.4018/IJQAETE.2016100103

Chen, H., Padilla, R. V., & Besarati, S. (2017). Supercritical Fluids and Their Applications in Power Generation. In L. Chen & Y. Iwamoto (Eds.), *Advanced Applications of Supercritical Fluids in Energy Systems* (pp. 369–402). Hershey, PA: IGI Global. doi:10.4018/978-1-5225-2047-4.ch012

Chen, L. (2017). Principles, Experiments, and Numerical Studies of Supercritical Fluid Natural Circulation System. In L. Chen & Y. Iwamoto (Eds.), *Advanced Applications of Supercritical Fluids in Energy Systems* (pp. 136–187). Hershey, PA: IGI Global. doi:10.4018/978-1-5225-2047-4.ch005

Clementi, F., Di Sciascio, G., Di Sciascio, S., & Lenci, S. (2017). Influence of the Shear-Bending Interaction on the Global Capacity of Reinforced Concrete Frames: A Brief Overview of the New Perspectives. In V. Plevris, G. Kremmyda, & Y. Fahjan (Eds.), *Performance-Based Seismic Design of Concrete Structures and Infrastructures* (pp. 84–111). Hershey, PA: IGI Global. doi:10.4018/978-1-5225-2089-4.ch004

Cortés-Polo, D., Calle-Cancho, J., Carmona-Murillo, J., & González-Sánchez, J. (2017). Future Trends in Mobile-Fixed Integration for Next Generation Networks: Classification and Analysis. *International Journal of Vehicular Telematics and Infotainment Systems, 1*(1), 33–53. doi:10.4018/IJVTIS.2017010103

Cui, X., Zeng, S., Li, Z., Zheng, Q., Yu, X., & Han, B. (2018). Advanced Composites for Civil Engineering Infrastructures. In K. Kumar & J. Davim (Eds.), *Composites and Advanced Materials for Industrial Applications* (pp. 212–248). Hershey, PA: IGI Global. doi:10.4018/978-1-5225-5216-1.ch010

Dalgıç, S., & Kuşku, İ. (2018). Geological and Geotechnical Investigations in Tunneling. In N. Ceryan (Ed.), *Handbook of Research on Trends and Digital Advances in Engineering Geology* (pp. 482–529). Hershey, PA: IGI Global. doi:10.4018/978-1-5225-2709-1.ch014 de la Varga, D., Soto, M., Arias, C. A., van Oirschot, D., Kilian, R., Pascual, A., & Álvarez, J. A. (2017). Constructed Wetlands for Industrial Wastewater Treatment and Removal of Nutrients. In Á. Val del Río, J. Campos Gómez, & A. Mosquera Corral (Eds.), *Technologies for the Treatment and Recovery of Nutrients from Industrial Wastewater* (pp. 202–230). Hershey, PA: IGI Global. doi:10.4018/978-1-5225-1037-6.ch008

del Valle-Zermeño, R., Chimenos, J. M., & Formosa, J. (2016). Flue Gas Desulfurization: Processes and Technologies. In T. Saleh (Ed.), *Applying Nanotechnology to the Desulfurization Process in Petroleum Engineering* (pp. 337–377). Hershey, PA: IGI Global. doi:10.4018/978-1-4666-9545-0.ch011

Delgado, J. M., Henriques, A. A., & Delgado, R. M. (2016). Structural Non-Linear Models and Simulation Techniques: An Efficient Combination for Safety Evaluation of RC Structures. In F. Miranda & C. Abreu (Eds.), *Handbook of Research on Computational Simulation and Modeling in Engineering* (pp. 540–584). Hershey, PA: IGI Global. doi:10.4018/978-1-4666-8823-0.ch018

Delgado, P. S., Arêde, A., Pouca, N. V., & Costa, A. (2016). Numerical Modeling of RC Bridges for Seismic Risk Analysis. In F. Miranda & C. Abreu (Eds.), *Handbook of Research on Computational Simulation and Modeling in Engineering* (pp. 457–481). Hershey, PA: IGI Global. doi:10.4018/978-1-4666-8823-0.ch015

Deng, Y., & Liu, S. (2016). Catalysis with Room Temperature Ionic Liquids Mediated Metal Nanoparticles. In H. Al-Megren & T. Xiao (Eds.), *Petrochemical Catalyst Materials, Processes, and Emerging Technologies* (pp. 285–329). Hershey, PA: IGI Global. doi:10.4018/978-1-4666-9975-5.ch011

Deperlioglu, O. (2018). Intelligent Techniques Inspired by Nature and Used in Biomedical Engineering. In U. Kose, G. Guraksin, & O. Deperlioglu (Eds.), *Nature-Inspired Intelligent Techniques for Solving Biomedical Engineering Problems* (pp. 51–77). Hershey, PA: IGI Global. doi:10.4018/978-1-5225-4769-3.ch003

Dias, G. L., Magalhães, R. R., Ferreira, D. D., & Vitoriano, F. A. (2016). The Use of a Robotic Arm for Displacement Measurements in a Cantilever beam. *International Journal of Manufacturing, Materials, and Mechanical Engineering*, *6*(3), 45–57. doi:10.4018/IJMMME.2016070104

Dimitratos, N., Villa, A., Chan-Thaw, C. E., Hammond, C., & Prati, L. (2016). Valorisation of Glycerol to Fine Chemicals and Fuels. In H. Al-Megren & T. Xiao (Eds.), *Petrochemical Catalyst Materials, Processes, and Emerging Technologies* (pp. 352–384). Hershey, PA: IGI Global. doi:10.4018/978-1-4666-9975-5.ch013

Dixit, A. (2018). Application of Silica-Gel-Reinforced Aluminium Composite on the Piston of Internal Combustion Engine: Comparative Study of Silica-Gel-Reinforced Aluminium Composite Piston With Aluminium Alloy Piston. In K. Kumar & J. Davim (Eds.), *Composites and Advanced Materials for Industrial Applications* (pp. 63–98). Hershey, PA: IGI Global. doi:10.4018/978-1-5225-5216-1.ch004

Drei, A., Milani, G., & Sincraian, G. (2016). Application of DEM to Historic Masonries, Two Case-Studies in Portugal and Italy: Aguas Livres Aqueduct and Arch-Tympana of a Church. In V. Sarhosis, K. Bagi, J. Lemos, & G. Milani (Eds.), *Computational Modeling of Masonry Structures Using the Discrete Element Method* (pp. 326–366). Hershey, PA: IGI Global. doi:10.4018/978-1-5225-0231-9.ch013

Dutta, S., Roy, P. K., & Nandi, D. (2016). Optimal Allocation of Static Synchronous Series Compensator Controllers using Chemical Reaction Optimization for Reactive Power Dispatch. *International Journal of Energy Optimization and Engineering*, *5*(3), 43–62. doi:10.4018/IJEOE.2016070103

Dutta, S., Roy, P. K., & Nandi, D. (2016). Quasi Oppositional Teaching-Learning based Optimization for Optimal Power Flow Incorporating FACTS. *International Journal of Energy Optimization and Engineering*, *5*(2), 64–84. doi:10.4018/IJEOE.2016040104

Eloy, S., Dias, M. S., Lopes, P. F., & Vilar, E. (2016). Digital Technologies in Architecture and Engineering: Exploring an Engaged Interaction within Curricula. In D. Fonseca & E. Redondo (Eds.), *Handbook of Research on Applied E-Learning in Engineering and Architecture Education* (pp. 368–402). Hershey, PA: IGI Global. doi:10.4018/978-1-4666-8803-2.ch017

Elsayed, A. M., Dakkama, H. J., Mahmoud, S., Al-Dadah, R., & Kaialy, W. (2017). Sustainable Cooling Research Using Activated Carbon Adsorbents and Their Environmental Impact. In T. Kobayashi (Ed.), *Applied Environmental Materials Science for Sustainability* (pp. 186–221). Hershey, PA: IGI Global. doi:10.4018/978-1-5225-1971-3.ch009

Ercanoglu, M., & Sonmez, H. (2018). General Trends and New Perspectives on Landslide Mapping and Assessment Methods. In N. Ceryan (Ed.), *Handbook of Research on Trends and Digital Advances in Engineering Geology* (pp. 350–379). Hershey, PA: IGI Global. doi:10.4018/978-1-5225-2709-1.ch010

Erinosho, M. F., Akinlabi, E. T., & Pityana, S. (2016). Enhancement of Surface Integrity of Titanium Alloy with Copper by Means of Laser Metal Deposition Process. In E. Akinlabi, R. Mahamood, & S. Akinlabi (Eds.), *Advanced Manufacturing Techniques Using Laser Material Processing* (pp. 60–91). Hershey, PA: IGI Global. doi:10.4018/978-1-5225-0329-3.ch004

Farag, H., & Kishida, M. (2016). Kinetic Models for Complex Parallel–Consecutive Reactions Assessment of Reaction Network and Product Selectivity. In H. Al-Megren & T. Xiao (Eds.), *Petrochemical Catalyst Materials, Processes, and Emerging Technologies* (pp. 330–351). Hershey, PA: IGI Global. doi:10.4018/978-1-4666-9975-5.ch012

Faroz, S. A., Pujari, N. N., Rastogi, R., & Ghosh, S. (2017). Risk Analysis of Structural Engineering Systems Using Bayesian Inference. In P. Samui, S. Chakraborty, & D. Kim (Eds.), *Modeling and Simulation Techniques in Structural Engineering* (pp. 390–424). Hershey, PA: IGI Global. doi:10.4018/978-1-5225-0588-4.ch014

Fernando, P. R., Hamigah, T., Disne, S., Wickramasingha, G. G., & Sutharshan, A. (2018). The Evaluation of Engineering Properties of Low Cost Concrete Blocks by Partial Doping of Sand with Sawdust: Low Cost Sawdust Concrete Block. *International Journal of Strategic Engineering*, *1*(2), 26–42. doi:10.4018/IJoSE.2018070103

Fragiadakis, M., Stefanou, I., & Psycharis, I. N. (2016). Vulnerability Assessment of Damaged Classical Multidrum Columns. In V. Sarhosis, K. Bagi, J. Lemos, & G. Milani (Eds.), *Computational Modeling of Masonry Structures Using the Discrete Element Method* (pp. 235–253). Hershey, PA: IGI Global. doi:10.4018/978-1-5225-0231-9.ch010

Gaines, T. W., Williams, K. R., & Wagener, K. B. (2016). ADMET: Functionalized Polyolefins. In H. Al-Megren & T. Xiao (Eds.), *Petrochemical Catalyst Materials, Processes, and Emerging Technologies* (pp. 1–21). Hershey, PA: IGI Global. doi:10.4018/978-1-4666-9975-5.ch001

Garg, H. (2016). Bi-Criteria Optimization for Finding the Optimal Replacement Interval for Maintaining the Performance of the Process Industries. In P. Vasant, G. Weber, & V. Dieu (Eds.), *Handbook of Research on Modern Optimization Algorithms and Applications in Engineering and Economics* (pp. 643–675). Hershey, PA: IGI Global. doi:10.4018/978-1-4666-9644-0.ch025

Gaspar, P. D., Dinho da Silva, P., Gonçalves, J. P., & Carneiro, R. (2016). Computational Modelling and Simulation to Assist the Improvement of Thermal Performance and Energy Efficiency in Industrial Engineering Systems: Application to Cold Stores. In F. Miranda & C. Abreu (Eds.), *Handbook of Research on Computational Simulation and Modeling in Engineering* (pp. 1–68). Hershey, PA: IGI Global. doi:10.4018/978-1-4666-8823-0.ch001

Ge, H., Tang, M., & Wen, X. (2016). Ni/ZnO Nano Sorbent for Reactive Adsorption Desulfurization of Refinery Oil Streams. In T. Saleh (Ed.), *Applying Nanotechnology to the Desulfurization Process in Petroleum Engineering* (pp. 216–239). Hershey, PA: IGI Global. doi:10.4018/978-1-4666-9545-0.ch007

Ghosh, S., Mitra, S., Ghosh, S., & Chakraborty, S. (2017). Seismic Reliability Analysis in the Framework of Metamodelling Based Monte Carlo Simulation. In P. Samui, S. Chakraborty, & D. Kim (Eds.), *Modeling and Simulation Techniques in Structural Engineering* (pp. 192–208). Hershey, PA: IGI Global. doi:10.4018/978-1-5225-0588-4.ch006

Gil, M., & Otero, B. (2017). Learning Engineering Skills through Creativity and Collaboration: A Game-Based Proposal. In R. Alexandre Peixoto de Queirós & M. Pinto (Eds.), *Gamification-Based E-Learning Strategies for Computer Programming Education* (pp. 14–29). Hershey, PA: IGI Global. doi:10.4018/978-1-5225-1034-5.ch002

Gill, J., Ayre, M., & Mills, J. (2017). Revisioning the Engineering Profession: How to Make It Happen! In M. Gray & K. Thomas (Eds.), *Strategies for Increasing Diversity in Engineering Majors and Careers* (pp. 156–175). Hershey, PA: IGI Global. doi:10.4018/978-1-5225-2212-6.ch008

Gopal, S., & Al-Hazmi, M. H. (2016). Advances in Catalytic Technologies for Selective Oxidation of Lower Alkanes. In H. Al-Megren & T. Xiao (Eds.), *Petrochemical Catalyst Materials, Processes, and Emerging Technologies* (pp. 22–52). Hershey, PA: IGI Global. doi:10.4018/978-1-4666-9975-5.ch002

Goyal, N., Ram, M., Bhardwaj, A., & Kumar, A. (2016). Thermal Power Plant Modelling with Fault Coverage Stochastically. *International Journal of Manufacturing, Materials, and Mechanical Engineering, 6*(3), 28–44. doi:10.4018/IJMMME.2016070103

Goyal, N., Ram, M., & Kumar, P. (2017). Welding Process under Fault Coverage Approach for Reliability and MTTF. In M. Ram & J. Davim (Eds.), *Mathematical Concepts and Applications in Mechanical Engineering and Mechatronics* (pp. 222–245). Hershey, PA: IGI Global. doi:10.4018/978-1-5225-1639-2.ch011

Gray, M., & Lundy, C. (2017). Engineering Study Abroad: High Impact Strategy for Increasing Access. In M. Gray & K. Thomas (Eds.), *Strategies for Increasing Diversity in Engineering Majors and Careers* (pp. 42–59). Hershey, PA: IGI Global. doi:10.4018/978-1-5225-2212-6.ch003

Guha, D., Roy, P. K., & Banerjee, S. (2016). Application of Modified Biogeography Based Optimization in AGC of an Interconnected Multi-Unit Multi-Source AC-DC Linked Power System. *International Journal of Energy Optimization and Engineering, 5*(3), 1–18. doi:10.4018/IJEOE.2016070101

Guha, D., Roy, P. K., & Banerjee, S. (2016). Grey Wolf Optimization to Solve Load Frequency Control of an Interconnected Power System: GWO Used to Solve LFC Problem. *International Journal of Energy Optimization and Engineering, 5*(4), 62–83. doi:10.4018/IJEOE.2016100104

Gupta, A. K., Dey, A., & Mukhopadhyay, A. K. (2016). Micromechanical and Finite Element Modeling for Composites. In S. Datta & J. Davim (Eds.), *Computational Approaches to Materials Design: Theoretical and Practical Aspects* (pp. 101–162). Hershey, PA: IGI Global. doi:10.4018/978-1-5225-0290-6.ch005

Guraksin, G. E. (2018). Internet of Things and Nature-Inspired Intelligent Techniques for the Future of Biomedical Engineering. In U. Kose, G. Guraksin, & O. Deperlioglu (Eds.), *Nature-Inspired Intelligent Techniques for Solving Biomedical Engineering Problems* (pp. 263–282). Hershey, PA: IGI Global. doi:10.4018/978-1-5225-4769-3.ch013

Hansman, C. A. (2016). Developing Mentoring Programs in Engineering and Technology Education. *International Journal of Quality Assurance in Engineering and Technology Education, 5*(2), 1–15. doi:10.4018/IJQAETE.2016040101

Hasan, U., Chegenizadeh, A., & Nikraz, H. (2016). Nanotechnology Future and Present in Construction Industry: Applications in Geotechnical Engineering. In A. Khitab & W. Anwar (Eds.), *Advanced Research on Nanotechnology for Civil Engineering Applications* (pp. 141–179). Hershey, PA: IGI Global. doi:10.4018/978-1-5225-0344-6.ch005

Hejazi, T., & Akbari, L. (2017). A Multiresponse Optimization Model for Statistical Design of Processes with Discrete Variables. In M. Ram & J. Davim (Eds.), *Mathematical Concepts and Applications in Mechanical Engineering and Mechatronics* (pp. 17–37). Hershey, PA: IGI Global. doi:10.4018/978-1-5225-1639-2.ch002

Hejazi, T., & Hejazi, A. (2017). Monte Carlo Simulation for Reliability-Based Design of Automotive Complex Subsystems. In M. Ram & J. Davim (Eds.), *Mathematical Concepts and Applications in Mechanical Engineering and Mechatronics* (pp. 177–200). Hershey, PA: IGI Global. doi:10.4018/978-1-5225-1639-2.ch009

Hejazi, T., & Poursabbagh, H. (2017). Reliability Analysis of Engineering Systems: An Accelerated Life Testing for Boiler Tubes. In M. Ram & J. Davim (Eds.), *Mathematical Concepts and Applications in Mechanical Engineering and Mechatronics* (pp. 154–176). Hershey, PA: IGI Global. doi:10.4018/978-1-5225-1639-2.ch008

Henao, J., & Sotelo, O. (2018). Surface Engineering at High Temperature: Thermal Cycling and Corrosion Resistance. In A. Pakseresht (Ed.), *Production, Properties, and Applications of High Temperature Coatings* (pp. 131–159). Hershey, PA: IGI Global. doi:10.4018/978-1-5225-4194-3.ch006

Huirache-Acuña, R., Alonso-Nuñez, G., Rivera-Muñoz, E. M., Gutierrez, O., & Pawelec, B. (2016). Trimetallic Sulfide Catalysts for Hydrodesulfurization. In T. Saleh (Ed.), *Applying Nanotechnology to the Desulfurization Process in Petroleum Engineering* (pp. 240–262). Hershey, PA: IGI Global. doi:10.4018/978-1-4666-9545-0.ch008

Ilori, O. O., Adetan, D. A., & Umoru, L. E. (2017). Effect of Cutting Parameters on the Surface Residual Stress of Face-Milled Pearlitic Ductile Iron. *International Journal of Materials Forming and Machining Processes*, *4*(1), 38–52. doi:10.4018/IJMFMP.2017010103

Imam, M. H., Tasadduq, I. A., Ahmad, A., Aldosari, F., & Khan, H. (2017). Automated Generation of Course Improvement Plans Using Expert System. *International Journal of Quality Assurance in Engineering and Technology Education*, *6*(1), 1–12. doi:10.4018/IJQAETE.2017010101

Injeti, S. K., & Kumar, T. V. (2018). A WDO Framework for Optimal Deployment of DGs and DSCs in a Radial Distribution System Under Daily Load Pattern to Improve Techno-Economic Benefits. *International Journal of Energy Optimization and Engineering*, *7*(2), 1–38. doi:10.4018/IJEOE.2018040101

Ishii, N., Anami, K., & Knisely, C. W. (2018). *Dynamic Stability of Hydraulic Gates and Engineering for Flood Prevention*. Hershey, PA: IGI Global. doi:10.4018/978-1-5225-3079-4

J., J., Chowdhury, S., Goyal, P., Samui, P., & Dalkiliç, Y. (2016). Determination of Bearing Capacity of Shallow Foundation Using Soft Computing. In P. Saxena, D. Singh, & M. Pant (Eds.), *Problem Solving and Uncertainty Modeling through Optimization and Soft Computing Applications* (pp. 292-328). Hershey, PA: IGI Global. doi:10.4018/978-1-4666-9885-7.ch014

Jagan, J., Gundlapalli, P., & Samui, P. (2016). Utilization of Classification Techniques for the Determination of Liquefaction Susceptibility of Soils. In S. Bhattacharyya, P. Banerjee, D. Majumdar, & P. Dutta (Eds.), *Handbook of Research on Advanced Hybrid Intelligent Techniques and Applications* (pp. 124–160). Hershey, PA: IGI Global. doi:10.4018/978-1-4666-9474-3.ch005

Jayapalan, S. (2018). A Review of Chemical Treatments on Natural Fibers-Based Hybrid Composites for Engineering Applications. In K. Kumar & J. Davim (Eds.), *Composites and Advanced Materials for Industrial Applications* (pp. 16–37). Hershey, PA: IGI Global. doi:10.4018/978-1-5225-5216-1.ch002

Jeet, K., & Dhir, R. (2016). Software Module Clustering Using Bio-Inspired Algorithms. In P. Vasant, G. Weber, & V. Dieu (Eds.), *Handbook of Research on Modern Optimization Algorithms and Applications in Engineering and Economics* (pp. 445–470). Hershey, PA: IGI Global. doi:10.4018/978-1-4666-9644-0.ch017

Joshi, S. D., & Talange, D. B. (2016). Fault Tolerant Control for a Fractional Order AUV System. *International Journal of Energy Optimization and Engineering*, *5*(2), 1–24. doi:10.4018/IJEOE.2016040101

Julião, D., Ribeiro, S., de Castro, B., Cunha-Silva, L., & Balula, S. S. (2016). Polyoxometalates-Based Nanocatalysts for Production of Sulfur-Free Diesel. In T. Saleh (Ed.), *Applying Nanotechnology to the Desulfurization Process in Petroleum Engineering* (pp. 426–458). Hershey, PA: IGI Global. doi:10.4018/978-1-4666-9545-0.ch014

Kamthan, P. (2016). On the Nature of Collaborations in Agile Software Engineering Course Projects. *International Journal of Quality Assurance in Engineering and Technology Education*, *5*(2), 42–59. doi:10.4018/IJQAETE.2016040104

Karaman, O., Celik, C., & Urkmez, A. S. (2016). Self-Assembled Biomimetic Scaffolds for Bone Tissue Engineering. In M. Bououdina (Ed.), *Emerging Research on Bioinspired Materials Engineering* (pp. 104–132). Hershey, PA: IGI Global. doi:10.4018/978-1-4666-9811-6.ch004

Karkalos, N. E., Markopoulos, A. P., & Dossis, M. F. (2017). Optimal Model Parameters of Inverse Kinematics Solution of a 3R Robotic Manipulator Using ANN Models. *International Journal of Manufacturing, Materials, and Mechanical Engineering*, *7*(3), 20–40. doi:10.4018/IJMMME.2017070102

Kesimal, A., Karaman, K., Cihangir, F., & Ercikdi, B. (2018). Excavatability Assessment of Rock Masses for Geotechnical Studies. In N. Ceryan (Ed.), *Handbook of Research on Trends and Digital Advances in Engineering Geology* (pp. 231–256). Hershey, PA: IGI Global. doi:10.4018/978-1-5225-2709-1.ch006

Khanh, D. V., Vasant, P. M., Elamvazuthi, I., & Dieu, V. N. (2016). Multi-Objective Optimization of Two-Stage Thermo-Electric Cooler Using Differential Evolution: MO Optimization of TEC Using DE. In F. Miranda & C. Abreu (Eds.), *Handbook of Research on Computational Simulation and Modeling in Engineering* (pp. 139–170). Hershey, PA: IGI Global. doi:10.4018/978-1-4666-8823-0.ch004

Kim, D., Hassan, M. K., Chang, S., & Bigdeli, Y. (2016). Nonlinear Vibration Control of 3D Irregular Structures Subjected to Seismic Loads. In P. Samui (Ed.), *Handbook of Research on Advanced Computational Techniques for Simulation-Based Engineering* (pp. 103–119). Hershey, PA: IGI Global. doi:10.4018/978-1-4666-9479-8.ch003

Knoflacher, H. (2017). The Role of Engineers and Their Tools in the Transport Sector after Paradigm Change: From Assumptions and Extrapolations to Science. In H. Knoflacher & E. Ocalir-Akunal (Eds.), *Engineering Tools and Solutions for Sustainable Transportation Planning* (pp. 1–29). Hershey, PA: IGI Global. doi:10.4018/978-1-5225-2116-7.ch001

Kose, U. (2018). Towards an Intelligent Biomedical Engineering With Nature-Inspired Artificial Intelligence Techniques. In U. Kose, G. Guraksin, & O. Deperlioglu (Eds.), *Nature-Inspired Intelligent Techniques for Solving Biomedical Engineering Problems* (pp. 1–26). Hershey, PA: IGI Global. doi:10.4018/978-1-5225-4769-3.ch001

Kostić, S. (2018). A Review on Enhanced Stability Analyses of Soil Slopes Using Statistical Design. In N. Ceryan (Ed.), *Handbook of Research on Trends and Digital Advances in Engineering Geology* (pp. 446–481). Hershey, PA: IGI Global. doi:10.4018/978-1-5225-2709-1.ch013

Kumar, A., Patil, P. P., & Prajapati, Y. K. (2018). *Advanced Numerical Simulations in Mechanical Engineering*. Hershey, PA: IGI Global. doi:10.4018/978-1-5225-3722-9

Kumar, G. R., Rajyalakshmi, G., & Manupati, V. K. (2017). Surface Micro Patterning of Aluminium Reinforced Composite through Laser Peening. *International Journal of Manufacturing, Materials, and Mechanical Engineering*, *7*(4), 15–27. doi:10.4018/IJMMME.2017100102

Kumari, N., & Kumar, K. (2018). Fabrication of Orthotic Calipers With Epoxy-Based Green Composite. In K. Kumar & J. Davim (Eds.), *Composites and Advanced Materials for Industrial Applications* (pp. 157–176). Hershey, PA: IGI Global. doi:10.4018/978-1-5225-5216-1.ch008

Kuppusamy, R. R. (2018). Development of Aerospace Composite Structures Through Vacuum-Enhanced Resin Transfer Moulding Technology (VERTMTy): Vacuum-Enhanced Resin Transfer Moulding. In K. Kumar & J. Davim (Eds.), *Composites and Advanced Materials for Industrial Applications* (pp. 99–111). Hershey, PA: IGI Global. doi:10.4018/978-1-5225-5216-1.ch005

Lemos, J. V. (2016). The Basis for Masonry Analysis with UDEC and 3DEC. In V. Sarhosis, K. Bagi, J. Lemos, & G. Milani (Eds.), *Computational Modeling of Masonry Structures Using the Discrete Element Method* (pp. 61–89). Hershey, PA: IGI Global. doi:10.4018/978-1-5225-0231-9.ch003

Loy, J., Howell, S., & Cooper, R. (2017). Engineering Teams: Supporting Diversity in Engineering Education. In M. Gray & K. Thomas (Eds.), *Strategies for Increasing Diversity in Engineering Majors and Careers* (pp. 106–129). Hershey, PA: IGI Global. doi:10.4018/978-1-5225-2212-6.ch006

Macher, G., Armengaud, E., Kreiner, C., Brenner, E., Schmittner, C., Ma, Z., ... Krammer, M. (2018). Integration of Security in the Development Lifecycle of Dependable Automotive CPS. In N. Druml, A. Genser, A. Krieg, M. Menghin, & A. Hoeller (Eds.), *Solutions for Cyber-Physical Systems Ubiquity* (pp. 383–423). Hershey, PA: IGI Global. doi:10.4018/978-1-5225-2845-6.ch015

Maghsoodlou, S., & Poreskandar, S. (2016). Controlling Electrospinning Jet Using Microscopic Model for Ideal Tissue Engineering Scaffolds. *International Journal of Chemoinformatics and Chemical Engineering*, *5*(2), 1–16. doi:10.4018/IJCCE.2016070101

Mahendramani, G., & Lakshmana Swamy, N. (2018). Effect of Weld Groove Area on Distortion of Butt Welded Joints in Submerged Arc Welding. *International Journal of Manufacturing, Materials, and Mechanical Engineering*, *8*(2), 33–44. doi:10.4018/IJMMME.2018040103

Maiti, S. (2016). Engineered Gellan Polysaccharides in the Design of Controlled Drug Delivery Systems. In M. Bououdina (Ed.), *Emerging Research on Bioinspired Materials Engineering* (pp. 266–293). Hershey, PA: IGI Global. doi:10.4018/978-1-4666-9811-6.ch010

Majumdar, J. D., Weisheit, A., & Manna, I. (2016). Laser Surface Processing for Tailoring of Properties by Optimization of Microstructure. In E. Akinlabi, R. Mahamood, & S. Akinlabi (Eds.), *Advanced Manufacturing Techniques Using Laser Material Processing* (pp. 121–171). Hershey, PA: IGI Global. doi:10.4018/978-1-5225-0329-3.ch006

Maldonado-Macías, A. A., García-Alcaraz, J. L., Hernández-Arellano, J. L., & Cortes-Robles, G. (2016). An Ergonomic Compatibility Perspective on the Selection of Advanced Manufacturing Technology: A Case Study for CNC Vertical Machining Centers. In G. Alor-Hernández, C. Sánchez-Ramírez, & J. García-Alcaraz (Eds.), *Handbook of Research on Managerial Strategies for Achieving Optimal Performance in Industrial Processes* (pp. 137–165). Hershey, PA: IGI Global. doi:10.4018/978-1-5225-0130-5.ch008

Mamaghani, I. H. (2016). Application of Discrete Finite Element Method for Analysis of Unreinforced Masonry Structures. In V. Sarhosis, K. Bagi, J. Lemos, & G. Milani (Eds.), *Computational Modeling of Masonry Structures Using the Discrete Element Method* (pp. 440–458). Hershey, PA: IGI Global. doi:10.4018/978-1-5225-0231-9.ch017

Mansor, M. R., Sapuan, S. M., Salim, M. A., Akop, M. Z., Musthafah, M. T., & Shaharuzaman, M. A. (2016). Concurrent Design of Green Composites. In D. Verma, S. Jain, X. Zhang, & P. Gope (Eds.), *Green Approaches to Biocomposite Materials Science and Engineering* (pp. 48–75). Hershey, PA: IGI Global. doi:10.4018/978-1-5225-0424-5.ch003

Mansouri, I., & Esmaeili, E. (2016). Nanotechnology Applications in the Construction Industry. In A. Khitab & W. Anwar (Eds.), *Advanced Research on Nanotechnology for Civil Engineering Applications* (pp. 111–140). Hershey, PA: IGI Global. doi:10.4018/978-1-5225-0344-6.ch004

Manzoor, A. (2016). MOOCs for Enhancing Engineering Education. In D. Fonseca & E. Redondo (Eds.), *Handbook of Research on Applied E-Learning in Engineering and Architecture Education* (pp. 204–223). Hershey, PA: IGI Global. doi:10.4018/978-1-4666-8803-2.ch010

Martin, A., Kalevaru, V. N., & Radnik, J. (2016). Palladium in Heterogeneous Oxidation Catalysis. In H. Al-Megren & T. Xiao (Eds.), *Petrochemical Catalyst Materials, Processes, and Emerging Technologies* (pp. 53–81). Hershey, PA: IGI Global. doi:10.4018/978-1-4666-9975-5.ch003

Melnyczuk, J. M., & Palchoudhury, S. (2016). Introduction to Bio-Inspired Hydrogel and Their Application: Hydrogels. In M. Bououdina (Ed.), *Emerging Research on Bioinspired Materials Engineering* (pp. 133–159). Hershey, PA: IGI Global. doi:10.4018/978-1-4666-9811-6.ch005

Mitra-Kirtley, S., Mullins, O. C., & Pomerantz, A. E. (2016). Sulfur and Nitrogen Chemical Speciation in Crude Oils and Related Carbonaceous Materials. In T. Saleh (Ed.), *Applying Nanotechnology to the Desulfurization Process in Petroleum Engineering* (pp. 53–83). Hershey, PA: IGI Global. doi:10.4018/978-1-4666-9545-0.ch002

Moalosi, R., Uziak, J., & Oladiran, M. T. (2016). Using Blended Learning Approach to Deliver Courses in An Engineering Programme. *International Journal of Quality Assurance in Engineering and Technology Education*, *5*(1), 23–39. doi:10.4018/IJQAETE.2016010103

Mohammadzadeh, S., & Kim, Y. (2017). Nonlinear System Identification of Smart Buildings. In P. Samui, S. Chakraborty, & D. Kim (Eds.), *Modeling and Simulation Techniques in Structural Engineering* (pp. 328–347). Hershey, PA: IGI Global. doi:10.4018/978-1-5225-0588-4.ch011

Mohanty, I., & Bhattacherjee, D. (2016). Artificial Neural Network and Its Application in Steel Industry. In S. Datta & J. Davim (Eds.), *Computational Approaches to Materials Design: Theoretical and Practical Aspects* (pp. 267–300). Hershey, PA: IGI Global. doi:10.4018/978-1-5225-0290-6.ch010

Mohebkhah, A., & Sarhosis, V. (2016). Discrete Element Modeling of Masonry-Infilled Frames. In V. Sarhosis, K. Bagi, J. Lemos, & G. Milani (Eds.), *Computational Modeling of Masonry Structures Using the Discrete Element Method* (pp. 200–234). Hershey, PA: IGI Global. doi:10.4018/978-1-5225-0231-9.ch009

Molina, G. J., Aktaruzzaman, F., Soloiu, V., & Rahman, M. (2017). Design and Testing of a Jet-Impingement Instrument to Study Surface-Modification Effects by Nanofluids. *International Journal of Surface Engineering and Interdisciplinary Materials Science*, *5*(2), 43–61. doi:10.4018/IJSEIMS.2017070104

Montalvan-Sorrosa, D., de los Cobos-Vasconcelos, D., & Gonzalez-Sanchez, A. (2016). Nanotechnology Applied to the Biodesulfurization of Fossil Fuels and Spent Caustic Streams. In T. Saleh (Ed.), *Applying Nanotechnology to the Desulfurization Process in Petroleum Engineering* (pp. 378–389). Hershey, PA: IGI Global. doi:10.4018/978-1-4666-9545-0.ch012

Montillet, J., Yu, K., Bonenberg, L. K., & Roberts, G. W. (2016). Optimization Algorithms in Local and Global Positioning. In P. Vasant, G. Weber, & V. Dieu (Eds.), *Handbook of Research on Modern Optimization Algorithms and Applications in Engineering and Economics* (pp. 1–53). Hershey, PA: IGI Global. doi:10.4018/978-1-4666-9644-0.ch001

Moreira, F., & Ferreira, M. J. (2016). Teaching and Learning Requirements Engineering Based on Mobile Devices and Cloud: A Case Study. In D. Fonseca & E. Redondo (Eds.), *Handbook of Research on Applied E-Learning in Engineering and Architecture Education* (pp. 237–262). Hershey, PA: IGI Global. doi:10.4018/978-1-4666-8803-2.ch012

Mukherjee, A., Saeed, R. A., Dutta, S., & Naskar, M. K. (2017). Fault Tracking Framework for Software-Defined Networking (SDN). In C. Singhal & S. De (Eds.), *Resource Allocation in Next-Generation Broadband Wireless Access Networks* (pp. 247–272). Hershey, PA: IGI Global. doi:10.4018/978-1-5225-2023-8.ch011

Mukhopadhyay, A., Barman, T. K., & Sahoo, P. (2018). Electroless Nickel Coatings for High Temperature Applications. In K. Kumar & J. Davim (Eds.), *Composites and Advanced Materials for Industrial Applications* (pp. 297–331). Hershey, PA: IGI Global. doi:10.4018/978-1-5225-5216-1.ch013

Náprstek, J., & Fischer, C. (2017). Dynamic Stability and Post-Critical Processes of Slender Auto-Parametric Systems. In V. Plevris, G. Kremmyda, & Y. Fahjan (Eds.), *Performance-Based Seismic Design of Concrete Structures and Infrastructures* (pp. 128–171). Hershey, PA: IGI Global. doi:10.4018/978-1-5225-2089-4.ch006

Nautiyal, L., Shivach, P., & Ram, M. (2018). Optimal Designs by Means of Genetic Algorithms. In M. Ram & J. Davim (Eds.), *Soft Computing Techniques and Applications in Mechanical Engineering* (pp. 151–161). Hershey, PA: IGI Global. doi:10.4018/978-1-5225-3035-0.ch007

Nazir, R. (2017). Advanced Nanomaterials for Water Engineering and Treatment: Nano-Metal Oxides and Their Nanocomposites. In T. Saleh (Ed.), *Advanced Nanomaterials for Water Engineering, Treatment, and Hydraulics* (pp. 84–126). Hershey, PA: IGI Global. doi:10.4018/978-1-5225-2136-5.ch005

Nogueira, A. F., Ribeiro, J. C., Fernández de Vega, F., & Zenha-Rela, M. A. (2018). Evolutionary Approaches to Test Data Generation for Object-Oriented Software: Overview of Techniques and Tools. In M. Khosrow-Pour, D.B.A. (Ed.), Incorporating Nature-Inspired Paradigms in Computational Applications (pp. 162-194). Hershey, PA: IGI Global. doi:10.4018/978-1-5225-5020-4.ch006

Nunes, J. F., Moreira, P. M., & Tavares, J. M. (2016). Human Motion Analysis and Simulation Tools: A Survey. In F. Miranda & C. Abreu (Eds.), *Handbook of Research on Computational Simulation and Modeling in Engineering* (pp. 359–388). Hershey, PA: IGI Global. doi:10.4018/978-1-4666-8823-0.ch012

Ogunlaja, A. S., & Tshentu, Z. R. (2016). Molecularly Imprinted Polymer Nanofibers for Adsorptive Desulfurization. In T. Saleh (Ed.), *Applying Nanotechnology to the Desulfurization Process in Petroleum Engineering* (pp. 281–336). Hershey, PA: IGI Global. doi:10.4018/978-1-4666-9545-0.ch010

Ong, P., & Kohshelan, S. (2016). Performances of Adaptive Cuckoo Search Algorithm in Engineering Optimization. In P. Vasant, G. Weber, & V. Dieu (Eds.), *Handbook of Research on Modern Optimization Algorithms and Applications in Engineering and Economics* (pp. 676–699). Hershey, PA: IGI Global. doi:10.4018/978-1-4666-9644-0.ch026

Osho, M. B. (2018). Industrial Enzyme Technology: Potential Applications. In S. Bharati & P. Chaurasia (Eds.), *Research Advancements in Pharmaceutical, Nutritional, and Industrial Enzymology* (pp. 375–394). Hershey, PA: IGI Global. doi:10.4018/978-1-5225-5237-6.ch017

Padmaja, P., & Marutheswar, G. (2017). Certain Investigation on Secured Data Transmission in Wireless Sensor Networks. *International Journal of Mobile Computing and Multimedia Communications*, *8*(1), 48–61. doi:10.4018/IJMCMC.2017010104

Paixão, S. M., Silva, T. P., Arez, B. F., & Alves, L. (2016). Advances in the Reduction of the Costs Inherent to Fossil Fuels' Biodesulfurization towards Its Potential Industrial Application. In T. Saleh (Ed.), *Applying Nanotechnology to the Desulfurization Process in Petroleum Engineering* (pp. 390–425). Hershey, PA: IGI Global. doi:10.4018/978-1-4666-9545-0.ch013

Palmer, S., & Hall, W. (2017). An Evaluation of Group Work in First-Year Engineering Design Education. In R. Tucker (Ed.), *Collaboration and Student Engagement in Design Education* (pp. 145–168). Hershey, PA: IGI Global. doi:10.4018/978-1-5225-0726-0.ch007

Panneer, R. (2017). Effect of Composition of Fibers on Properties of Hybrid Composites. *International Journal of Manufacturing, Materials, and Mechanical Engineering*, *7*(4), 28–43. doi:10.4018/IJMMME.2017100103

Parker, J. (2016). Hubble's Expanding Universe: A Model for Quality in Technology Infused engineering and Technology Education. *International Journal of Quality Assurance in Engineering and Technology Education*, *5*(2), 16–29. doi:10.4018/IJQAETE.2016040102

Paul, S., & Roy, P. (2018). Optimal Design of Power System Stabilizer Using a Novel Evolutionary Algorithm. *International Journal of Energy Optimization and Engineering*, *7*(3), 24–46. doi:10.4018/IJEOE.2018070102

Pavaloiu, A. (2018). Artificial Intelligence Ethics in Biomedical-Engineering-Oriented Problems. In U. Kose, G. Guraksin, & O. Deperlioglu (Eds.), *Nature-Inspired Intelligent Techniques for Solving Biomedical Engineering Problems* (pp. 219–231). Hershey, PA: IGI Global. doi:10.4018/978-1-5225-4769-3.ch010

Peña, F. (2016). A Semi-Discrete Approach for the Numerical Simulation of Freestanding Blocks. In V. Sarhosis, K. Bagi, J. Lemos, & G. Milani (Eds.), *Computational Modeling of Masonry Structures Using the Discrete Element Method* (pp. 416–439). Hershey, PA: IGI Global. doi:10.4018/978-1-5225-0231-9.ch016

Penchovsky, R., & Traykovska, M. (2016). Synthetic Approaches to Biology: Engineering Gene Control Circuits, Synthesizing, and Editing Genomes. In M. Bououdina (Ed.), *Emerging Research on Bioinspired Materials Engineering* (pp. 323–351). Hershey, PA: IGI Global. doi:10.4018/978-1-4666-9811-6.ch012

Pieroni, A., & Iazeolla, G. (2016). Engineering QoS and Energy Saving in the Delivery of ICT Services. In P. Vasant & N. Voropai (Eds.), *Sustaining Power Resources through Energy Optimization and Engineering* (pp. 208–226). Hershey, PA: IGI Global. doi:10.4018/978-1-4666-9755-3.ch009

Pioro, I., Mahdi, M., & Popov, R. (2017). Application of Supercritical Pressures in Power Engineering. In L. Chen & Y. Iwamoto (Eds.), *Advanced Applications of Supercritical Fluids in Energy Systems* (pp. 404–457). Hershey, PA: IGI Global. doi:10.4018/978-1-5225-2047-4.ch013

Plaksina, T., & Gildin, E. (2017). Rigorous Integrated Evolutionary Workflow for Optimal Exploitation of Unconventional Gas Assets. *International Journal of Energy Optimization and Engineering*, *6*(1), 101–122. doi:10.4018/IJEOE.2017010106

Puppala, A. J., Bheemasetti, T. V., Zou, H., Yu, X., Pedarla, A., & Cai, G. (2016). Spatial Variability Analysis of Soil Properties using Geostatistics. In P. Samui (Ed.), *Handbook of Research on Advanced Computational Techniques for Simulation-Based Engineering* (pp. 195–226). Hershey, PA: IGI Global. doi:10.4018/978-1-4666-9479-8.ch008

Ramdani, N., & Azibi, M. (2018). Polymer Composite Materials for Microelectronics Packaging Applications: Composites for Microelectronics Packaging. In K. Kumar & J. Davim (Eds.), *Composites and Advanced Materials for Industrial Applications* (pp. 177–211). Hershey, PA: IGI Global. doi:10.4018/978-1-5225-5216-1.ch009

Ramesh, M., Garg, R., & Subrahmanyam, G. V. (2017). Investigation of Influence of Quenching and Annealing on the Plane Fracture Toughness and Brittle to Ductile Transition Temperature of the Zinc Coated Structural Steel Materials. *International Journal of Surface Engineering and Interdisciplinary Materials Science*, *5*(2), 33–42. doi:10.4018/IJSEIMS.2017070103

Razavi, A. M., & Ahmad, R. (2016). Agile Software Development Challenges in Implementation and Adoption: Focusing on Large and Distributed Settings – Past Experiences, Emergent Topics. In I. Ghani, D. Jawawi, S. Dorairaj, & A. Sidky (Eds.), *Emerging Innovations in Agile Software Development* (pp. 175–207). Hershey, PA: IGI Global. doi:10.4018/978-1-4666-9858-1.ch010

Reccia, E., Cecchi, A., & Milani, G. (2016). FEM/DEM Approach for the Analysis of Masonry Arch Bridges. In V. Sarhosis, K. Bagi, J. Lemos, & G. Milani (Eds.), *Computational Modeling of Masonry Structures Using the Discrete Element Method* (pp. 367–392). Hershey, PA: IGI Global. doi:10.4018/978-1-5225-0231-9.ch014

Ro, H. K., & McIntosh, K. (2016). Constructing Conducive Environment for Women of Color in Engineering Undergraduate Education. In U. Thomas & J. Drake (Eds.), *Critical Research on Sexism and Racism in STEM Fields* (pp. 23–48). Hershey, PA: IGI Global. doi:10.4018/978-1-5225-0174-9.ch002

Rodulfo-Baechler, S. M. (2016). Dual Role of Perovskite Hollow Fiber Membrane in the Methane Oxidation Reactions. In H. Al-Megren & T. Xiao (Eds.), *Petrochemical Catalyst Materials, Processes, and Emerging Technologies* (pp. 385–430). Hershey, PA: IGI Global. doi:10.4018/978-1-4666-9975-5.ch014

Rudolf, S., Biryuk, V. V., & Volov, V. (2018). Vortex Effect, Vortex Power: Technology of Vortex Power Engineering. In V. Kharchenko & P. Vasant (Eds.), *Handbook of Research on Renewable Energy and Electric Resources for Sustainable Rural Development* (pp. 500–533). Hershey, PA: IGI Global. doi:10.4018/978-1-5225-3867-7.ch021

Sah, A., Bhadula, S. J., Dumka, A., & Rawat, S. (2018). A Software Engineering Perspective for Development of Enterprise Applications. In A. Elçi (Ed.), *Handbook of Research on Contemporary Perspectives on Web-Based Systems* (pp. 1–23). Hershey, PA: IGI Global. doi:10.4018/978-1-5225-5384-7.ch001

Sahoo, P., & Roy, S. (2017). Tribological Behavior of Electroless Ni-P, Ni-P-W and Ni-P-Cu Coatings: A Comparison. *International Journal of Surface Engineering and Interdisciplinary Materials Science, 5*(1), 1–15. doi:10.4018/IJSEIMS.2017010101

Sahoo, S. (2018). Laminated Composite Hypar Shells as Roofing Units: Static and Dynamic Behavior. In K. Kumar & J. Davim (Eds.), *Composites and Advanced Materials for Industrial Applications* (pp. 249–269). Hershey, PA: IGI Global. doi:10.4018/978-1-5225-5216-1.ch011

Sahu, H., & Hungyo, M. (2018). Introduction to SDN and NFV. In A. Dumka (Ed.), *Innovations in Software-Defined Networking and Network Functions Virtualization* (pp. 1–25). Hershey, PA: IGI Global. doi:10.4018/978-1-5225-3640-6.ch001

Saikia, P., Bharadwaj, S. K., & Miah, A. T. (2016). Peroxovanadates and Its Bio-Mimicking Relation with Vanadium Haloperoxidases. In M. Bououdina (Ed.), *Emerging Research on Bioinspired Materials Engineering* (pp. 197–219). Hershey, PA: IGI Global. doi:10.4018/978-1-4666-9811-6.ch007

Saladino, R., Botta, G., & Crucianelli, M. (2016). Advances in Nanotechnology Transition Metal Catalysts in Oxidative Desulfurization (ODS) Processes: Nanotechnology Applied to ODS Processing. In T. Saleh (Ed.), *Applying Nanotechnology to the Desulfurization Process in Petroleum Engineering* (pp. 180–215). Hershey, PA: IGI Global. doi:10.4018/978-1-4666-9545-0.ch006

Saleh, T. A., Danmaliki, G. I., & Shuaib, T. D. (2016). Nanocomposites and Hybrid Materials for Adsorptive Desulfurization. In T. Saleh (Ed.), *Applying Nanotechnology to the Desulfurization Process in Petroleum Engineering* (pp. 129–153). Hershey, PA: IGI Global. doi:10.4018/978-1-4666-9545-0.ch004

Saleh, T. A., Shuaib, T. D., Danmaliki, G. I., & Al-Daous, M. A. (2016). Carbon-Based Nanomaterials for Desulfurization: Classification, Preparation, and Evaluation. In T. Saleh (Ed.), *Applying Nanotechnology to the Desulfurization Process in Petroleum Engineering* (pp. 154–179). Hershey, PA: IGI Global. doi:10.4018/978-1-4666-9545-0.ch005

Salem, A. M., & Shmelova, T. (2018). Intelligent Expert Decision Support Systems: Methodologies, Applications, and Challenges. In T. Shmelova, Y. Sikirda, N. Rizun, A. Salem, & Y. Kovalyov (Eds.), *Socio-Technical Decision Support in Air Navigation Systems: Emerging Research and Opportunities* (pp. 215–242). Hershey, PA: IGI Global. doi:10.4018/978-1-5225-3108-1.ch007

Samal, M. (2017). FE Analysis and Experimental Investigation of Cracked and Un-Cracked Thin-Walled Tubular Components to Evaluate Mechanical and Fracture Properties. In P. Samui, S. Chakraborty, & D. Kim (Eds.), *Modeling and Simulation Techniques in Structural Engineering* (pp. 266–293). Hershey, PA: IGI Global. doi:10.4018/978-1-5225-0588-4.ch009

Samal, M., & Balakrishnan, K. (2017). Experiments on a Ring Tension Setup and FE Analysis to Evaluate Transverse Mechanical Properties of Tubular Components. In P. Samui, S. Chakraborty, & D. Kim (Eds.), *Modeling and Simulation Techniques in Structural Engineering* (pp. 91–115). Hershey, PA: IGI Global. doi:10.4018/978-1-5225-0588-4.ch004

Santhanakumar, M., Adalarasan, R., & Rajmohan, M. (2016). An Investigation in Abrasive Waterjet Cutting of Al6061/SiC/Al2O3 Composite Using Principal Component Based Response Surface Methodology. *International Journal of Manufacturing, Materials, and Mechanical Engineering*, *6*(4), 30–47. doi:10.4018/IJMMME.2016100103

Sareen, N., & Bhattacharya, S. (2016). Cleaner Energy Fuels: Hydrodesulfurization and Beyond. In T. Saleh (Ed.), *Applying Nanotechnology to the Desulfurization Process in Petroleum Engineering* (pp. 84–128). Hershey, PA: IGI Global. doi:10.4018/978-1-4666-9545-0.ch003

Sarhosis, V. (2016). Micro-Modeling Options for Masonry. In V. Sarhosis, K. Bagi, J. Lemos, & G. Milani (Eds.), *Computational Modeling of Masonry Structures Using the Discrete Element Method* (pp. 28–60). Hershey, PA: IGI Global. doi:10.4018/978-1-5225-0231-9.ch002

Sarhosis, V., Oliveira, D. V., & Lourenco, P. B. (2016). On the Mechanical Behavior of Masonry. In V. Sarhosis, K. Bagi, J. Lemos, & G. Milani (Eds.), *Computational Modeling of Masonry Structures Using the Discrete Element Method* (pp. 1–27). Hershey, PA: IGI Global. doi:10.4018/978-1-5225-0231-9.ch001

Satyam, N. (2016). Liquefaction Modelling of Granular Soils using Discrete Element Method. In P. Samui (Ed.), *Handbook of Research on Advanced Computational Techniques for Simulation-Based Engineering* (pp. 381–441). Hershey, PA: IGI Global. doi:10.4018/978-1-4666-9479-8.ch015

Sawant, S. (2018). Deep Learning and Biomedical Engineering. In U. Kose, G. Guraksin, & O. Deperlioglu (Eds.), *Nature-Inspired Intelligent Techniques for Solving Biomedical Engineering Problems* (pp. 283–296). Hershey, PA: IGI Global. doi:10.4018/978-1-5225-4769-3.ch014

Sezgin, H., & Berkalp, O. B. (2018). Textile-Reinforced Composites for the Automotive Industry. In K. Kumar & J. Davim (Eds.), *Composites and Advanced Materials for Industrial Applications* (pp. 129–156). Hershey, PA: IGI Global. doi:10.4018/978-1-5225-5216-1.ch007

Shah, M. Z., Gazder, U., Bhatti, M. S., & Hussain, M. (2018). Comparative Performance Evaluation of Effects of Modifier in Asphaltic Concrete Mix. *International Journal of Strategic Engineering*, *1*(2), 13–25. doi:10.4018/IJoSE.2018070102

Shah, V. S., Shah, H. R., & Samui, P. (2016). Application of Meta-Models (MPMR and ELM) for Determining OMC, MDD and Soaked CBR Value of Soil. In S. Bhattacharyya, P. Banerjee, D. Majumdar, & P. Dutta (Eds.), *Handbook of Research on Advanced Hybrid Intelligent Techniques and Applications* (pp. 454–482). Hershey, PA: IGI Global. doi:10.4018/978-1-4666-9474-3.ch015

Sharma, N., & Kumar, K. (2018). Fabrication of Porous NiTi Alloy Using Organic Binders. In K. Kumar & J. Davim (Eds.), *Composites and Advanced Materials for Industrial Applications* (pp. 38–62). Hershey, PA: IGI Global. doi:10.4018/978-1-5225-5216-1.ch003

Sharma, T. K. (2016). Application of Shuffled Frog Leaping Algorithm in Software Project Scheduling. In P. Saxena, D. Singh, & M. Pant (Eds.), *Problem Solving and Uncertainty Modeling through Optimization and Soft Computing Applications* (pp. 225–238). Hershey, PA: IGI Global. doi:10.4018/978-1-4666-9885-7.ch011

Shivach, P., Nautiyal, L., & Ram, M. (2018). Applying Multi-Objective Optimization Algorithms to Mechanical Engineering. In M. Ram & J. Davim (Eds.), *Soft Computing Techniques and Applications in Mechanical Engineering* (pp. 287–301). Hershey, PA: IGI Global. doi:10.4018/978-1-5225-3035-0.ch014

Shmelova, T. (2018). Stochastic Methods for Estimation and Problem Solving in Engineering: Stochastic Methods of Decision Making in Aviation. In S. Kadry (Ed.), *Stochastic Methods for Estimation and Problem Solving in Engineering* (pp. 139–160). Hershey, PA: IGI Global. doi:10.4018/978-1-5225-5045-7.ch006

Shukla, R., Anapagaddi, R., Singh, A. K., Allen, J. K., Panchal, J. H., & Mistree, F. (2016). Integrated Computational Materials Engineering for Determining the Set Points of Unit Operations for Production of a Steel Product Mix. In S. Datta & J. Davim (Eds.), *Computational Approaches to Materials Design: Theoretical and Practical Aspects* (pp. 163–191). Hershey, PA: IGI Global. doi:10.4018/978-1-5225-0290-6.ch006

Siero González, L. R., & Romo Vázquez, A. (2017). Didactic Sequences Teaching Mathematics for Engineers With Focus on Differential Equations. In M. Ramírez-Montoya (Ed.), *Handbook of Research on Driving STEM Learning With Educational Technologies* (pp. 129–151). Hershey, PA: IGI Global. doi:10.4018/978-1-5225-2026-9.ch007

Singh, R., & Dutta, S. (2018). Visible Light Active Nanocomposites for Photocatalytic Applications. In K. Kumar & J. Davim (Eds.), *Composites and Advanced Materials for Industrial Applications* (pp. 270–296). Hershey, PA: IGI Global. doi:10.4018/978-1-5225-5216-1.ch012

Singh, R., & Lou, H. H. (2016). Safety and Efficiency Enhancement in LNG Terminals. In H. Al-Megren & T. Xiao (Eds.), *Petrochemical Catalyst Materials, Processes, and Emerging Technologies* (pp. 164–176). Hershey, PA: IGI Global. doi:10.4018/978-1-4666-9975-5.ch007

Sözbilir, H., Özkaymak, Ç., Uzel, B., & Sümer, Ö. (2018). Criteria for Surface Rupture Microzonation of Active Faults for Earthquake Hazards in Urban Areas. In N. Ceryan (Ed.), *Handbook of Research on Trends and Digital Advances in Engineering Geology* (pp. 187–230). Hershey, PA: IGI Global. doi:10.4018/978-1-5225-2709-1.ch005

Stanciu, I. (2018). Stochastic Methods in Microsystems Engineering. In S. Kadry (Ed.), *Stochastic Methods for Estimation and Problem Solving in Engineering* (pp. 161–176). Hershey, PA: IGI Global. doi:10.4018/978-1-5225-5045-7.ch007

Strebkov, D., Nekrasov, A., Trubnikov, V., & Nekrasov, A. (2018). Single-Wire Resonant Electric Power Systems for Renewable-Based Electric Grid. In V. Kharchenko & P. Vasant (Eds.), *Handbook of Research on Renewable Energy and Electric Resources for Sustainable Rural Development* (pp. 449–474). Hershey, PA: IGI Global. doi:10.4018/978-1-5225-3867-7.ch019

Subburaman, D., Jagan, J., Dalkiliç, Y., & Samui, P. (2016). Reliability Analysis of Slope Using MPMR, GRNN and GPR. In F. Miranda & C. Abreu (Eds.), *Handbook of Research on Computational Simulation and Modeling in Engineering* (pp. 208–224). Hershey, PA: IGI Global. doi:10.4018/978-1-4666-8823-0.ch007

Sun, J., Wan, S., Lin, J., & Wang, Y. (2016). Advances in Catalytic Conversion of Syngas to Ethanol and Higher Alcohols. In H. Al-Megren & T. Xiao (Eds.), *Petrochemical Catalyst Materials, Processes, and Emerging Technologies* (pp. 177–215). Hershey, PA: IGI Global. doi:10.4018/978-1-4666-9975-5.ch008

Tüdeş, Ş., Kumlu, K. B., & Ceryan, S. (2018). Integration Between Urban Planning and Natural Hazards For Resilient City. In N. Ceryan (Ed.), *Handbook of Research on Trends and Digital Advances in Engineering Geology* (pp. 591–630). Hershey, PA: IGI Global. doi:10.4018/978-1-5225-2709-1.ch017

Tyukhov, I., Rezk, H., & Vasant, P. (2016). Modern Optimization Algorithms and Applications in Solar Photovoltaic Engineering. In P. Vasant & N. Voropai (Eds.), *Sustaining Power Resources through Energy Optimization and Engineering* (pp. 390–445). Hershey, PA: IGI Global. doi:10.4018/978-1-4666-9755-3.ch016

Ulamis, K. (2018). Soil Liquefaction Assessment by Anisotropic Cyclic Triaxial Test. In N. Ceryan (Ed.), *Handbook of Research on Trends and Digital Advances in Engineering Geology* (pp. 631–664). Hershey, PA: IGI Global. doi:10.4018/978-1-5225-2709-1.ch018

Umar, M. A., Tenuche, S. S., Yusuf, S. A., Abdulsalami, A. O., & Kufena, A. M. (2016). Usability Engineering in Agile Software Development Processes. In I. Ghani, D. Jawawi, S. Dorairaj, & A. Sidky (Eds.), *Emerging Innovations in Agile Software Development* (pp. 208–221). Hershey, PA: IGI Global. doi:10.4018/978-1-4666-9858-1.ch011

Üzüm, O., & Çakır, Ö. A. (2016). A Bio-Inspired Phenomena in Cementitious Materials: Self-Healing. In M. Bououdina (Ed.), *Emerging Research on Bioinspired Materials Engineering* (pp. 220–246). Hershey, PA: IGI Global. doi:10.4018/978-1-4666-9811-6.ch008

Valente, M., & Milani, G. (2017). Seismic Assessment and Retrofitting of an Under-Designed RC Frame Through a Displacement-Based Approach. In V. Plevris, G. Kremmyda, & Y. Fahjan (Eds.), *Performance-Based Seismic Design of Concrete Structures and Infrastructures* (pp. 36–58). Hershey, PA: IGI Global. doi:10.4018/978-1-5225-2089-4.ch002

Vasant, P. (2018). A General Medical Diagnosis System Formed by Artificial Neural Networks and Swarm Intelligence Techniques. In U. Kose, G. Guraksin, & O. Deperlioglu (Eds.), *Nature-Inspired Intelligent Techniques for Solving Biomedical Engineering Problems* (pp. 130–145). Hershey, PA: IGI Global. doi:10.4018/978-1-5225-4769-3.ch006

Vergara, D., Lorenzo, M., & Rubio, M. (2016). On the Use of Virtual Environments in Engineering Education. *International Journal of Quality Assurance in Engineering and Technology Education*, *5*(2), 30–41. doi:10.4018/IJQAETE.2016040103

Verrollot, J., Tolonen, A., Harkonen, J., & Haapasalo, H. J. (2018). Challenges and Enablers for Rapid Product Development. *International Journal of Applied Industrial Engineering*, *5*(1), 25–49. doi:10.4018/IJAIE.2018010102

Wagner, C., & Ryan, C. (2016). Physical and Digital Integration Strategies of Electronic Device Supply Chains and Their Applicability to ETO Supply Chains. In R. Addo-Tenkorang, J. Kantola, P. Helo, & A. Shamsuzzoha (Eds.), *Supply Chain Strategies and the Engineer-to-Order Approach* (pp. 224–245). Hershey, PA: IGI Global. doi:10.4018/978-1-5225-0021-6.ch011

Wang, Z., Wu, P., Lan, L., & Ji, S. (2016). Preparation, Characterization and Desulfurization of the Supported Nickel Phosphide Catalysts. In H. Al-Megren & T. Xiao (Eds.), *Petrochemical Catalyst Materials, Processes, and Emerging Technologies* (pp. 431–458). Hershey, PA: IGI Global. doi:10.4018/978-1-4666-9975-5.ch015

Yardimci, A. G., & Karpuz, C. (2018). Fuzzy Rock Mass Rating: Soft-Computing-Aided Preliminary Stability Analysis of Weak Rock Slopes. In N. Ceryan (Ed.), *Handbook of Research on Trends and Digital Advances in Engineering Geology* (pp. 97–131). Hershey, PA: IGI Global. doi:10.4018/978-1-5225-2709-1.ch003

Zhang, L., Ding, S., Sun, S., Han, B., Yu, X., & Ou, J. (2016). Nano-Scale Behavior and Nano-Modification of Cement and Concrete Materials. In A. Khitab & W. Anwar (Eds.), *Advanced Research on Nanotechnology for Civil Engineering Applications* (pp. 28–79). Hershey, PA: IGI Global. doi:10.4018/978-1-5225-0344-6.ch002

Zindani, D., & Kumar, K. (2018). Industrial Applications of Polymer Composite Materials. In K. Kumar & J. Davim (Eds.), *Composites and Advanced Materials for Industrial Applications* (pp. 1–15). Hershey, PA: IGI Global. doi:10.4018/978-1-5225-5216-1.ch001

Zindani, D., Maity, S. R., & Bhowmik, S. (2018). A Decision-Making Approach for Material Selection of Polymeric Composite Bumper Beam. In K. Kumar & J. Davim (Eds.), *Composites and Advanced Materials for Industrial Applications* (pp. 112–128). Hershey, PA: IGI Global. doi:10.4018/978-1-5225-5216-1.ch006

About the Contributors

Angelia Melani Adrian is an Associate Professor at the Informatics Engineering Department, Universitas Katolik De La Salle Manado, Indonesia. She received her Master degree in Computer Science and Ph.D. degree in Industrial Management from National Taiwan University of Science and Technology. Her research interest is related to Medical Data Mining, Big Data and Metaheuristic Algorithm. So far, Dr. Adrian has published several journals indexed by SCI.

Mourad Belgasmia, Lecturer and Thesis Adviser joined the team of the Civil Engineering department of university of Bejaia (Algeria) on December 2001 as a lecturer (assistant professor), till 2008. Since 2008 till now with the team of the Civil Engineering department of university of Setif 1 (Algeria) he spends two years (from 2004 to 2006) at the laboratory of structural and continuum Mechanics at Swiss Federal Institute of Technology in Lausanne, Switzerland, working then on seismic evaluation of existing structures and new structures using Pushover Analysis. And at several occasion since for shorter periods the work was mainly on approximate method for seismic evaluation of buildings this research is documented in two substantial internal reports. His responsibilities during this period in the laboratory of structural and continuum Mechanics at Swiss Federal Institute of Technology(EPFL) Switzerland (Lausanne) From July 2004 to July 2006 is Assistant of the implementation of new method in z-soil software. His teaching experience is Strength of materials, Finite element method, Continuum Mechanics, Steel Construction, Computer language programming, Structural Dynamic, Pre-stressed concrete. His Professional path 2015 – Present Lecturer & Thesis Adviser, Civil Engineering Department, Setif1 University 2008 – 2015 Lecturer (assistant professor), Civil Engineering Department, Setif1 University 2001 – 2008 Lecturer (assistant professor), Civil Engineering Department, Bejaia University 2004 – 2006 Visiting Researcher, Laboratory of Structural and Continuum Mechanics, Swiss Federal Institute of Technology, Lausanne (Prof. Thomas Zimmermann). His research directions are Seismic expertise, Dynamic of structures, Object oriented programming, XFEM extend finite element method. Mourad Belgasmia is (co) author of 14 papers in refereed international journals and more than 20 international conferences.

Lyes Bennamoun is a research professional at University of New Brunswick, Fredericton, New Brunswick, Canada, since 2013. His main axes of research are: mathematical modeling of the woody biomass supply chain management, heat treatment including convective and microwave drying and pyrolysis of biomass. From 2011 to 2013, Dr. Bennamoun was working at the Department of Applied Chemistry, University of Liège, Belgium, as foreign fellow, where he performed experimental and modeling studying related to convective and conductive drying of thermos-sensitive and rheologically complex products, such as wastewater sludge and yeast. In 2008, Dr. Bennamoun was postdoctoral fellow at Food Research and Development Centre, St-Hyacinthe, Quebec, Canada. His main study in the research centre was the characterization and extraction of fibers from canola residues. Dr. Bennamoun is also working on the thermal conversion of solar energy and its applications. He has published 81 papers in international journals, edited books and international conferences, and he is member of the editorial board of 9 international journals linked to renewable and green energy, energy conversion and processing.

Mounir Bouassida is a professor of civil engineering at the National Engineering School of Tunis (ENIT) of the University of Tunis El Manar where he earned his B.S., M.S., Ph.D., and doctorate of sciences diplomas, all in civil engineering. He is the director of the Research Laboratory in Geotechnical Engineering and has supervised 20 Ph.D. and 30 master of science graduates. His research focuses on soil improvement techniques and behavior of soft clays. Dr. Bouassida is the (co)author of 90 papers in refereed international journals; 130 conference papers, including 20 keynote lectures; and three books. He is a member of the editorial committees of journals Ground Improvement (ICE), Geotechnical Geological Engineering, Infrastructure Innovative Solutions, and International Journal of Geomechanics (ASCE). He is also an active reviewer in several international journals. As a 2006 Fulbright scholar, Bouassida elaborated a novel methodology for the design of foundations on reinforced soil by columns. He was awarded the 2006 S. Prakash Prize for Excellence in the practice of geotechnical engineering. In 2008, Bouassida launched a Tunisian consulting office in geotechnical engineering, SIMPRO. He is a co-developer of the software Columns 1.01 used for designing column-reinforced foundations. Prof. Bouassida held the office of the vice president of ISSMGE for Africa (2005–2009). He benefited from several grants as a visiting professor in the USA, France, Belgium, Australia and Vietnam. Recently, Prof. Bouassida became an appointed member of the ISSMGE board (2017-2021).

Wafy Bouassida is a geotechnical engineer at Simpro consulting, graduated in Civil Engineering from the National engineering School of Tunis (ENIT) and MS from the University of Sherbrooke (Québec, Canada). He is a PhD fellow at Research Laboratory in Geotechnical Engineering (ENIT).

Arif Djunaidy holds BEng degree in Electrical Engineering from Institut Teknologi Sepuluh Nopember (ITS), Indonesia; and MSc as well as PhD degrees in Computer Science from University of Manchester, UK. He published some articles in referred journals and contributed to some international conferences. His research interests include data management, data and knowledge engineering, data mining, and soft computing. He is affiliated as a university Professor in Department of Information Systems, Faculty of Information and Communication Technology, Institut Teknologi Sepuluh Nopember (ITS) at Surabaya, Indonesia. He has been a member of Institute of Electrical and Electronics Engineers (IEEE) Computer Society since 1992, Association for Computing Machinary (ACM) since 2002, and Association for Information Systems Indonesia (AISINDO) since 2015.

Souhir Ellouze is Assistant Prof. at National engineering school of Sfax. Graduated in Civil Engineering from the National engineering school of Tunis (BS, MS and PhD all supervised by Prof. M. Bouassida). She is a member of the Research Laboratory in Geotechnical Engineering (ENIT).

Nadeem Faisal, B.Tech (Mechanical Engineering, ITM University, Gwalior, India), M.E. (Design of Mechanical Equipment, Birla Institute of Technology, Mesra, India). He has over 1 year of Industrial experience. His areas of interests are Optimization, Material Science, Product and Process Design, CAD/CAM/CAE and Rapid Prototyping. He has 2 books, 8 Book Chapter, 3 SCI Indexed international journals to his credit.

Md. Farrukh, B.Tech (Civil Engineering, Maulana Azad College of Engineering & Technology, Patna, India), M.E.(Structural Engineering, Birla Institute of Technology, Mesra, India). He has over 1 year of industrial experience. His interest lies in Concrete Technology, Steel Structures, etc. He has 2 book chapters under his credit.

Aboubaker Gherbi is a PhD student in the department of civil engineering at Constantine University. His research interests include the design of additional viscous dampers, and the behavior of structures under wind and seismic loads.

Kaushik Kumar, B.Tech (Mechanical Engineering, REC (Now NIT), Warangal), MBA (Marketing, IGNOU) and Ph.D (Engineering, Jadavpur University), is presently an Associate Professor in the Department of Mechanical Engineering, Birla Institute of Technology, Mesra, Ranchi, India. He has 16 years of Teaching & Research and over 11 years of industrial experience in a manufacturing unit of Global repute. His areas of teaching and research interest are Quality Management Systems, Optimization, Non-conventional machining, CAD / CAM, Rapid Prototyping and Composites. He has 9 Patents, 16 Books, 12 Edited Books, 38 Book Chapters, 124 international Journal publications, 18 International and 8 National Conference publications to his credit. He is on the editorial board and review panel of 7 International and 1 National Journals of repute. He has been felicitated with many awards and honours.

Mavis Sika Okyere (neé Nyarko) is a pipeline integrity engineer at Ghana National Gas Company. She is an expert in risk-based assessment, pipeline integrity, corrosion monitoring, and cathodic protection design. She has experience with subsea structural engineering, piping & pipeline engineering principles as applied to both onshore & offshore conditions. Mavis studied MSc. Gas Engineering and Management at University of Salford, United Kingdom and BSc. Civil Engineering at Kwame Nkrumah University of Science and Technology, Ghana. She worked with LUDA Development Ltd, Bluecrest College, INTECSEA/Worleyparsons Atlantic Ltd, Technip, Ussuya Ghana Ltd, and Ghana Highway Authority. She has published in several books and journals, and is a member of many National and International bodies such as the Institute of Gas Engineers and Managers (IGEM).

Bougherra Souhaib is a PhD student in the department of civil engineering at Constantine University. His research interests include the design of dashpots for modelization of soil structure interaction.

Moussaoui Sabah joined the team of the Civil Engineering department of university of Bejaia (Algeria) on December 2003 as a lecturer (assistant professor), till 2008, Since 2008 till now she is with the team of the Civil Engineering department of university of Setif 1 (Algeria). His teaching experience is Strength of materials, Finite element method, Computer language programming, Structural Dynamic. Moussaoui Sabah is (co) author 4 papers in refereed international journals and 03 International conferences.

Kong Fah Tee, BEng (Hons), PhD, PGCert (HE), MBA, DIC, FHEA, is a Reader in Infrastructure Engineering at the University of Greenwich. His research effort has been focused on structural system identification and health monitoring, structural reliability and failure analysis, experimental stress analysis, fatigue, fracture mechanics and structural dynamics. He has published a book and over 130 refereed international journal and conference papers. He has been awarded research grants with the total amount of over £1m. He is currently serving as an Editor-in-Chief of the International Journal of Forensic Engineering (Inderscience), editorial board members for over 20 international journals and organizing committee/international scientific committee for over 35 international conferences. He was awarded an International Research Collaboration Award from the University of Sydney and invited as a foreign expert to the Nanjing University of Aeronautics and Astronautics. He has been invited as an expert witness and a forensic engineering consultant to industry. He is a regular reviewer of papers, books, research proposals as well as an external examiner of research degrees. He has given a short course, a keynote speech and over 30 seminars and conference presentations.

Amalia Utamima is a lecturer in Faculty of Communication and Information Technology, Institut Teknologi Sepuluh Nopember. She got her master degree from NTUST, Taiwan and the bachelor of computing degree from ITS, Indonesia. Her research interests are in optimization, algorithm development, and computational intelligence. Currently, she has several journal and conference papers indexed by Scopus.

Index

G

H

I

K

L

M

N

O

P

R

S

T

V

W

www.ingramcontent.com/pod-product-compliance
Lightning Source LLC
LaVergne TN
LVHW081252011026
849162LV00007BA/1548